SALTERS ADVANCED CHEMISTRY

Chemical *Storylines* A2

D0587967

Central Team

Derek Denby

Chris Otter

Kay Stephenson

WITHDRAWN

www.heinemann.co.uk

✓ Free online support
✓ Useful weblinks
✓ 24 hour online ordering

01865 888080

Lincoln College
2004761

ACHIEVEMENT
– In Exclusive Partnership

Heinemann is an imprint of Pearson Education Limited, a company incorporated in England and Wales, having its registered office at Edinburgh Gate, Harlow, Essex, CM20 2JE. Registered company number 872828

www.heinemann.co.uk

Heinemann is a registered trademark of Pearson Education Limited

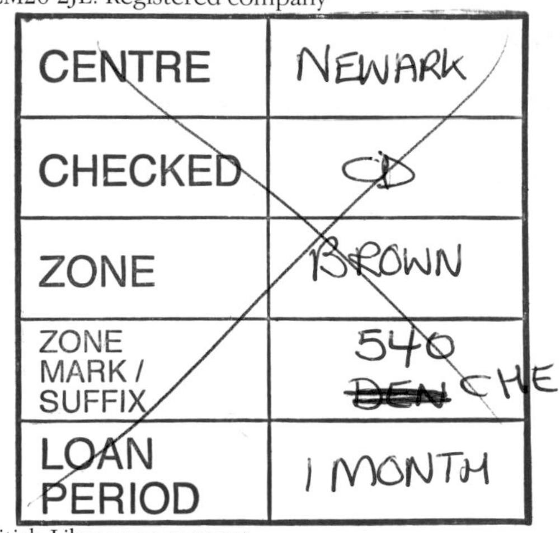

Text © University of York 2009

First published 1994
Second edition published 2000
This edition published 2009

13 12 11 10 09
10 9 8 7 6 5 4 3 2

British Library Cataloguing in Publication Data is available from the British Library on request.

ISBN 978 0 435631 48 2

Copyright notice
All rights reserved. No part of this publication may be reproduced in any form or by any means (including photocopying or storing it in any medium by electronic means and whether or not transiently or incidentally to some other use of this publication) without the written permission of the copyright owner, except in accordance with the provisions of the Copyright, Designs and Patents Act 1988 or under the terms of a licence issued by the Copyright Licensing Agency, Saffron House, 6–10 Kirby Street, London EC1N 8TS (www.cla.co.uk). Applications for the copyright owner's written permission should be addressed to the publisher.

Edited by Tony Clappison
Designed, produced, illustrated and typeset by Wearset Limited, Boldon, Tyne and Wear
Original illustrations © Pearson Education Limited 2009
Cover design by Wearset Limited, Boldon, Tyne and Wear
Picture research by Wearset Limited, Boldon, Tyne and Wear
Cover photo/illustration © NASA/Science Photo Library
Printed in China (CTPS/02)

Websites
The websites used in this book were correct and up-to-date at the time of publication. It is essential for tutors to preview each website before using it in class to ensure that the URL is still accurate, relevant and appropriate. We suggest that tutors bookmark useful websites and consider enabling students to access them through the school/college intranet.

2004761
£17.99

CONTENTS

ACKNOWLEDGEMENTS

Every effort has been made to contact copyright holders of material reproduced in this book. Any omissions will be rectified in subsequent printings if notice is given to the publishers.

Thanks are due to the following for permission to reproduce copyright material:

p3 T Nigel Cattlin/Alamy; **p3 B** Peter Evans 2003; **p4** dmac/Alamy; **p5** Dr Teofilus Tonnisson/Interspectrum OU; **p6** John McLean/Science Photo Library; **p9 L** Arco Images GmbH/Alamy; **p9 R** Bayer Corporate History & Archives; **p10 L** Bayer Corporate History & Archives; **p10 TR** Arco Images GmbH/Alamy; **p10 BR** Bayer Corporate History & Archives; **p11 L** Laguna Design/Science Photo Library; **p11 R** vario images GmbH & Co. KG/Alamy; **p12 T** Prof. P. Motta/Dept. of Anatomy/ University 'La Sapienza', Rome/Science Photo Library; **p12 B** US National Library of Medicine/Science Photo Library; **p15** Photo Library International/Science Photo Library; **p18** Pasquale Sorrentino/Science Photo Library; **p19 T** G. Muller, Struers GmbH/Science Photo Library; **p19 B** Zephyr/Science Photo Library; **p20** Hagley Museum and Library; **p21** National Archives and Records Administration; **p22** NILIT; **p23 TL** Charles D. Winters/Science Photo Library; **p23 BL** Hank Morgan/ Science Photo Library; **p23 R** Philippe Plailly/Eurelios/Science Photo Library; **p24 L** Arno Massee/Science Photo Library; **p24 R** Steve Horrell/Science Photo Library; **p25 T** National Media Museum/SSPL; **p25 B** hanapix/Alamy; **p26 T** David Ball/Alamy; **p26 M** Andrej Crcek/ Alamy; **p26 B** Angela Hampton Picture Library/Alamy; **p27** Onne van der Wal/Corbis; **p28 T** Alex Bartel/Science Photo Library; **p28 B** Science Photo Library; **p30 L** Content Mine International/Alamy; **p30 TR** Oleksiy Maksymenko/Alamy; **p30 R** Jack Sullivan/Alamy; **p31** Francoise Sauze/Science Photo Library; **p35 L** Lea Paterson/Science Photo Library; **p35 R** Damien Lovegrove/Science Photo Library; **p38** Lea Paterson/Science Photo Library; **p40 L** Alfred Pasieka/Science Photo Library; **p40 TR** Pasieka/Science Photo Library; **p40 BR** Dr Mark J. Winter/Science Photo Library; **p41 R** Science Photo Library; **p41 L** Ted Spiegel/Corbis; **p42 L** Ian Hooton/Science Photo Library; **p42 R** Div. of Computer Research & Technology, National Institute of Health/ Science Photo Library; **p44** iStockphoto/Joe Gough; **p45** BSIP Chassenet/Science Photo Library; **p48 TLL** Science Photo Library; **p48 TLR** Science Photo Library; **p48 BR** A. Barrington Brown/Science Photo Library; **p48 R** Omikron/Science Photo Library; **p49 L** Science Source/Science Photo Library; **p49 R** Mike Agliolo/Science Photo Library; **p53 T** TEK Image/Science Photo Library; **p53 M** David Parker/ Science Photo Library; **p53 B** The Illustrated London News Picture Library; **p54** David Parker/Science Photo Library; **p58 TL** www.white-windmill.co.uk/Alamy; **p58 MBL** Capstone Global Library Ltd/MM Studios; **p58 ML** PhotoDisc. 1999/Photolink; **p58 MTL** Zephyr/Science Photo Library; **p58 BR** Fancy. Punchstock; **p58 TR** Royal Armouries, Leeds; **p58 BL** www.baa.com/photolibrary; **p57** Bildarchiv Monheim GmbH/Alamy; **p61** Uwe Niggemeier; **p64** Sam Ogden/Science Photo Library; **p65 L** Heini Schneebeli/Science Photo Library; **p65 T** Sheffield Industrial Museums Trust; **p65 B** Phil King/Alamy; **p67 T** Chris Rout/ Alamy; **p67 M** Sonia Clark; **p67 B** vario images GmbH & Co. KG/Alamy; **p68** Aberdeen Foundaries; **p69 L** Science Museum/SSPL; **p69 R** Paul Rapson/Science Photo Library; **p70 L** Crown Copyright Courtesy of CSL/Science Photo Library; **p70 R** Andrew Lambert Photography/ Science Photo Library; **p72** Holmes Garden Photos/Alamy; **p75 T** Giacomo Pirozzi/Panos; **p75 B** Jim Gipe/AGSTOCKUSA/Science Photo Library; **p77 T** Shiela Terry/Science Photo Library; **p77 B** Jeff Greenberg/Alamy; **p80 L** Nigel Cattlin/Alamy; **p80 R** Nigel Cattlin/ Alamy; **p81** Digital Vision/Robert Harding World Imagery. Jim Reed; **p83 L** The Soil Association; **p83 R** Adam Hart-Davis/Science Photo Library; **p84 TL** Nigel Cattlin/Alamy; **p84 BL** Cotswold Photo Library/ Alamy; **p84 R** Andrew Tiley/Alamy; **p85 L** Mary Evans Picture Library; **p85 TR** The Print Collector/Alamy; **p85 BR** David J. Green – tools/ Alamy; **p86** Tim Graham/Alamy; **p87 TL** Emilio Segre Visual Archives/ American Institute of Physics/Science Photo Library; **p87 BL** Science Photo Library; **p87 R** DreamPictures/Getty Images; **p89 T** Dave Ellison/ Alamy; **p89 B** Nigel Cattlin/Alamy; **p91** Bill Bachman/Alamy; **p92 TL** Geoff Kidd/Science Photo Library; **p92 BL** Vaughan Fleming/Science Photo Library; **p92 TR** Nigel Cattlin/Science Photo Library; **p92 MR** Alistair Petrie/Alamy; **p92 BR** blickwinkel/Alamy; **p93** Images of Africa Photobank/Alamy; **p95** Maximilian Stock Ltd/Science Photo Library; **p97 T** Inga Leksina/Alamy; **p97 B** Don Ainley; **p98 L** Brian Atkinson/ Alamy; **p98 R** blickwinkel/Alamy; **p100 L** Robert Harding Picture Library Limited/Alamy; **p100 R** John Kershaw/Alamy; **p101 T** Mary Evans Picture Library; **p101 B** Mary Evans Picture Library; **p102 L** Vaughan Fleming/Science Photo Library; **p102 R** Roger Wood/Corbis; **p103** Mark Sykes/Alamy; **p104** Science Photo Library; **p105** Steve Welsh/Alamy; **p106 L** M. Flynn/Alamy; **p106 TR** Vincent van Gogh, *A Wheatfield, with Cypresses* © The National Gallery, London; **p106 MR** Vincent van Gogh, *A Wheatfield, with Cypresses* © The National Gallery, London; **p106 BR** Vincent van Gogh, *A Wheatfield, with Cypresses* © The National Gallery, London; **p107** Giovanni Battista Cima de Conegliano, *The Incredulity of Saint Thomas* © The National Gallery, London; **p108** Giovanni Battista Cima de Conegliano, *The Incredulity of Saint Thomas* © The National Gallery, London; **p109** Giovanni Battista Cima de Conegliano, *The Incredulity of Saint Thomas* © The National Gallery, London; **p111** Giovanni Battista Cima de Conegliano, *The Incredulity of Saint Thomas* © The National Gallery, London; **p112 T** Dept. of Physics, Imperial College/Science Photo Library; **p112 B** Science Photo Library; **p114 TL** Mary Evans Picture Library; **p114 BL** Dr Tony Travis; **p114 R** The London Art Archive/Alamy; **p115 T** Arco Images GmbH/Alamy; **p115 B** SDC Colour Museum; **p118** Dr Tony Travis; **p119** PhotoDisc 1993/Photolink; **p120** Science Museum/SSPL; **p122** Philippe Psaila/Science Photo Library; **p123** WoodyStock/Alamy; **p125 L** PhotoDisc/StockTrek; **p125 R** PhotoDisc/StockTrek; **p126 L** PhotoDisc/StockTrek; **p126 R** Tom van Sant, Geosphere Project/ Planetary Visions/Science Photo Library; **p127** Mary Evans Picture Library/MARY EVANS ILN PICTURES; **p130 TL** The Salters' Institute; **p130 BL** Steve Allen/Getty Images; **p130 R** The Maldon Crystal Salt Co Ltd; **p132 T** PhotoDisc 1996/PhotoLink Tomi; **p132 B** Jan Hinsch/ Science Photo Library; **p133** Digital Vision; **p138 L** Brand X Pictures 2001/Morey Milbradt; **p138 TR** PhotoDisc. 1998/Seide Preis; **p138 BR** nagelestock.com/Alamy; **p139** Scottish Natural Heritage; **p140** L. Newman & A. Flowers/Science Photo Library; **p143** Corbis 2003/Corbis; **p146** MedioImages/MedioImages Alamy; **p147** PhotoDisc/StockTrek; **p149** ©2008 J. Aspinall/Lion Laboratories Limited; **p150 L** ©2008 J. Aspinall/Lion Laboratories Limited; **p150 TR** ©2008 J. Aspinall/Lion Laboratories Limited; **p150 BR** ©2008 J. Aspinall/Lion Laboratories Limited; **p151** Dr. Mark J. Winter/Science Photo Library; **p152** PHOTOTAKE Inc./Alamy; **p153** Jochen Tack/Alamy; **p154** Michael P. Gadomski/Science Photo Library; **p159 R** PHOTOTAKE Inc./Alamy; **p159 L** Peter O'Brien, University of York; **p161 L** Nigel Cattlin/Alamy; **p161 TR** Science Museum/SSPL; **p161 BRL** Dr P. Marazzi/Science Photo Library; **p161 BRR** Dr P. Marazzi/Science Photo Library; **p164 T** Kwangshin Kim/Science Photo Library; **p164 B** Dr Kari Lounatmaa/ Science Photo Library; **p165** Maximilian Stock Ltd/Science Photo Library.

The author and publisher would like to thank the following individuals and organisations for permission to reproduce photographs:

p31 fig 30: Figure 1 from Mohr *et al. Initiation of shape-memory effect by inductive heating of magnetic nanoparticles in thermoplastic polymers. PNAS 2006;103:3540–3545, Copyright (2006) National Academy of Sciences, U.S.A.*; **p34 fig 1**: Artwork adapted from the U.S. Department of Energy Human Genome Program; **p63 fig 6:** Corus UK Limited; **p75 fig 1:** Data from the US Census Bureau; **p76:** Key findings from Quality Low Input Food Project taken from an article 'Organic food really is better for you', published in *The Times* newspaper, 29th October 2007; **p96:** Text from 'Carbonate synthesis by solid-base catalyzed reaction of disubstituted ureas and carbonate' by S.P. Gupte, A.B. Shivarkar and R.V. Chaudhari in *Chemical Communication* 2620–2621, 2001, www.rsc.org/publishing/ journals/CC/article.asp?doi=b107947f; **p147 fig 38:** 'Will a sea change turn up the heat?' by F. Pearce in *New Scientist* 30 November 1996, www.newscientist.com; **p153:** 'The pill of life: statin benefits last a decade' published in *The Times* newspaper, 11th October 2007.

CONTRIBUTORS

The following people have contributed to the development of *Chemical Storylines A2* (Third Edition) for the Salters Advanced Chemistry Project.

Editor

Chris Otter (Project Director) — University of York Science Education Group (UYSEG)

Associate Editors

Adelene Cogill — Idsall School, Shifnal
Frank Harriss — Formerly Malvern College
Dave Newton — Greenhead College, Huddersfield
Gill Saville — Dover Grammar School for Boys
Kay Stephenson — CLEAPSS
David Waistnidge — King Edward VI College, Totnes
Ashley Wheway — Formerly Oakham School

Acknowledgement

We would like to thank the following for their advice and contribution to the development of these materials:

Sandra Wilmott (Project Administrator) — University of York Science Education Group (UYSEG)

Sponsors

We are grateful for sponsorship from the Salters' Institute, which has continued to support the Salters Advanced Chemistry Project and has enabled the development of these materials.

Dedication

This publication is dedicated to the memory of Don Ainley, a valued contributor to the development of the Salters Advanced Chemistry Project over the years.

The Third Edition Salters Advanced Chemistry course materials draw heavily upon the previous two editions and the work of all contributors, including the following:

First Edition

Central Team

George Burton	Cranleigh School and University of York
Margaret Ferguson (1990–1991)	King Edward VI School, Louth
John Holman (Project Director)	Watford Grammar School and University of York
Gwen Pilling	University of York
David Waddington (Chairman of Steering Committee)	University of York

Associate Editors

Malcolm Churchill	Wycombe High School
Derek Denby	John Leggott Sixth Form College, Scunthorpe
Frank Harriss	Malvern College
Miranda Stephenson	Chemical Industry Education Centre
Brian Ratcliff	OCR (formerly Long Road Sixth Form College, Cambridge)
Ashley Wheway	Oakham School

Second Edition

Central Team

John Lazonby	University of York
Gwen Pilling (Project Director)	University of York
David Waddington	University of York

Associate Editors

Derek Denby	John Leggot College, Scunthorpe
John Dexter	The Trinity School, Nottingham
Margaret Ferguson	Lews Castle School, Stornoway
Frank Harriss	Malvern College
Gerald Keeling	Oundle School
Dave Newton	Greenhead College, Huddersfield
Brian Ratcliff	OCR (formerly Long Road Sixth Form College, Cambridge)
Mike Shipton	Oxted School, Surrey (formerly Reigate College)
Terri Vine	Loreto College (formerly Epsom and Ewell School)

INTRODUCTION FOR STUDENTS

The Salters Advanced Chemistry course for AS and A2 is made up of thirteen teaching modules. *Chemical Storylines A2* forms the backbone of the eight A2 teaching modules. There is a separate book covering *Chemical Ideas* and a *Support Pack* containing activities to accompany the A2 teaching modules.

Each teaching module is driven by the storyline. You work through each storyline, making 'excursions' to activities and chemical ideas at appropriate points.

The storylines are broken down into numbered sections. You will find that there are **assignments** at intervals. These are designed to help you through each storyline and to check your understanding, and they are best done as you go along.

Excursions to Activities

As you work through each storyline, you will find that there are references to particular **activities**. Each activity is referred to at that point in the storyline to which it most closely relates. Activities are numbered to correspond with the relevant section of each storyline.

Excursions to Chemical Ideas

As you work through the storylines, you will also find that there are references to sections in *Chemical Ideas*. These sections cover the chemical principles that are needed to understand that particular part of the storyline, and you will probably need to study that section of the *Chemical Ideas* book before you can go much further.

As you study *Chemical Ideas* you will find **problems** relating to each section. These are designed to check and consolidate your understanding of the chemical principles involved.

Building up the Chemical Ideas

Salters Advanced Chemistry has been planned so that you build up your understanding of chemical ideas gradually. For example, the idea of chemical equilibrium is introduced in a simple, qualitative way in 'The Atmosphere' module. A more detailed, quantitative treatment is given in the A2 teaching modules 'Agriculture and Industry' and 'The Oceans'.

Sections in *Chemical Ideas* cover chemical principles that may be needed in more than one module of the course. *Chemical Ideas* covers both AS and A2 content, so those sections that were met for the first time at AS are marked clearly – some of these sections may be revisited at A2. The context of the chemistry for a particular module is dealt with in the storyline itself and in related activities. *Chemical Storylines* features coloured boxes carrying extra background chemistry (green boxes), and case studies or in-depth information about certain aspects of the storyline (purple boxes).

How much do you need to remember?

The specification for OCR Chemistry B (Salters) defines what you have to remember. Each teaching module includes one or more 'Check your knowledge and understanding' activities. These can be used to check that you have mastered all the required knowledge, understanding and skills for the module. Each 'Check your knowledge and understanding' activity lists whether a topic is covered in *Chemical Ideas*, *Chemical Storylines* or in the associated activities.

Salters Advanced Chemistry Project

NewScientist **Reinforce your learning and keep up to date with recent developments in science by taking advantage of Heinemann's unique partnership with New Scientist. Visit www.heinemann.co.uk/newscientistmagazine for guidance and subscription discounts.**

WHAT'S IN A MEDICINE?

Why a module on 'What's in a Medicine?'?

This module introduces the pharmaceutical industry, which not only produces new and more effective medicines but is a net exporter of medicinal products and so contributes to the financial health of the UK. Through a study of aspirin, the module illustrates many of the principal activities involved in the development of a medicine. Finally, it considers the problems of development and safety testing.

During the module you will study the application of instrumental methods for determining the structure of molecules and practise the extraction of a natural product, organic synthesis and the use of test-tube reactions to identify functional groups.

The chemistry of alcohols, phenols, aldehydes, ketones and carboxylic acids is studied in some detail. You will see how alcohols can react with carboxylic acids to produce esters, and that esters can also be made from phenols. You will also develop an understanding of acids and bases.

Overview of chemical principles

In this module you will learn more about ideas introduced in earlier modules in this course:

- the interaction of radiation with matter (**Elements of Life** and **The Atmosphere**)
- atom economy (**Elements from the Sea**)
- mass spectrometry (**Elements of Life**)
- infrared spectroscopy (**Polymer Revolution**)
- alcohols (**Developing Fuels** and **Polymer Revolution**).

You will also learn new ideas about:

- molecular structure determination
- medicine manufacture and testing
- acids and bases
- phenols
- carboxylic acids
- esters
- aldehydes and ketones.

WHAT'S IN A MEDICINE?

WM1 *The development of modern ideas about medicines*

This module is about medicines and how the pharmaceutical industry works. Many pharmaceuticals are complex compounds, but in this module we focus on the chemistry of a simple and familiar substance – aspirin.

The active ingredients of **medicines** are **drugs** – substances that alter the way your body works. If your body is already working normally the drug will not be beneficial, and if the drug throws the body a long way off balance it may even be a poison. When your body is working wrongly a medicine prevents things getting worse and can help to bring about a cure – for example when you take aspirin or penicillin. Not all drugs are medicines – alcohol and nicotine are not medicines, but they certainly are drugs. Some drugs, such as opium, may or may not be medicines depending on your state of health.

The study of drugs and their action is called **pharmacology**; the art and science of making and dispensing medicines is called **pharmacy**.

People have been using medicines for thousands of years – most of that time with no idea how they worked. Their effectiveness was discovered by trial and error and sometimes there were disastrous mistakes.

Today's medicines are increasingly designed to have specific effects – this is becoming easier as we learn more about the body's chemistry and begin to understand the intricate detail of the complex molecules from which we are made.

Work at this level comes into the field of **molecular pharmacology** and you will gain some insight into this in a later module, **Medicines by Design**.

▲ **Figure 1** Feverfew has been used since ancient times for the treatment of migraine – research in the 1970s confirmed that it was an effective medicine for this disorder.

WM2 *Medicines from nature*

Modern pharmacy has its origins in folklore, and the history of medicine abounds with herbal and folk remedies. Many of these can be explained in present-day terms and the modern pharmaceutical industry investigates 'old wives' tales' to see if they lead to important new medicines.

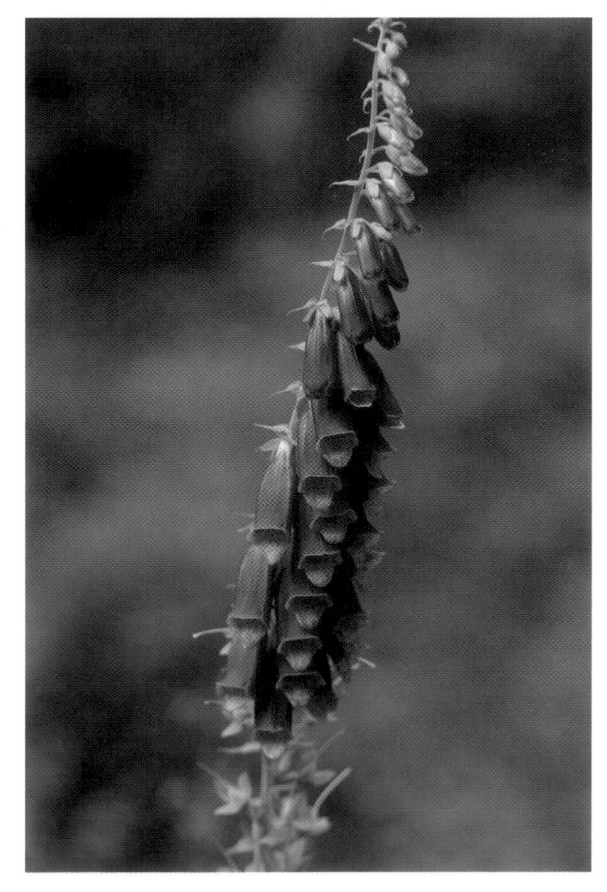

▲ **Figure 2** Foxgloves contain the compound digitalis, which is active against heart disease.

Medicines from willow bark

In 400 BC, Hippocrates recommended a brew of willow leaves to ease the pain of childbirth, and in 1763 the Reverend Edward Stone, an English clergyman living in Chipping Norton, Oxfordshire, used a willow bark brew to reduce fevers.

▲ **Figure 3** Extracts from willow trees have been used in medicine for thousands of years.

'Culpeper's Herbal'

Nicholas Culpeper (1616–54) rose to fame in the wave of enthusiasm for astrological botany that swept England in the seventeenth century. He believed that plants were 'owned' by certain planets, stars, etc. He also believed that the celestial bodies were the causes of diseases, and that illness could be cured by administering a plant 'owned' by an opposing body, or sometimes by a sympathetic body.

Extract from Culpeper's Herbal (published in 1640) about the willow tree

The decoction, made by boiling 1 oz (28 g) of willow bark in 1 pt (568 ml) of water until the mixture measures 1 pt (568 ml), is given in doses of 1–2 fl oz (28–56 ml) for fevers, diarrhoea and dysentery. The powdered root can be taken in sweetened water in doses of one teaspoonful. An infusion of the leaves – 1 oz (28 g) to 1 pt (568 ml) of boiling water – is a useful digestive tonic.

The substance salicin, extracted in the recipe you have just read, has no pharmacological effect by itself. The body converts it by hydrolysis and oxidation into the active chemical, *salicylic acid*. (The acid is named after the Latin name for willow, *Salix*.)

In **Activity WM2.1** you can produce some of the fever-curing chemical salicylic acid from willow bark.

Salicin can be oxidised to salicylic acid because it contains a primary alcohol group. You can remind yourself about the structure of alcohols, and their oxidation, in **Chemical Ideas 13.2**.

Chemical Ideas 13.7 tells you about the oxidation products of primary and secondary alcohols – aldehydes and ketones.

Activity WM2.2 will help you to understand the mechanism of an aldehyde reaction.

WM3 *Identifying the active chemical in willow bark*

How can we find out the chemical structure of compounds like salicylic acid? One way is to use chemical reactions, and in this section you will learn how chemical tests reveal the presence of particular functional groups in salicylic acid.

SOME –OH GROUP CHEMISTRY

A knowledge of some relatively simple test-tube experiments can often be used effectively in the identification of unknown substances. For example, you may already know about the use of bromine solutions for detecting double bonds between carbon atoms in alkenes (see **Chemical Ideas 12.2**).

Three chemical tests are particularly helpful in providing clues about the structure of salicylic acid:

1. An aqueous solution of the compound is weakly acidic.
2. Salicylic acid reacts with alcohols (such as ethanol) to produce compounds called esters. Esters have strong odours, often of fruit or flowers.
3. A neutral solution of iron(III) chloride turns an intense pink colour when salicylic acid is added.

Tests 1 and 2 are characteristic of *carboxylic acids* (compounds containing the –COOH functional group); test 3 indicates the presence of a **phenol** group (an –OH group attached to a benzene ring).

Before you read any further, you need to find out about the chemistry of compounds containing these functional groups.

Chemical Ideas 13.3 tells you about the structure of carboxylic acids and some compounds related to them.

In **Chemical Ideas 13.4** you compared the behaviour of the –OH group in an alcohol, a phenol and a carboxylic acid.

Activity WM3 allows you to investigate the behaviour of –OH groups in phenols and carboxylic acids.

WM4 *Instrumental analysis*

Although chemical tests provide evidence for the presence of carboxylic acid and phenol groups in salicylic acid, instrumental techniques are today's most efficient research tools. In this section you will learn about three frequently used instrumental techniques:

- mass spectrometry (m.s.)
- infrared (i.r.) spectroscopy
- nuclear magnetic resonance (n.m.r.) spectroscopy.

Making use of infrared spectroscopy

One of the very first things that would be done with any unidentified, new substance is to record its infrared (i.r.) spectrum. Figure 5 shows the i.r. spectrum of salicylic acid.

You can remind yourself about infrared spectroscopy in **Chemical Ideas 6.4**.

An i.r. spectrum measures the extent to which electromagnetic radiation in part of the infrared region is transmitted through a sample of a substance. The frequency ranges that are absorbed provide important clues about the **functional groups** that are present. The functional groups absorb at similar frequencies in many different compounds, so an absorption pattern provides a kind of *fingerprint* of the molecule.

The i.r. spectrum of salicylic acid shows clear evidence of the presence of C=O and O–H groups.

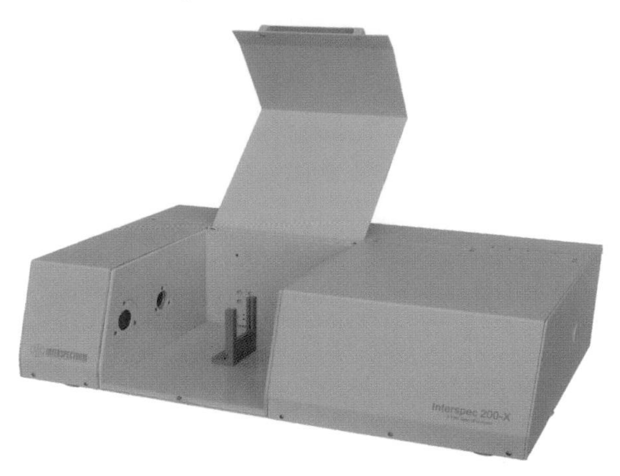

▲ **Figure 4** An infrared spectrometer – the spectrum can be viewed on a computer screen or as hard copy.

Assignment 1

Examine the i.r. spectrum of salicylic acid shown in Figure 5. Compare the absorptions marked by an asterisk (*) with the characteristic absorption bands of the different functional groups listed in **Chemical Ideas 6.4**, and suggest which groupings could be responsible. (Information about i.r. absorptions is also listed in the **Data Sheets** (Table 21).)

Do these groupings correspond to what you know of the formula for this compound?

Evidence from n.m.r. spectroscopy

A second instrumental technique that could be applied to an unidentified compound is nuclear magnetic resonance (n.m.r.) spectroscopy.

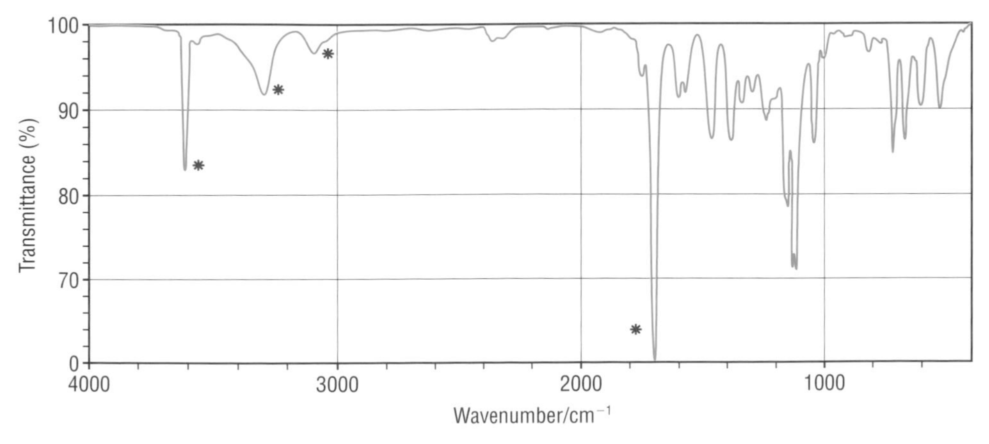

▲ **Figure 5** Infrared spectrum of salicylic acid (in the gas phase).

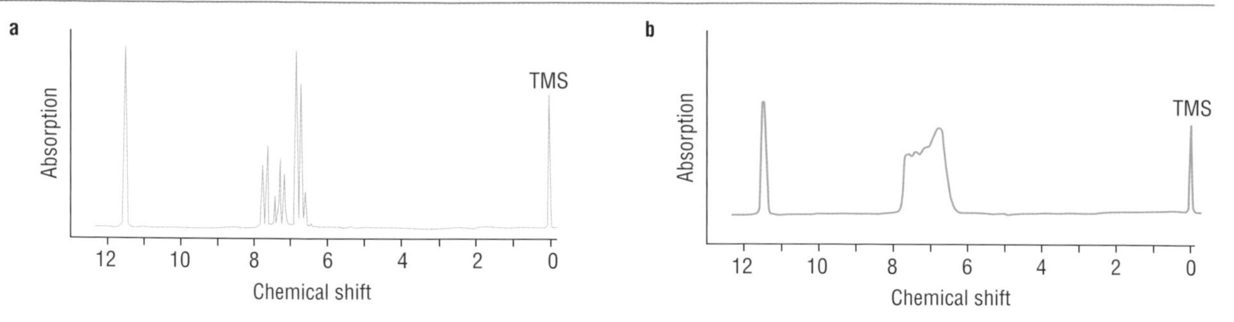

▲ **Figure 6** Proton n.m.r. spectra of salicylic acid – **a** high-resolution, **b** low-resolution. The signal labelled TMS is made by a reference compound called tetramethylsilane.

This investigates the different chemical environments in which the nuclei of one particular element are situated. Most often this element is hydrogen – the nucleus of a hydrogen atom consists of just one proton. It is the ^1_1H isotope that is utilised. The proton n.m.r. spectrum for salicylic acid is shown in Figure 6 – it shows that salicylic acid contains

- one proton in a –COOH environment
- one proton in a phenolic –OH environment
- four protons attached to a benzene ring.

In this case, the n.m.r. spectrum is quite complicated because all six hydrogen atom nuclei are in different environments in the molecule, and five of them give signals that are close together. Although n.m.r. spectroscopy is of limited value in this case, it generally provides a powerful technique for determining the structure of many organic compounds.

The n.m.r. spectrum in Figure 6a is a high-resolution spectrum. The spectrum in Figure 6b represents a low-resolution spectrum – information is lost in such a spectrum, but it is much simpler and the positions of the signals still tell us about the environments of the hydrogen atoms. You will learn more about n.m.r. spectroscopy in **Medicines by Design**.

Assignment 2

Examine the structure of 2-hydroxybenzoic acid shown here.

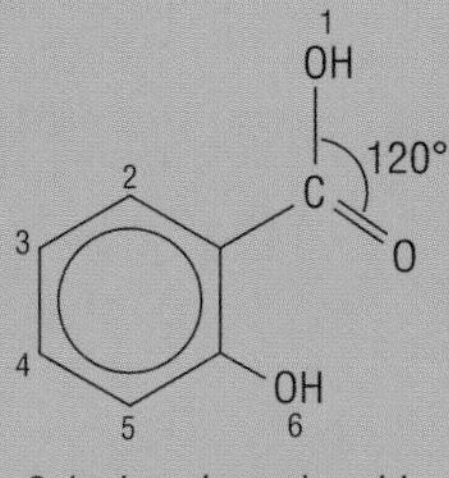

2-hydroxybenzoic acid

Explain why the following hydrogen atom nuclei are in different environments within the molecule:

a hydrogen atoms 1 and 6

b hydrogen atoms 3 and 4.

The mass spectrum of salicylic acid

A combination of i.r. and n.m.r. spectroscopy shows that salicylic acid has an –OH group and a –COOH group both attached to a benzene ring – in other words, a better name for salicylic acid is hydroxybenzoic acid.

However, there are three possible isomeric hydroxybenzoic acids: 2-hydroxybenzoic acid, 3-hydroxybenzoic acid and 4-hydroxybenzoic acid. A decision about which isomer salicylic acid is can be made by analysis of the **mass spectrum** of the compound.

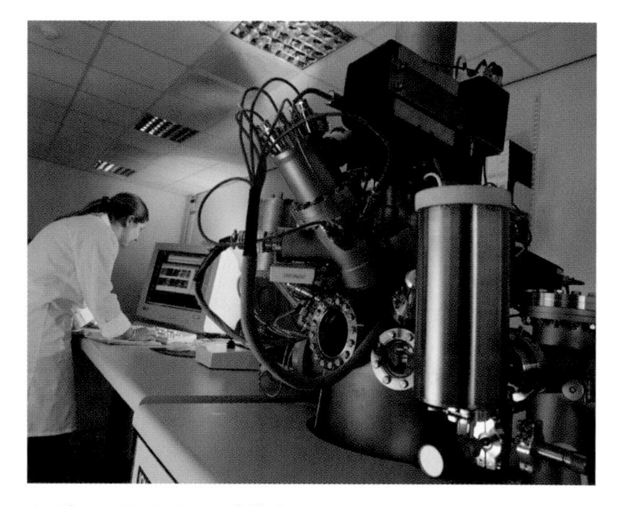

▲ **Figure 7** A time-of-flight mass spectrometer.

In **Chemical Ideas 2.1** you saw how information about the relative abundance of isotopes of an element can be obtained from a mass spectrum. **Chemical Ideas 6.5** tells you how mass spectrometry is used to find the structure of compounds.

USING MASS SPECTRA TO DETERMINE MOLECULAR STRUCTURES

A mass spectrum shows signals that correspond to positively charged ions formed from the parent compound, and also to **fragment ions** into which the parent compound has broken down.

Figure 8 shows the mass spectrum of salicylic acid. The signal at mass 138 is from the parent ion, called the **molecular ion**. Modern machines, such as the one in Figure 7, give very accurate mass values for the signals in the spectrum. For salicylic acid, the high-resolution spectrum (Figure 8a) gives a mass of 138.0317 for the molecular ion. This confirms that the substance has an empirical formula of $C_7H_6O_3$.

The way in which a parent ion breaks down – called its **fragmentation pattern** – is characteristic of that compound. In this case, comparison with a database of known mass spectra identifies salicylic acid as 2-hydroxybenzoic acid. Thinking about the way that the 3-hydroxybenzoic acid and the 4-hydroxybenzoic acid isomers would break down also leads to the conclusion that these isomers could not form some of the fragments observed in the mass spectrum of salicylic acid. For example, the signal at mass 120 could be formed only from breakdown of 2-hydroxybenzoic acid.

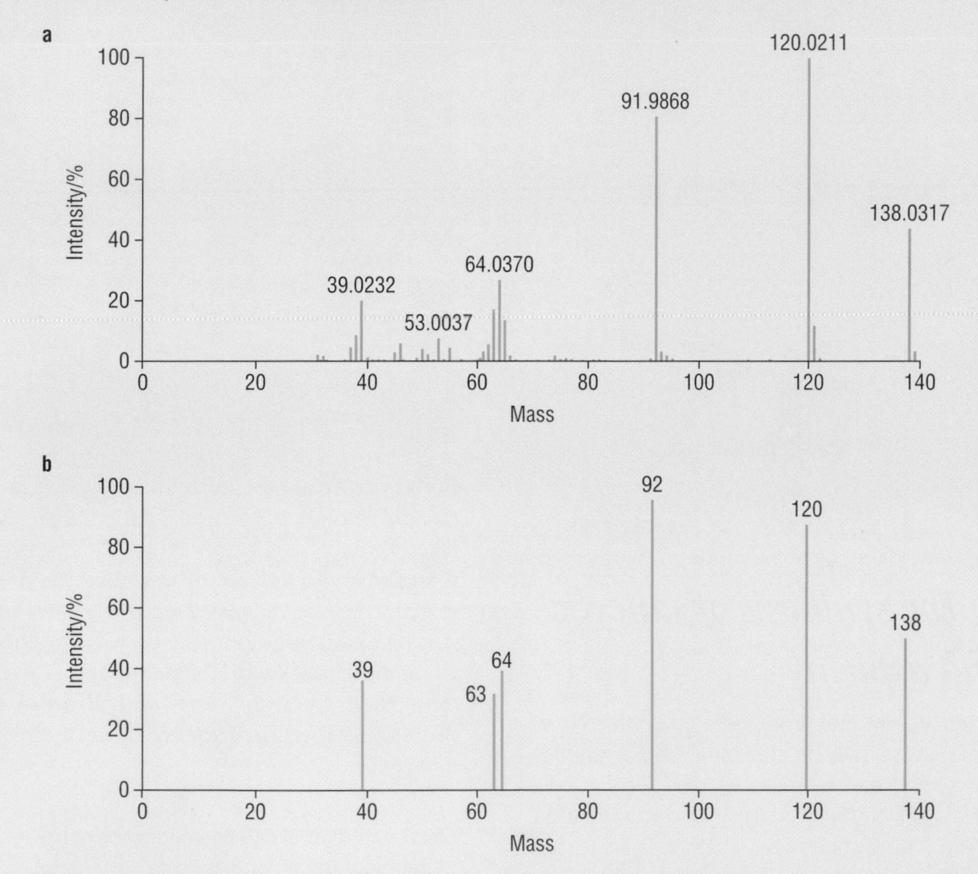

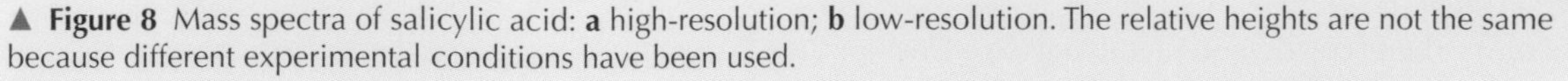

▲ **Figure 8** Mass spectra of salicylic acid: **a** high-resolution; **b** low-resolution. The relative heights are not the same because different experimental conditions have been used.

Activity WM4 uses both mass spectra and i.r. spectra to determine the structure of organic compounds. It shows how chemists can use fragmentation patterns in mass spectra to deduce or confirm a molecular structure.

Drawing the evidence together

- Chemical tests showed the presence of a phenolic –OH group and a carboxylic acid –COOH group.
- Infrared spectroscopy showed that the –OH and –C=O groups were certainly present.
- Nuclear magnetic resonance spectroscopy confirmed that there were hydrogen atoms in three types of environment: –COOH, –OH attached directly to a benzene ring and –H attached directly to a benzene ring.

- Mass spectrometry showed that salicylic acid was the same compound as that stored in the database as 2-hydroxybenzoic acid. The structure of 2-hydroxybenzoic acid is

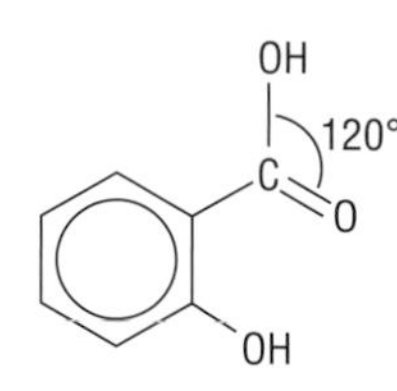

2-hydroxybenzoic acid

- The mass spectrum fragmentation pattern showed that the structure could not be

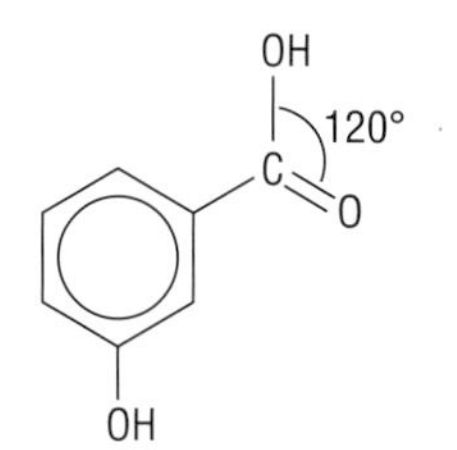

3-hydroxybenzoic acid

or

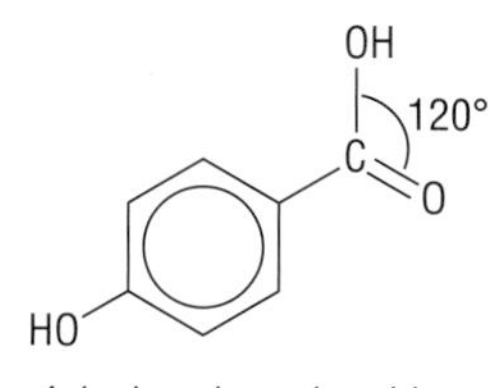

4-hydroxybenzoic acid

WM5 *The synthesis of salicylic acid and aspirin*

Medicines that are 'natural products' – those that come directly from plants – may be difficult to obtain when needed. The supply may be seasonal, may depend on weather conditions and may be liable to contamination. Collecting plants from their natural habitat is not environmentally acceptable.

Chemists, therefore, do not want to rely on willow trees as their source of 2-hydroxybenzoic acid. Once the chemical structure of the active compound in a plant is known, chemists can instead begin to search for ways of producing it artificially.

Simple inorganic substances, such as aluminium chloride, can be synthesised directly from their elements, but larger, more complex molecules cannot be made directly in this way. Instead, chemists search for a compound that is already known, which has a similar structure to the required compound that can be modified.

At the end of the nineteenth century, the compound phenol was already well known in the pharmaceutical industry – it has germicidal properties. It was also readily available as a product from heating coal in gas-works. Its molecular structure differs from that of 2-hydroxybenzoic acid by only one functional group. The problem in synthesis is to introduce this extra group in the right position – without disrupting the rest of the molecule.

Assignment 3

Compare the structural formulae of the starting and finishing compounds:

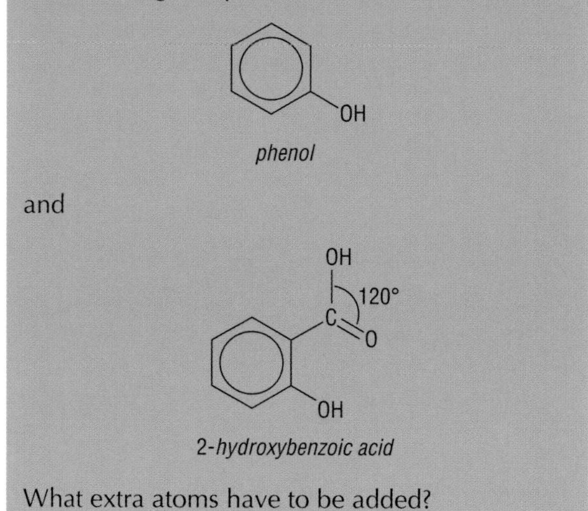

What extra atoms have to be added?

In this particular case, carbon dioxide can be combined directly with phenol to give 2-hydroxybenzoic acid by careful control of the conditions. This general method is known as the Kolbe synthesis (details can be found in many organic chemistry textbooks) and an industrial version of this addition reaction was developed by the German chemist Felix Hoffmann. This is an early example of 'green chemistry' with an **atom economy** of 100% because there are no leaving atoms or molecules.

You can read about atom economy in **Chemical Ideas 15.8**.

Thus, synthetic 2-hydroxybenzoic acid of reliable purity became available and was marketed by the chemical company Bayer.

Synthetic 2-hydroxybenzoic acid was widely used for curing fevers and suppressing pain, but reports began to accumulate of irritating effects on the mouth, gullet and stomach. Clearly the new wonder medicine had unpleasant side-effects. Chemists had a new problem – could they modify the structure to reduce the irritating effects, while still retaining the beneficial ones?

Hoffmann prepared a range of compounds by making slight modifications to the structure of 2-hydroxybenzoic acid. His father was a sufferer from

▲ **Figure 9** Meadowsweet (*Spiraea ulmaria*), from which salicylic acid was first extracted in 1835 – aspirin got its name from 'a' for acetyl (an older word for ethanoyl) and 'spirin' for *spirsaüre* (the German word for salicylic acid).

chronic rheumatism and Hoffmann tried out each of the new preparations on him to test its effects. This was a bit more primitive than the modern testing of medicines. It is not recorded what Hoffmann senior thought of all this, but he survived long enough for his son to prepare, in 1898, a derivative that was as effective as 2-hydroxybenzoic acid and had less unpleasant side effects.

The effective product was 2-ethanoxybenzoic acid (sometimes called 2-ethanoylhydroxybenzoic acid or acetylsalicylic acid) – this is now known as *aspirin*:

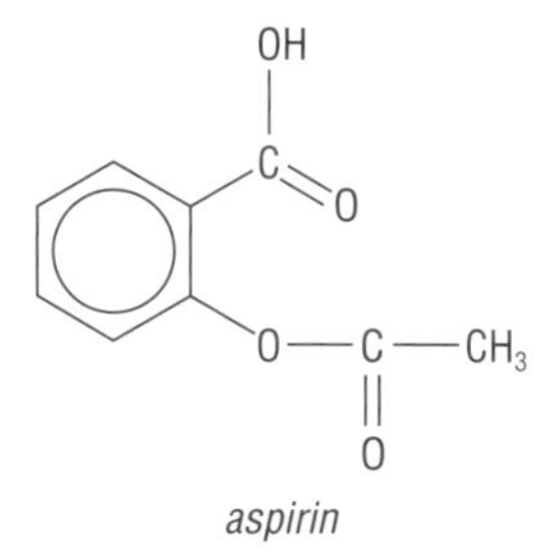

aspirin

The problem was that aspirin is not very soluble in water. It was first available as a powder in sachets – Bayer then decided to pellet the powder and aspirin became the first medicine to be sold as tablets.

Aspirin belongs to a class of compounds known as **esters**, and Hoffmann used the process of **esterification** to produce aspirin. **Chemical Ideas 13.5** introduces you to the structure of esters. You will find out more about esters and esterification later in the course in the **Materials Revolution** module.

In **Activity WM5.1** you can convert 2-hydroxybenzoic acid into aspirin.

The technique of thin-layer chromatography used in activity **WM5.1** is described in **Chemical Ideas 7.3** – as well as **Chemical Ideas Appendix 1: Experimental Techniques** (Technique 4).

In **Activity WM5.2** you can see how combinatorial chemistry can be used to produce a range of related esters quickly and in a convenient form.

In **Activity WM5.3** you can practise classifying organic reactions by types.

▲ **Figure 10** Felix Hoffmann – said to be the first person to synthesise aspirin in a chemically pure and stable form in 1897.

WM6 *Delivering the product*

Protecting the discovery

Developing a new medicine costs an enormous amount of money. The selling price charged by the pharmaceutical company must be sufficient not only to cover the costs of production and marketing, but also to recover the development costs. If other companies could simply copy the medicine, they would be able to sell it at a much lower price. This is where *patents* become important.

When a pharmaceutical company discovers a new medicine, it takes out patents to protect the discovery. Patents apply in only one country, so several patents must be taken out to prevent companies in other

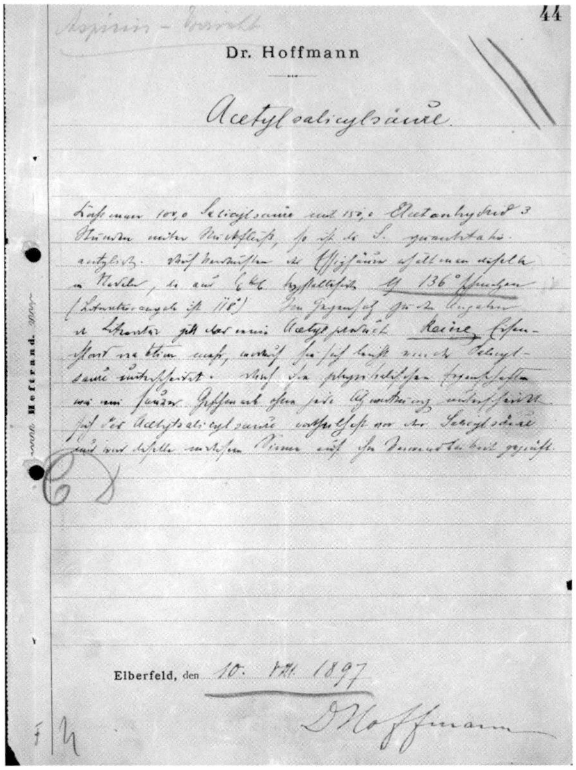

▲ **Figure 11** Felix Hoffmann's lab notes.

countries manufacturing the medicine. When the medicine is approved, the pharmaceutical company markets it under a trade name – or *brand name*. Patents last only for a specific amount of time, but while the patent is in force no other company can manufacture the medicine in that country.

Pharmaceuticals are usually complex compounds with long and unwieldy chemical names. For convenience they are known by shorter, trivial names – these are called 'generic' names in the pharmaceutical business. So most pharmaceuticals have three names:

- a chemical name
- a generic name
- a brand name.

An example is the compound 2-(4-(2-methylpropyl)-phenyl)propanoic acid, known by the generic name *ibuprofen* and marketed, for example, as Brufen (by Knoll) or simply supplied as Ibuprofen.

By the time a patent runs out, the company that discovered the medicine will hopefully have sold enough of it to cover its development costs. Afterwards, any company can produce and sell the medicine – usually under its own, new brand name. That's why there are so many ibuprofen tablets around.

When the Bayer company in Germany first decided to market 2-ethanoylhydroxybenzoic acid in 1899, it was not a new compound so the compound itself could not be protected by patents. However, the company

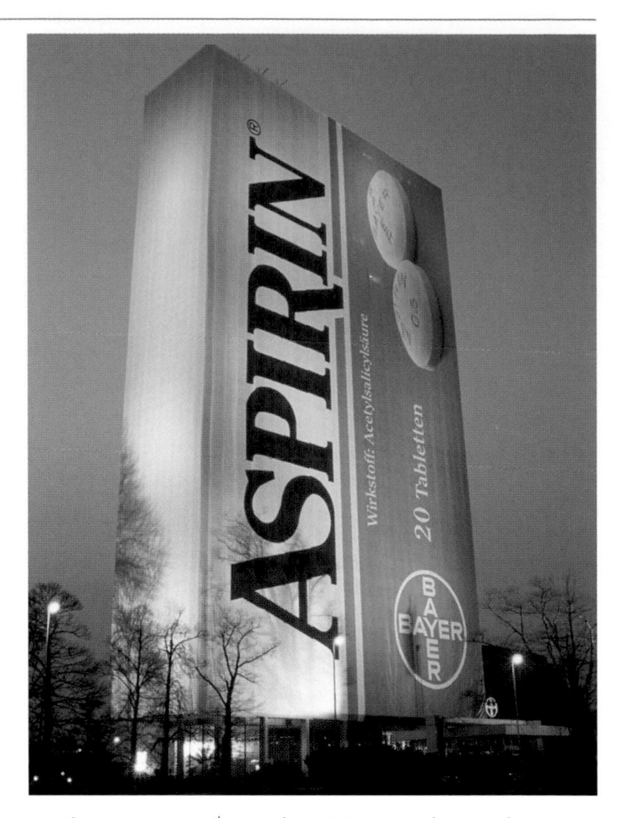

▲ **Figure 12** On 6th March 1999, 30 professional mountaineers draped the 120-metre-high building that is Bayer's headquarters in fabric made of a woven polyester coated with PVC using 32 zip fasteners. It was in celebration of the 100th anniversary of Bayer registering the trademark Aspirin®. *The Guinness Book of Records* has this as the biggest aspirin box in the world!

patented the process in which it was made and also sought copyright for the trade name *Aspirin* in as many countries as possible.

The trade protection lasted until the First World War, when other countries were no longer able to obtain aspirin from Germany. American firms were restrained by the patent agreements from producing it

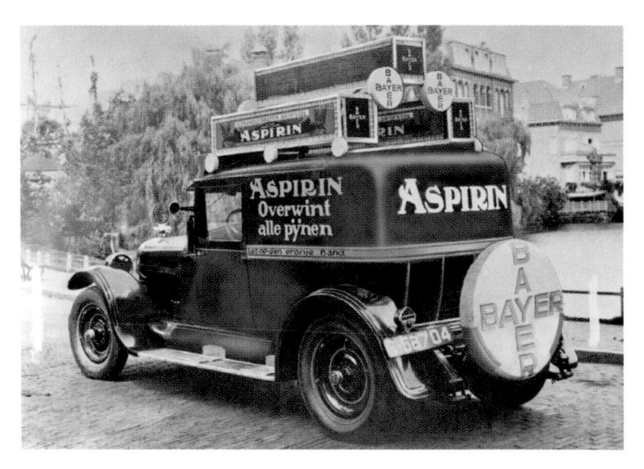

▲ **Figure 13** Early advertising for aspirin in the Netherlands – the slogan on the vehicle means 'Aspirin conquers every pain'.

and UK chemists were busy with the war effort, but an Australian pharmacist, George Nicholas, developed a process for producing a 'soluble' form of the medicine, and marketed this under the new trade name *Aspro*. As part of war reparations, the German rights to the trade name Aspirin were given up in the UK and Commonwealth countries.

Because there are no development costs, the new brands can be made more cheaply than the original one. Currently, the Government encourages doctors to prescribe some medicines by their generic name. The pharmacist is then free to dispense the cheapest brand – which helps to keep National Health Service costs down.

▲ **Figure 14** If all the aspirin tablets produced each year were laid end to end, they would make a path to the Moon and back.

Different ways of buying aspirin

There are some 200 analgesic (pain-relieving) formulations worldwide that contain aspirin. There are so many because:

- The medicine may come in different forms – e.g. solids, soluble substances and syrups.
- Other compounds may be present to help relieve other symptoms that occur along with the one being treated.
- There may be other substances present to help the action of the principal compound.
- For all these different formulations, and aspirin itself, there are many companies each producing their own brand-name equivalent.

Manufacturers need to analyse samples from each batch of a medicine to ensure that it has been properly blended and that each tablet contains exactly the stated amount of active ingredient.

More critically, because these medicines are so easily available and widely used, many households keep some permanently in stock. These are not always stored safely and are sometimes kept unlabelled.

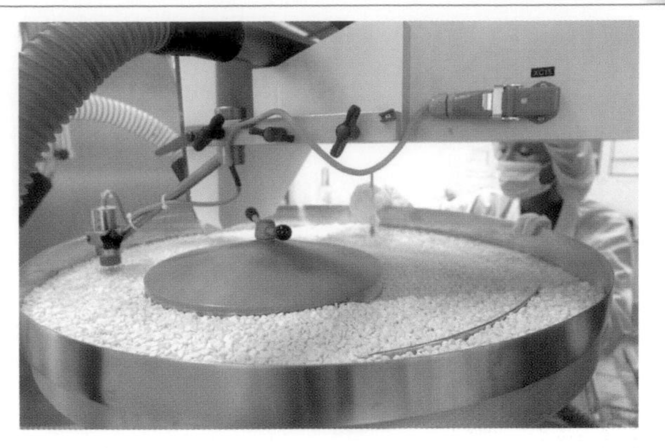

▲ **Figure 15** Tablet production today – batches of 6–10 tonnes can be processed at a time.

Each year many thousands of cases of accidental poisoning occur. Hospital analysts need techniques for establishing quickly which compounds – and how much of them – are present in tablets, so that the correct treatment can be given.

In **Activity WM6** you can perform an aspirin assay – in other words an experiment to find out the amount of aspirin in a medicine. This is an acid–base titration, as described in **Chemical Ideas Appendix 1: Experimental Techniques** (Technique 9).

Chemical Ideas 8.1 describes acids and bases in more detail.

The safety of aspirin

People tend to think of aspirin as a safe medicine because it is so familiar. But like all medicines, it is only safe if taken in the recommended dose. The lethal dose of aspirin is 30 g for an adult of average size. A typical aspirin tablet contains 0.3 g (300 mg) of aspirin, so 100 tablets could be a lethal dose – unpleasant symptoms would be experienced with far fewer tablets than this. The recommended dose of aspirin is no more than 12 tablets a day, and it is not recommended at all for children under 12 years old.

Aspirin has been prohibited from being sold in packets containing more than 16 tablets (each of 300 mg) since 1998, unless you consult a pharmacist.

WM7 *The miraculous medicine*

The life cycle of modern medicines is often short because medical and pharmaceutical research offers better remedies all the time. Aspirin is so cheap that it is generally bought by patients rather than through an NHS prescription. The drug has such miraculous properties that over 1 trillion tablets have been consumed since it

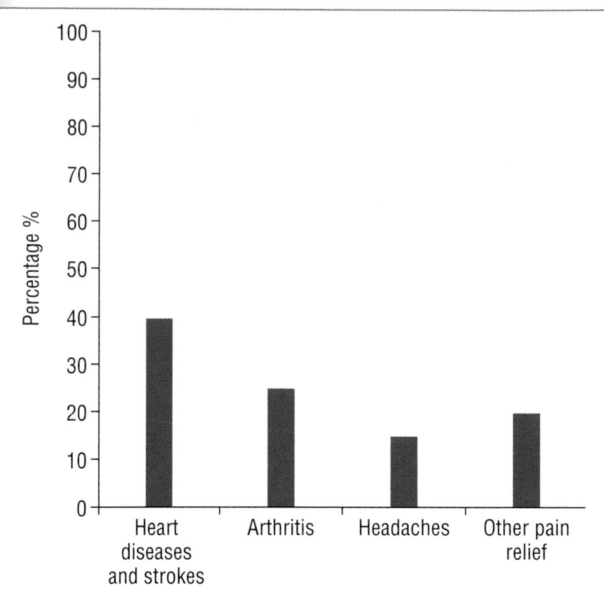

was first launched on the market. Every year about 50 billion tablets are consumed in the UK.

Aspirin is now over 100 years old and new uses continue to emerge. For example, the discovery of its potential in treating heart disease came from observations by doctors on their patients. By careful analysis, Dr Laurence Craven in the US noticed that his male patients who took aspirin suffered fewer heart attacks. Similar work is being carried out to observe its effects on certain cancers, diabetes, deep vein thrombosis and Alzheimer's disease, among others. This research is being carried out by doctors and universities using statistical analysis of medical records, and not by pharmaceutical companies, who have nothing to gain by it.

◀ **Figure 16** Current uses of aspirin.

SIR JOHN VANE

Sir John Vane, of the Wellcome Research Laboratories in Beckenham, shared the Nobel Prize in Medicine in 1982 for his work on a group of hormones called *prostaglandins*.

Prostaglandins are formed in the body from unsaturated carboxylic acids by the action of COX enzymes. They are released from tissue when it suffers disease or stress in order to maintain its normal function, and so they act to defend the cells against sudden change.

Different prostaglandins help to control reactions in different parts of the body. For example, one type of prostaglandin is involved in the prevention of peptic ulcers in the stomach lining, whereas another keeps the kidneys healthy. However, the presence of prostaglandins can cause inflammation, fever and pain and this can be relieved by aspirin. Sir John's work helped to show that aspirin reduces the production of prostaglandins and is thus able to act as a painkiller.

Another prostaglandin induces thickening of the blood. Aspirin greatly reduces the 'stickiness' of the platelets in the blood so that they do not clump together and form clots. It is now the most effective drug in the treatment of strokes. In fact, aspirin has an effect wherever prostaglandins, whether beneficial or not, are being produced. There are different types of COX enzymes and pharmaceutical companies are now producing drugs that are specific for only one type. This should remove one of the major side effects of aspirin – long-term use is thought to cause damage to the stomach lining.

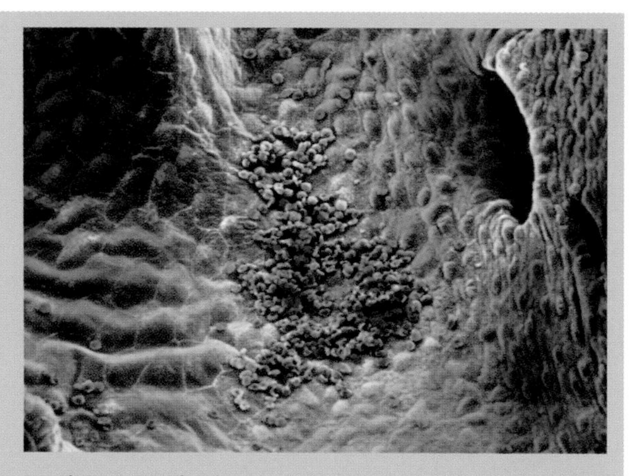

▲ **Figure 17** Platelets sticking together in the early stages of blood clot formation (magnification about ×550).

▲ **Figure 18** Sir John Vane.

Sir John has written about his start in chemistry:

'At the age of 12, my parents gave me a chemistry set for Christmas and experimentation soon became a consuming passion in my life. At first, I was able to use a Bunsen burner attached to my mother's gas stove, but the use of the kitchen as a laboratory came to an abrupt end when a minor explosion involving hydrogen sulfide spattered the newly painted decor and changed the colour from blue to dirty green!

'Shortly afterwards, my father, who ran a small company making portable buildings, erected a wooden shed for me in the garden, fitted with bench, gas and water. This became my first real laboratory, and my chemical experimentation rapidly expanded into new fields.'

WM8 *Development and safety testing of medicines*

A great deal of time, money and effort is spent by the pharmaceutical industry in discovering and developing new medicines. The following account comes from *An A to Z of British Medicines Research*, published by the Association of the British Pharmaceutical Industry (ABPI). When developing potential new drugs, the following questions (among others) need answering:

- Is the new product safe?
- Does the new product work?
- Is the new product better than the standard existing treatment?

There follows a summary of the essential stages in drug development. This process is often costly and time-consuming, but is necessary to ensure that the right drugs are developed.

THE STAGES IN THE MEDICINES RESEARCH AND DEVELOPMENT PROCESS

Terms such as 'discovery research', 'development' and 'phase I, II and III clinical trials' are used by the pharmaceutical industry. It is important that you understand these terms, so you are not misled into believing that progress is more advanced than it really is. The timescale for developing a new medicine is surprisingly long and the average for all medicines is about 15 years. The many stages involved in developing a conventional medicine are shown on page 14. However, the procedures for biological products (cytokines, growth factors, gene therapy, etc.) differ in a number of respects and are covered by their own requirements and regulations.

● *Discovery research* – steps 1–6 in Figure 19

Discovery research relates to the activities of chemists, biologists and pharmacologists who extract, synthesise and test new molecules. For every new medicine that reaches the patient, many thousands of other molecules fall by the wayside. A compound having the desired activity is identified. This is called a *lead* compound. Once the useful activity for a molecule has been identified, chemists optimise its structure by making many close variations (called *analogues*), to try to maximise the desired effects. Combinatorial chemistry is a common method in the synthesis of new molecules and analogues. (See **Activity WM5.2**.)

Once this is done, the improved molecule enters the development stage.

● *Development research* – steps 7–8 in Figure 19

Before a new potential medicine can be given to humans, much work has to be done to determine if it is acceptably safe, if it is sufficiently stable and how it is likely to be absorbed and excreted by the body. It is also necessary to prepare a dosage form that suits specific medical needs – such as injection, capsule, a tablet, aerosol or suppository.

Once all this has been completed, a long and complex process of clinical studies begins. These are generally divided into three distinct phases.

● *Phase I trials* – step 9 in Figure 19

Phase I trials are the first time the new substance is administered to humans, usually in studies of healthy, informed volunteers conducted under the close supervision of a qualified doctor. The purpose is to determine if the new compound is tolerated and behaves in the way predicted by all the previous experimental investigations. Initial doses will be the lowest possible consistent with obtaining the required information, but may gradually be raised to the expected therapeutic dose level. If the compounds under investigation are particularly powerful, as in cancer treatment for example, it may be that people who actually have the condition will take part in these trials.

Once the data from volunteers are available, an application has to be made to the Medicines and Health Care Products Regulatory Agency (MHRA) for a

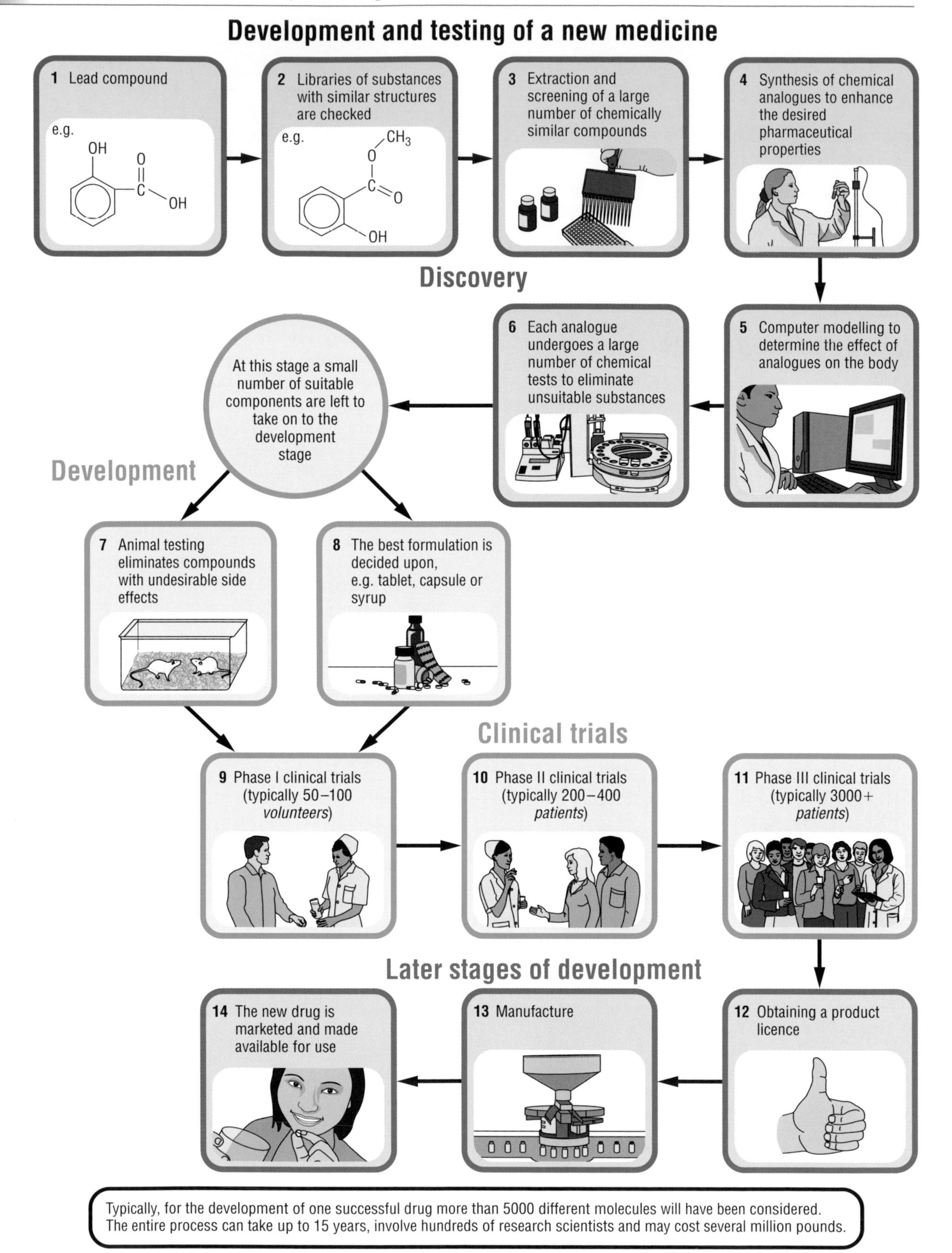

▲ **Figure 19** Development and testing of a new medicine.

certificate to conduct clinical trials. The information is reviewed by independent medical and scientific experts, who make their recommendation on whether further trials can start or whether more information is required. If a certificate is granted, a new medicine will then pass through two further phases of clinical trials before the company can seek a licence for its more widespread use.

- *Phase II trials* – step 10 in Figure 19

Phase II trials are the first in which the illness is actually treated. Different dose levels may be given to different patient groups to establish whether the compound is suitable for further study or should be abandoned. Patient numbers in these trials are usually small (typically 200–400 patients).

- *Phase III trials* – step 11 in Figure 19

Phase III trials only follow encouraging results from the Phase II study. As this stage, the new medicine is likely to be compared with a 'dummy' medication, called a *placebo*, and possibly with another medicine already used for the disease under investigation, to provide a reference standard. Patients are allocated randomly to one of the groups and neither doctor nor patient knows which preparation is being given during the trials. When the code is broken, a positive result would be indicated by an improvement in those patients who received the real medication compared with those on a placebo. Phase III trials usually involve much larger patient groups so that the results can be analysed statistically (typically 3000+ patients). If the medicine proves successful and well tolerated at this stage, the way is open for a product licence application to be made that includes all aspects of the data generated on the new medicine and runs to many volumes.

- *Later stages of development* – steps 12–14 in Figure 19

Of 10 to 15 compounds reaching Phase I studies, only one is likely to survive through to licensing. Also, the timescales for the above studies are very variable. For example, if a new compound is an antibiotic for urinary tract infections, a positive result will be apparent in each patient within a few days as the infection is eradicated. However, for chronic diseases, such as multiple sclerosis, AIDS, arthritis or some forms of cancer, the trial may last for more than a year and involve long-term follow-up to ensure that any observed positive effects are of lasting value.

Development of aspirin

Note that the medicine you have been looking at closely in this module, aspirin, did not go through all the safety testing described here, because it was developed long before all these safety procedures were established. But aspirin has been known about for so long and so widely that its use is accepted by most people – provided the recommended dosage isn't exceeded.

Which medicine to develop?

You have seen how long the process takes, and therefore how expensive it is, to develop a medicine from discovery right through to marketing. A firm will have wasted millions of pounds if it produces a medicine that does not meet a perceived need by the medical professionals or the public. Very careful decisions have to be made at several stages to ensure that the medicine is both safe and commercially viable.

Activity WM8 gives you a chance to discuss the steps involved in developing and safety testing a new medicine.

▲ **Figure 20** Pharmaceutical research has to be carried out under conditions of safety and cleanliness.

WM9 *Summary*

In this module you have learned about some of the chemistry associated with the pharmaceutical industry using one familiar and important medicine – aspirin – as an example. You saw how the analytical techniques of mass spectrometry and infrared spectroscopy allow us to identify the compound that is responsible for its pharmacological activity.

You were introduced briefly to another analytical technique, called nuclear magnetic resonance spectroscopy, which is also useful in determining the structure of molecules. You will find out more about this technique in the **Medicines by Design** module.

A study of aspirin led you to find out about a new series of organic compounds called esters. You saw how esters can be formed in the reaction of an alcohol with a carboxylic acid. Esters can also be made from phenols. This led to a more general study of compounds that contain the –OH group.

Activity WM9.1 will help you to check your knowledge and understanding of the organic names and structures met in this module.

This module also introduced you to a new experimental technique – you learned how to use thin-layer chromatography to identify the components of a mixture. You also used acid–base chemistry to assay a compound.

Knowledge of the chemical reactions of organic functional groups gives us the power to construct molecules of compounds that are essential for our well-being from readily available starting materials. You experienced something of the scale, complexity and costs involved in the production of a medicine for mass use.

Activity WM9.2 will help you to check your knowledge and understanding of this module.

MATERIALS REVOLUTION

Why a module on 'Materials Revolution'?

The main theme of this module is the development of materials with the specific properties to meet particular needs. The module picks up the thread from **Polymer Revolution** and continues the story of polymers and polymerisation – this time concentrating on condensation polymers.

An understanding of the relationship between the properties of materials and their structure and bonding is the key to designing new materials. This will require you to consider the factors that affect the properties of materials in more detail and will allow you to revise earlier work on bonding, structure and intermolecular bonds.

During the module you will revisit and extend some of the organic chemistry you met in earlier modules – the chemistry of alcohols, carboxylic acids and esters – as well as learning about two new series of compounds – amines and amides.

Disposing of polymers in landfill sites is no longer acceptable on environmental and economic grounds. Three possible solutions involve recycling, incineration or the use of degradable polymers. The advantages and disadvantages of each are discussed.

Overview of chemical principles

In this module you will learn more about ideas introduced in earlier modules in this course:

- polymers and polymerisation (**Polymer Revolution**)
- alcohols (**Developing Fuels** and **What's in a Medicine?**)
- carboxylic acids and esters (**What's in a Medicine?**)
- intermolecular bonds (**Polymer Revolution**)
- the relationship between structure and bonding and properties (**Polymer Revolution**).

You will also learn new ideas about:

- condensation polymerisation
- reactions of amines and amides
- the use of acyl chlorides to make amides
- the effect of temperature changes on polymers
- disposal of polymers.

MR

MATERIALS REVOLUTION

MR1 *Designing materials*

When people first started to make tools, clothing and other articles to help them to live more comfortably, they used naturally occurring materials. The material they chose for a particular item took advantage of that material's natural properties. For example, wood was used to make bowls, spoons and handles for tools because it is easily shaped and strong. Wool was used to make clothing because it is a good thermal insulator and it can be spun into threads for weaving or knitting into a fabric. The range of items that could be made, and their usefulness, was limited by the fact that the natural materials that were being used to make them did not necessarily have the ideal properties for the object that was being made.

As our understanding of methods for making things developed, and our ability to investigate materials on a molecular level improved, new developments allowed people to produce goods with improved properties. This area of chemistry overlaps with other science and engineering disciplines and is called materials science. It includes the science of how materials behave, coupled with improving the performance of those materials. Changes in the way we can make metallic products provide a good example of the developments in materials that have been made throughout history. Around 3500 years ago people discovered how to make alloys, such as bronze – an alloy of copper and tin.

Bronze tools were made, replacing stone tools that quickly became blunt. Metal tools made woodworking on a large scale possible, allowing people to produce larger buildings and boats. Bronze has greater strength and hardness than either copper or tin alone. This is because a pure metal is made from atoms of equal sizes, which pack together in a regular arrangement of rows. When a force is applied to the metal, the rows of atoms can slide over each other, allowing the shape to be distorted. In an alloy, atoms of a second metal are introduced and these are of a different size (represented as blue in Figure 2c). These atoms reduce slip between layers, as shown in Figure 2.

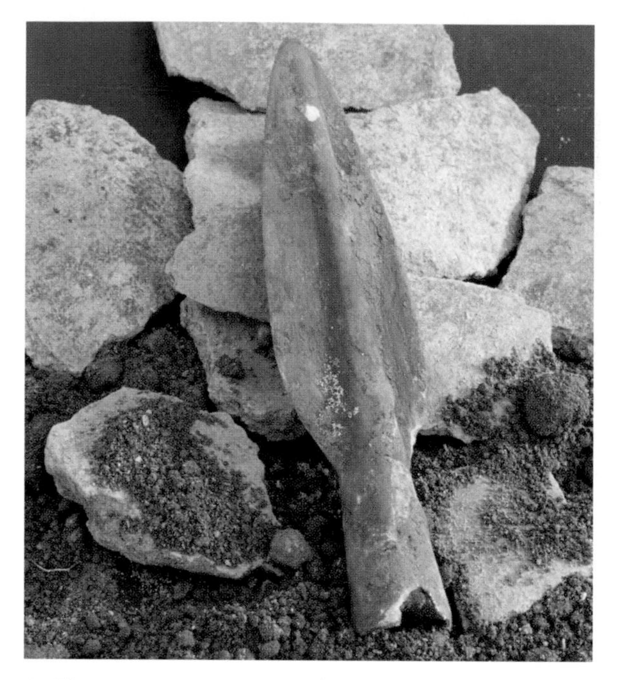

▲ **Figure 1** Bronze-age arrowhead found during excavations at a settlement dating from around 1500 BC, near present-day Naples, Italy.

Because of the greater strength and hardness of bronze over its constituent metals, it can be used for making objects such as gear wheels, propellers and bearings, which require a material that will not wear down. Nowadays we can make even more versatile alloys called superalloys, many of which were originally developed in the aerospace industry. Superalloys have

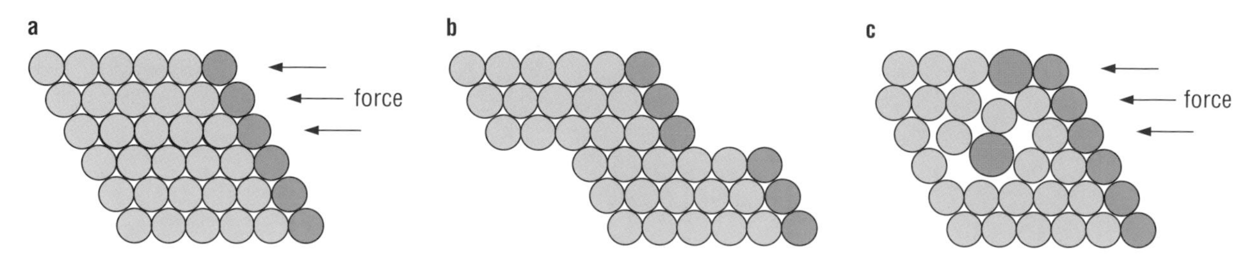

▲ **Figure 2** Diagrams **a** and **b** show atoms in a metal, with **a** being before and **b** after slip has taken place. The dark orange circles represent the end of a row of atoms. Diagram **c** shows the arrangement of atoms in an alloy. The blue circles are the larger atoms of a metal added to make an alloy.

excellent strength and corrosion resistance, even at elevated temperatures, and can be used to make goods that take advantage of these properties, such as turbine blades for jet engines (Figure 3).

It is also possible to make shape memory metals. These are metals that can be deformed, but then return to their original shape again. Shape memory metals such as Nitinol (an alloy containing nickel and titanium) can be used to make spectacle frames, braces for teeth and stents (see Figure 4).

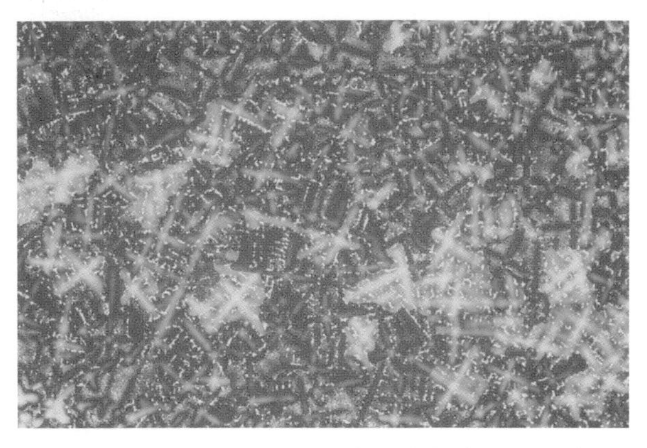

Figure 3 Light micrograph of a polished section of a cobalt and chromium-based 'superalloy' used in the construction of turbine blades for jet engines.

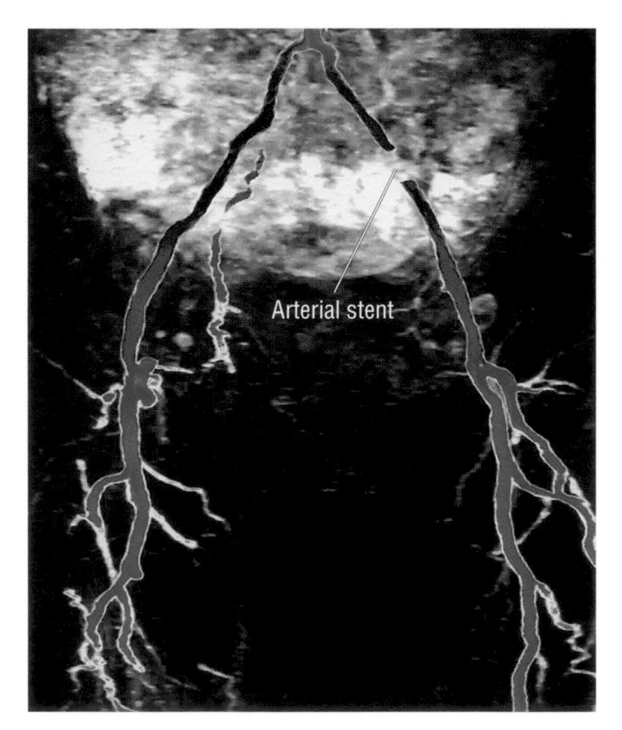

Figure 4 MRI scan of an arterial stent, made of Nitinol, in an artery in a patient's groin. This operation widened the artery from its previously blocked state, restoring the blood flow.

MR2 *Making and breaking polymers*

Making polymers

In the **Polymer Revolution** module you found out about polymers made from unsaturated molecules by addition reactions (where the monomer molecules join together to form the polymer as the only product). In this module you will be introduced to a range of polymers made by **condensation polymerisation**. This is a process in which a large number of monomer molecules react to form a long-chain polymer and a second product consisting of small molecules – this is often water or hydrogen chloride. Nylons and polyesters are examples of condensation polymers.

Nylons

Wallace Carothers joined the US chemical company DuPont in 1928. He led a team investigating the production of polymers that might be used as fibres. This was at a time when scientists were beginning to understand more about the structure of polymers, so Carothers had a scientific basis for his work.

Before you read this section, it will be helpful for you to revise your previous work on carboxylic acids in **Chemical Ideas 13.3** and **13.4**.

You will also need to know about **amines** and **amides**, two important families of nitrogen-containing organic compounds – you can read about these in **Chemical Ideas 13.8**.

It was already known that wool and silk have protein structures, and are polymers involving the **peptide linkage** –CONH–. Chemists had also begun to discover that many natural fibres are composed of molecules that are very long and narrow – like the fibres themselves. Figure 5 (page 20) shows the structure of part of a protein chain in a silk fibre.

Carothers did not make his discoveries by accident – he set about systematically trying to create new polymers. In one series of experiments he decided to try to make synthetic polymers in which the polymer molecules were built up in a similar way to the protein chains in silk and wool. Instead of using amino acids (the starting materials for proteins), Carothers began with **amines** and **carboxylic acids**.

Amines are organic compounds that contain the $-NH_2$ functional group. When an $-NH_2$ group reacts with the –COOH group in a carboxylic acid, an **amide** group –CONH– is formed. A molecule of water is eliminated in the process – this is an example of a *condensation* reaction.

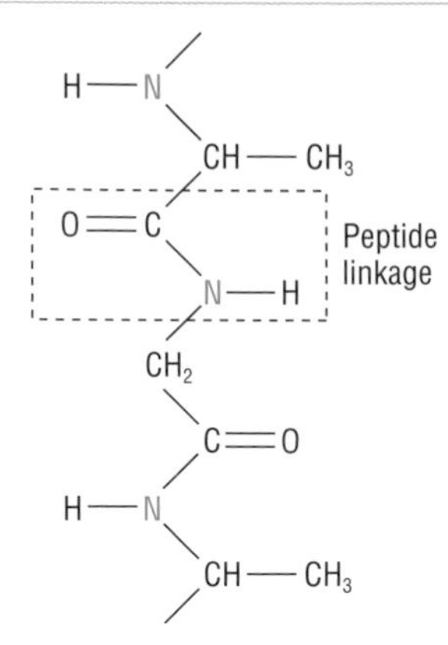

Figure 5 Part of a protein chain in silk.

Carothers used *di*amines and *di*carboxylic acids that contained reactive groups in *two* places in their molecules, so they could link together to form a chain. In this way he was able to make polymers in which monomer units were linked together by amide groups. The process is called **condensation polymerisation** because the individual steps are condensation reactions (Figure 6).

Examples of a diamine and a dicarboxylic acid that can be made to polymerise in this way are

$$H_2NCH_2CH_2CH_2\ CH_2CH_2CH_2NH_2$$
1,6-diaminohexane

and

$$HOOCCH_2CH_2CH_2CH_2COOH$$
hexanedioic acid

Because the group linking the monomer groups together is an amide group, these polymers are called **polyamides**. More commonly they are known as **nylons**.

The –CONH– group is also found in proteins where it joins amino acids together. You can see this in the structure of the silk protein chain in Figure 5. In this case, the secondary amide group is given the special name **peptide group**.

The industrial preparation of nylon from a diamine and a dicarboxylic acid is quite slow. It is easier to demonstrate the process in the laboratory if an **acyl chloride** derivative of the acid is used:

1,6-diaminohexane and decanedioyl dichloride react readily and the equation is

$$nH_2N(CH_2)_6NH_2 + nClCO(CH_2)_8COCl \rightarrow$$
1,6diaminohexane *decanedioyl dichloride*

$$-(NH(CH_2)_6\text{–}NH\text{–}CO\text{–}(CH_2)_8CO)_n^- + 2nHCl$$

Carothers discovered nylon in the spring of 1935 and by 1938 the first product using nylon, 'Dr West's Miracle Toothbrush', appeared. Nylon stockings were seen for the first time in 1939 (Figure 7). Most of the nylon produced at this time was used instead of silk for parachute material (Figure 8), so nylon stockings did not become generally available in Britain until the end of the Second World War.

Figure 7 Advertising nylon stockings in Los Angeles when the polymer was first being used, 1939–40.

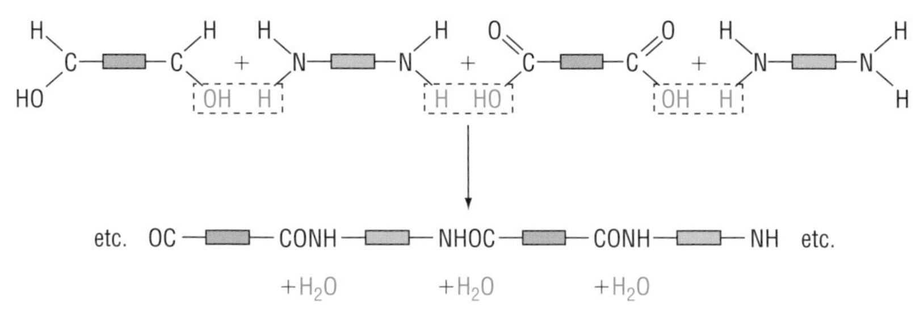

Figure 6 Formation of a polyamide.

▲ **Figure 8** Production of nylon for stockings took second place to its use for parachutes in the Second World War.

NAMING NYLONS

A nylon is named according to the number of carbon atoms in the monomers. If two monomers are used, then the first digit indicates the number of carbon atoms in the diamine and the second digit indicates the number of carbon atoms in the acid. So nylon-6,6 is made from 1,6-diaminohexane and hexanedioic acid. Nylon-6,10 is made from 1,6-diaminohexane and decanedioic acid:

1,6-diamino*hexane* + *hex*anedioic acid – nylon-6,6
1,6-diamino*hexane* + *dec*anedioic acid – nylon-6,10

It is also possible to make nylon from a single monomer containing an amine group at one end and an acid group at the other. For example, nylon-6 is $-(NH-(CH_2)_5-CO)_n-$ and is made from molecules of $H_2N(CH_2)_5COOH$.

Assignment 1

a Name the nylons that contain the following repeating units:
 i $-HN-(CH_2)_6-NHCO-(CH_2)_6-CO-$
 ii $-HN-(CH_2)_9-NHCO-(CH_2)_7-CO-$
 iii $-HN-(CH_2)_4-NHCO-(CH_2)_2-CO-$

b Write out the repeating units and give the names for the polymers formed from the following molecules:
 i $HOOC(CH_2)_5NH_2$
 ii $H_2N(CH_2)_5NH_2$ and $HOOC(CH_2)_5COOH$.

c Nylon-5,10 can be formed by the reaction of a diamine with a diacyl dichloride.
 i Write down the structures of the two monomers.
 ii What small molecule is lost when the monomers react?
 iii Draw the structure of the repeating unit in nylon-5,10.

In **Activity MR2.1** you can match the names, monomers and repeat units of some nylons.

If you did not make a nylon in an earlier chemistry or science course, **Activity MR2.2** allows you to make some nylon-6,10.

In **Activity MR2.3** you can take some nylon molecules apart again.

Hard choices – right decisions

The development of polyamides, with their potential to replace silk as well as to open up entirely new markets, required some hard choices. Many different types of nylon had been prepared. Wallace Carothers favoured nylon-5,10, because the material was easy to process and would be appropriate for detailed studies of the relationships between polymer structures and properties. A principal drawback would have been the high cost of the decanedioic acid needed and this could have limited the widespread use of the polymer.

Elmer Bolton, the director of chemical research at DuPont, held a different view. He believed that a better choice would be nylon-6,6. Its physical properties would allow more applications than nylon-5,10. In addition, the starting materials have six carbon atoms each (1,6-diaminohexane and hexanedioic acid) and they could both be made from readily available and cheap chemicals – for example, benzene. Even so, a synthesis of the diamine from benzene had first to be developed, as did techniques to process the polymer, which has a higher melting point than nylon-5,10.

Despite these initial difficulties, Bolton foresaw greater commercial possibilities for nylon-6,6. The success of nylon depended on the combination of Carother's pioneering research and Bolton's brilliant ideas about its development. This coming together of basic science and applied technology must occur in industry if a product is to be successful.

Nylon machine parts

You know that nylon is widely used as a fibre, but it is also a very important *engineering plastic* – a material that can be used in place of a metal in things like machine parts.

Its usefulness arises from its excellent combination of strength, toughness, rigidity and abrasion resistance, as well as its lack of chemical reactivity in many environments. As an engineering material, it is far superior to poly(ethene) or poly(propene). The polymer chains in nylon need only be about half as long as HDPE chains to show the same strength. The more powerful intermolecular bonds that act between nylon chains are the source of this increased strength.

In nylon there is hydrogen bonding between adjacent polymer chains. In poly(ethene) and poly(propene) there are only the weaker, instantaneous dipole–induced dipole attractive bonds between adjacent chains.

At this point it will be helpful to look again at **Chemical Ideas 5.3** and **5.4** to check your understanding of the intermolecular bonds responsible for nylon's strength. You might also like to revisit **Chemical Ideas 3.1** (the section on bond polarity and electronegativities).

In and out of fashion

Towards the end of the 1970s, people's tastes were moving away from nylon clothes and back towards the look and softer feel of natural fibres.

One of the problems with nylon fibres is that they are *hydrophobic* – they repel water. The nylon fabrics produced did not absorb moisture, and did not allow water vapour to escape through the weave. This made them rather sweaty and uncomfortable to wear.

Chemical companies were facing a big downturn in the demand for their nylon, bringing major financial losses, but the high cost of developing a completely new polymer on the scale necessary to replace nylon was too high. To make things worse, the machines that had been specifically designed for making nylon fabrics could not be used for cotton or other natural fibres.

ICI's answer to this problem was to redevelop their nylon so that it bore a much closer resemblance to natural fibres. The first steps were to slim the thickness of the nylon filament to the equivalent of cotton, and to add a delustrant to the fibre to reduce the shiny appearance. The major breakthrough came when they developed a process for changing the shape and texture of the nylon yarn. They were able to create a large number of loops along the nylon filaments by blowing bundles of them apart with high-pressure air. The new fibre was called Tactel – when the fibre is woven, the loops give the material a softness and texture similar to cotton.

Further refinement of the Tactel family of yarns has led to fabrics that are waterproof but that 'breathe'. For example, very fine yarns have been developed that allow water *vapour* to escape through the weave but do not allow *liquid* water in. Material made from these yarns is ideal for lightweight raincoats and ski-wear.

The solution to the problem of falling demand for nylon was not chemical – it did not involve the creation of a new polymer to suit the new fashion – but *technological*. It involved discovering new ways of handling the existing polymer to produce materials of the type people wanted.

Nylon molecules are made from amines – you can find out more about the chemistry of amines in **Activity MR2.4**.

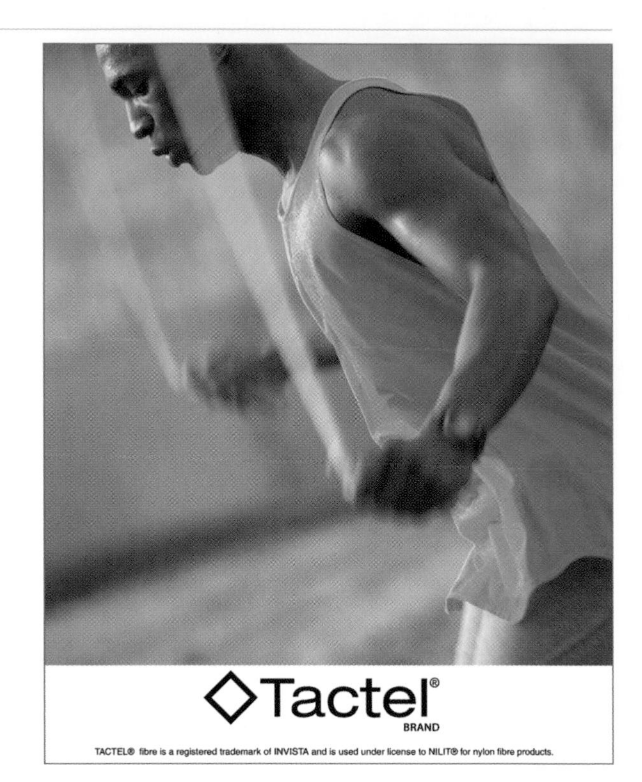

Figure 9 Tactel fabrics 'breathe' but remain waterproof.

Polyesters

The development of **polyesters** follows a similar story to that of nylon. A brilliant idea of UK chemist Rex Whinfield, in the 1940s, led to an incredible journey developing materials undreamt of by their originators.

Whinfield knew that Carothers had tried to produce polyesters, but felt that Carothers had not used the acid most likely to give the desired properties for the polymer. Whinfield used 1,4-benzenedicarboxylic acid (terephthalic acid) and reacted it with ethane-1,2-diol. The structure of the polyester formed is

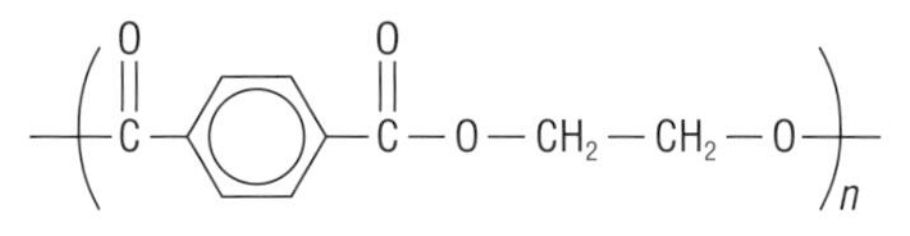

The 'old' name for this polyester is polyethylene terephthalate, often known as PET.

You can revise the formation and reactions of esters in **Chemical Ideas 13.5**.

The polyester is produced as small granules – these are melted and squeezed through fine holes (see Figure 10). The resulting filaments are spun to form a continuous fibre. The fibre, known as Terylene or Dacron, is widely used in making clothes – such as suits, shirts and skirts – and for filling anoraks and duvets because it gives good heat insulation. The polyester can also be made into sheets for X-ray films.

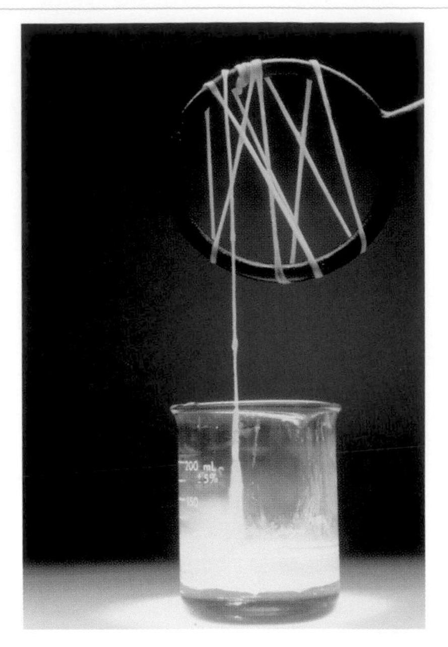

▲ **Figure 10** Birth of a continuous fibre of nylon.

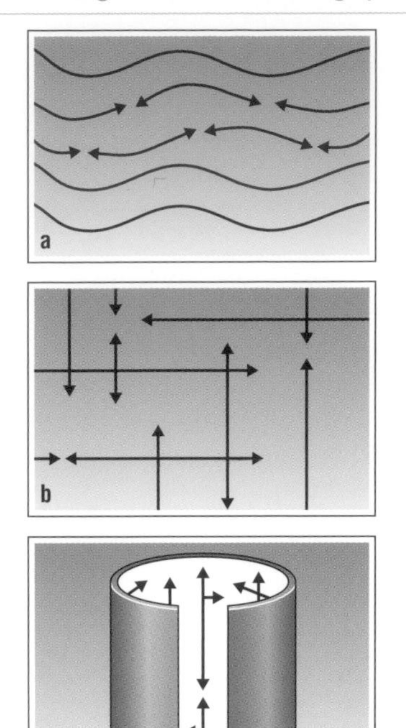

▲ **Figure 12** **a** Yarn – the molecules are mostly in one dimension (Terylene/Dacron); **b** Film – the molecules are aligned in two dimensions; **c** Bottles – the molecules are aligned in three dimensions.

A newer use of PET is for packaging. The granules of the polyester are heated to about 240 °C and further polymerisation takes place – a process known as *curing*. When the polymer is then stretched and moulded, the molecules are orientated in three dimensions. The plastic has great strength and is impermeable to gases – no wonder PET is widely used for bottling carbonated drinks (Figure 11).

The versatility of the polymer arises because the molecules can be aligned in one, two or three dimensions (Figure 12), giving rise to different properties.

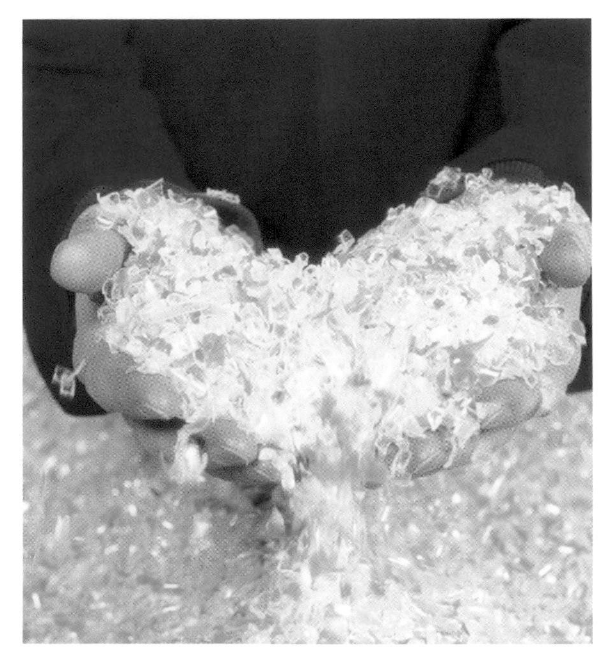

▲ **Figure 11** Granules of PET that are heated and moulded into bottles.

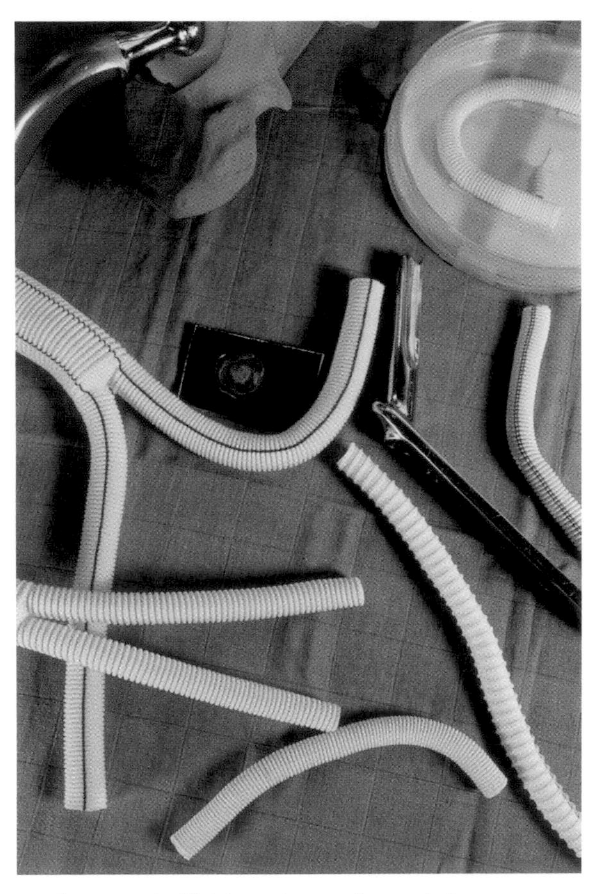

▲ **Figure 13** Artificial arteries can be made from polyester and nylon.

Assignment 2

a Draw out full structural formulae for the two monomers used to make the polyester PET.

b What types of intermolecular bonds will exist between the polymer chains in PET?

Breaking polymers

Disappearing in the body

Since the 1980s, threads made from a special polyester have been used by surgeons to stitch together the sides of wounds. What is special about the thread is that it 'disappears' as the wound heals.

The polyester is made from monomers that contain both a hydroxyl group (–OH) and a carboxylic acid group (–COOH), so only one type of monomer is needed.

The monomer used may be lactic acid (2-hydroxypropanoic acid) or glycolic acid (2-hydroxyethanoic acid) or a mixture of the two.

The polymers form strong threads, but water in the body slowly hydrolyses the ester bonds. The products of the hydrolysis (lactic acid or glycolic acid) are non-toxic. Indeed, lactic acid is a normal breakdown product of glucose in the body.

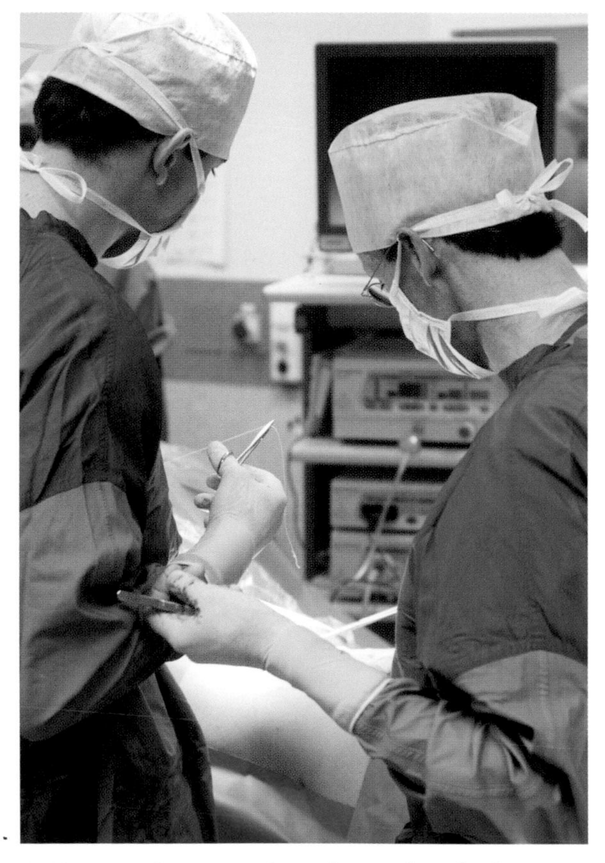

▲ **Figure 14** Surgeons using polyester thread – this hydrolyses slowly in the body to harmless compounds after the wound has healed.

Assignment 3

In **Polymer Revolution** you met a dissolving polymer, poly(ethenol), that is used to make laundry bags in hospitals. It can be used to make soluble threads for surgical stitches. Polyester thread made from lactic acid is also used for stitches.

Compare the mechanism by which the two polymers dissolve, and suggest why the polyester is more suited to use inside the body.

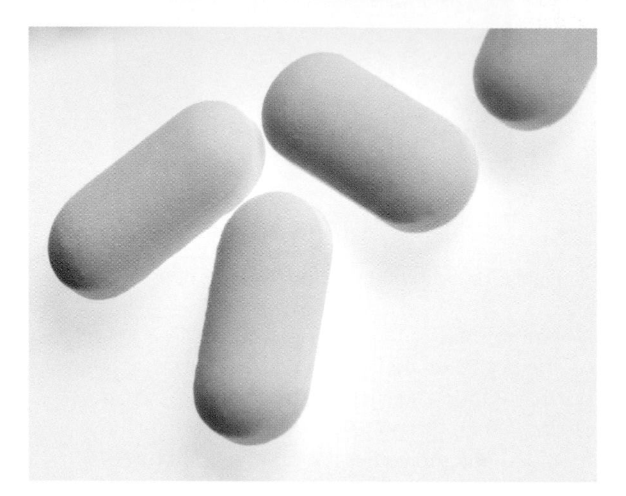

▲ **Figure 15** Tablets have coatings that dissolve slowly. For example, some aspirin tablets contain fine crystals of aspirin coated in a polymer (ethyl cellulose). This allows the aspirin to pass through the stomach and into the gut, where it is released.

Another use for the degradable polyesters is the controlled delivery of a medicine. The medicine is dispersed throughout a tablet of the polymer, and this is then implanted in a suitable part of the body. The medicine is released at a rate determined by the rate of hydrolysis of the polyester.

The rate of hydrolysis of the polymer is very important. It is affected by the relative molecular mass of the polymer and its crystallinity. Tuning and balancing these features in polyesters made from lactic acid and glycolic acid can achieve a precise rate of hydrolysis.

Seeing things differently

When Logie Baird invented the first television, it was a large piece of equipment filled with valves and transistors. The cathode-ray tube fires a single beam of electrons across the screen very quickly. The electrons hit tiny pixels one after the other, making each glow in turn – it all happens so quickly that you don't notice the flickering. Because the television used a cathode-ray tube to produce its pictures, the TV box had to be very deep to allow a large enough distance between the tube and the screen for a good picture to be formed.

▲ **Figure 16** Inside an old television with a cathode-ray tube.

▲ **Figure 17** LCD TVs on a factory assembly line.

Many modern televisions are flat-screen models that use liquid crystal display (LCD) technology.

These televisions are slim enough to be wall mounted. How has the technology changed to allow this development?

In your studies of chemistry so far, you will have been introduced to the idea that there are three states of matter – solid, liquid and gas. You will have considered particle models for these states of matter, like those shown in Figure 18. You will also have used these particle models to help you to explain some of the properties of materials that are in each of these physical states.

'Liquid crystals' are materials that have a physical state somewhere between liquid and solid, causing them to have unusual properties. For example, a pure solid chemical has a fixed melting point, but a liquid crystal shows 'double melting' behaviour. The particles in a solid are arranged in a highly ordered structure, where they can occupy only certain positions, and each particle has a specific orientation compared to the other particles. If a solid is heated to above its melting temperature and it becomes a liquid, the orderly arrangement of particles is lost completely. However, if the substance melts to a liquid crystalline phase, only the positional order of the particles is lost, but most of the particles remain in the same orientation as they had in the solid. The fact that the material retains some of the organisation of its particles that was present in the solid state means that the material shows properties between that of a solid and a liquid. The orientation of the particles in the liquid crystal phase can be affected by the presence of an electric field.

How does this help to give a TV picture? In a typical LCD device, a liquid crystal layer is placed between two pieces of polarised material that are oriented at right angles. Polarised material only allows light waves through it if they are travelling in one specific direction. If two pieces of polarised material are held at right

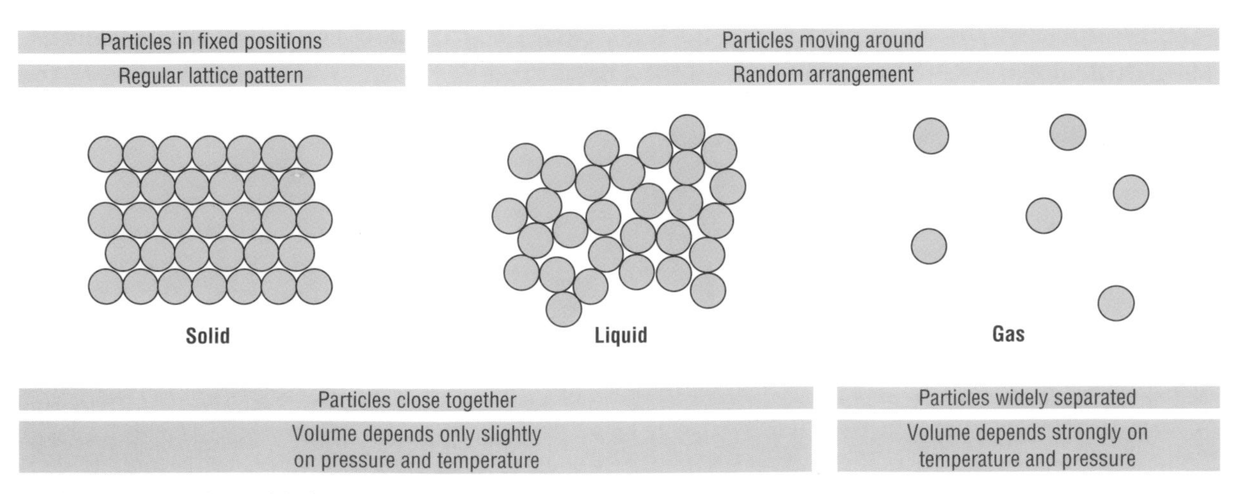

▲ **Figure 18** Particle models for solids, liquids and gases.

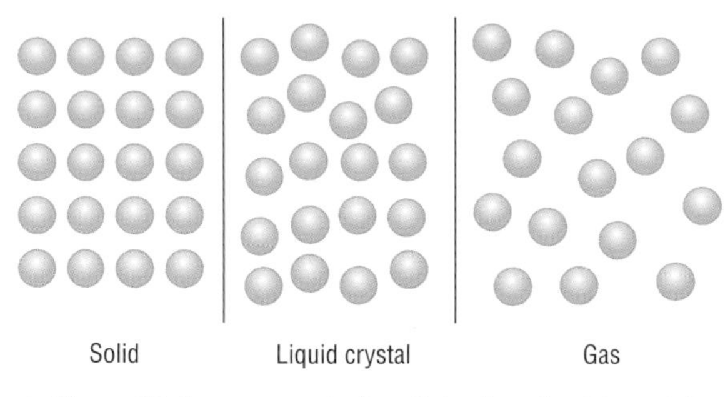

▲ **Figure 19** Arrangement of particles in a liquid crystal.

angles, no light can get through. The particle alignment in the liquid crystal layer of the device is chosen so that the particles reorient the light that has passed through the first polariser, allowing it to be transmitted through the second polariser. The device therefore appears transparent. When an electric field is applied to the liquid crystal layer, the particles realign themselves. In this state the particles do not reorient light, so the light polarised at the first polariser is absorbed at the second polariser, and the device loses transparency. In this way, the electric field can be used to make a pixel switch between transparent or opaque on command. LCD systems use the same technique with colour filters to generate red, green and blue pixels. Similar principles can be used to make other liquid crystal-based optical devices.

You will investigate the formation of a liquid crystal in **Activity MR2.5**.

You can make sure you understand the work on condensation polymers and their hydrolysis products in this section by doing **Activity MR2.6**.

MR3 *Reuse or refuse?*

Is there a problem?

Within the European Union, over 2 billion tonnes of waste are reportedly produced each year. While plastics account for only about 7% by volume, this is still a vast amount – most of it is household waste.

Until recently, much of our waste has been disposed of by dumping it in landfill sites – and these are getting harder to find nowadays. Landfill tax is set to rise significantly (the 2010 figure is set at 200% of the 1995 figure) and plastics are notorious for not being readily decomposed when buried. There is an ongoing search for other means of disposal. Recycling sounds attractive degradable plastics even more so.

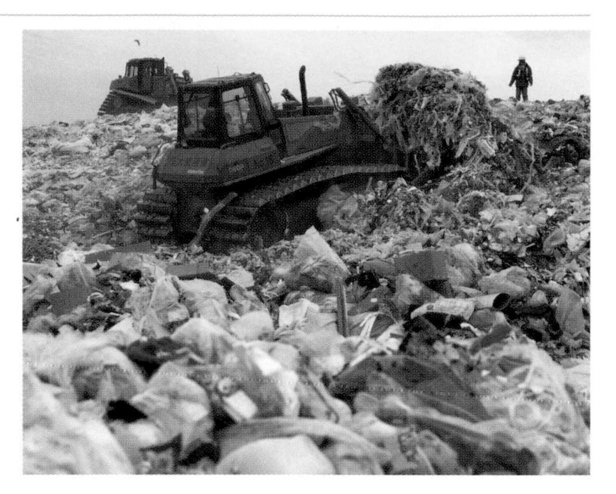

▲ **Figure 20** Much of our plastic waste goes into landfill sites like this.

Recycling plastics

The obvious answer is to **recycle** the polymers, because most of them are thermoplastics and can be reworked without decomposition. However, there are many problems in trying to reuse domestic waste. For one thing, the plastics have to be sorted (Figure 21) and generally this is a very expensive process. A better source of plastics for recycling is the waste from

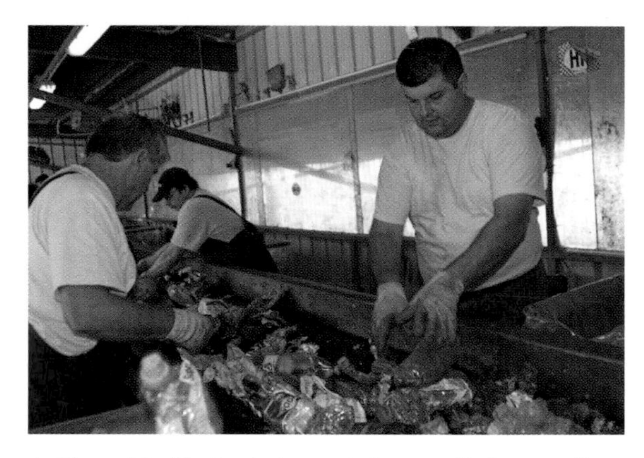

▲ **Figure 21** Plastics being sorted manually for recycling.

▲ **Figure 22** This fleece top has been produced from plastic drinks bottles made from PET.

factories where plastic articles are fabricated. Often only one type of plastic is produced in a factory, reducing the need to sort the waste before recycling it.

Another approach is to recycle plastics chemically by converting them back to their monomers and repolymerising. This is only practicable where there is enough high-quality single-material waste available. Some polyesters and polyamides are being recycled in this way. Recycling reduces carbon emissions by reducing the amount of polymer production and increases the number of useful phases in the life cycle of the polymer.

A third possibility is to crack the polymer and break it into smaller molecules. These small molecules can then be used as feedstocks in the chemical industry. Small, trial cracking plants are being built that can be used on a relatively small scale to prepare feedstock for existing chemical processes.

Figure 24 summarises these options.

▲ **Figure 23** The sails of replica tall ship *Surprise* (featured in the 2003 film *Master and Commander*) are woven from 125,000 soft drinks bottles made from PET.

You can find out how the principles of green chemistry are used in polymer manufacture, recycling and disposal of polymers in **Chemical Ideas 15.10**.

Is burning an option?

Simply burning plastics prior to disposal of the residual ash is becoming unacceptable in terms of its impact on the environment – it is also a waste of valuable energy.

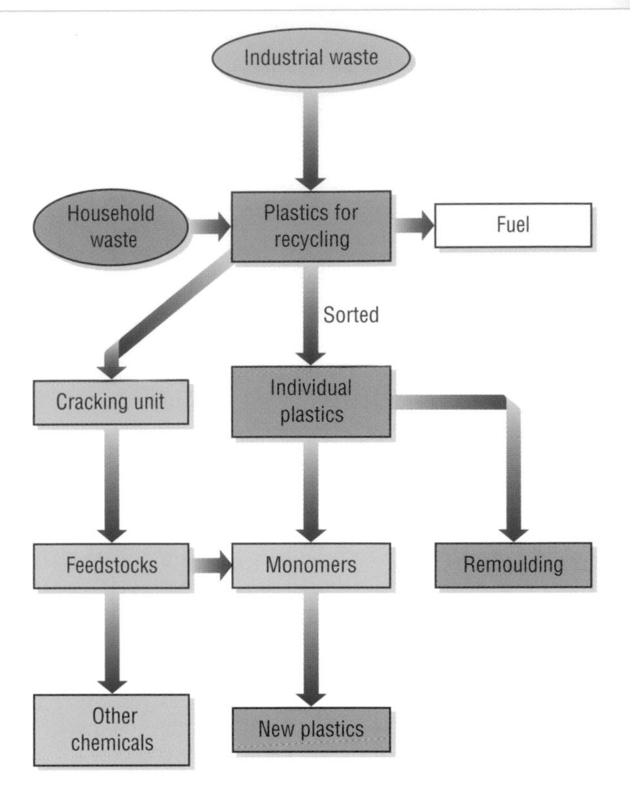

▲ **Figure 24** Summarising ways in which waste plastics can be used to save energy and oil – it is estimated that 4% of world oil resources are used for plastic production.

But **incinerators** can be effective and a great effort is being made to develop furnaces that trap harmful emissions, such as toxic gases and smoke, and use the energy to generate heat and power.

Degradable plastics

Most plastics are not degradable, because decomposer organisms (bacteria and fungi) do not have the enzymes needed to break them down. There are three important categories of degradable plastic:

- **biopolymers** – made by living organisms and broken down by bacteria
- **synthetic biodegradable plastics** – broken down by bacteria
- **photodegradable plastics** – broken down by sunlight.

Biopolymers

Poly(hydroxybutanoate), PHB, is a natural polyester made by certain bacteria. The formula for PHB is:

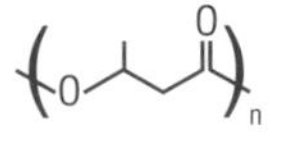

When the nutrient glucose is in short supply, the bacteria break down the PHB instead, digesting the polymer to survive. So the polymer can be made into plastic articles and when these are no longer needed, the plastic can be broken down by the bacteria that helped to make it.

The plastic has superior properties to other plastics, such as poly(propene) and poly(styrene), but it is costly – about 10 times the price of these conventional polymers. This has prevented the material from being commercially successful.

However, the story does not finish here. PHB is a member of a series of related compounds known as the poly(hydroxyalkanoate)s that all have similar structures (sometimes called PHAs – but don't confuse them with the PHA, poly(hydroxamide) on page 30). PHB has a methyl side group on its polymer chain – other members of the series have different side groups. PHB has packaging and horticultural applications.

Work is ongoing to produce a range of PHAs using plants. Think of that! Plastics not being made from oil, but by photosynthesis of carbon dioxide and water. Then, when they are not wanted, they are readily decomposed back to carbon dioxide and water – a truly renewable plastic. One of the problems with PHB is that it is brittle. Recent work has produced PHB that is made stronger by adding nanostructures of silicates.

Synthetic biodegradable plastics

Some plastic bags are made of poly(ethene) that has starch granules encapsulated in it (Figure 25). The starch is digested by microorganisms in the soil when the plastic bag is buried. The bag then breaks up into very small pieces of leftover poly(ethene) that have a large surface area and biodegrade more quickly.

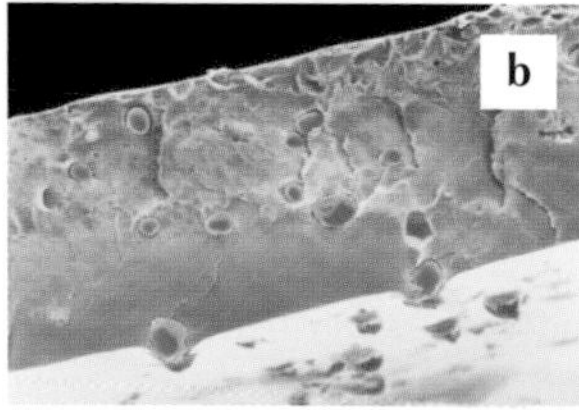

▲ **Figure 25** **a** Protecting plants from frost is one use for biodegradable plastics. **b** Plastic sheeting seen in close up (×400) – starch granules, coloured orange, are embedded in the plastic.

Photodegradable plastics

Carbonyl groups (C=O) absorb radiation in the wavelength range 270–360 nm (about 10^{15} Hz frequency). This corresponds to light in the near ultraviolet region of the spectrum. These groups can be incorporated into polymer chains to act as 'energy traps'. The trapped energy causes fission of bonds in the neighbourhood of the carbonyl group, and the polymer chain breaks down into short fragments that can then biodegrade. For example, carbonyl groups have been incorporated in the process of manufacture into poly(ethene) chains.

You will find out more about the life cycle of a polymer in **Activity MR3**.

MR4 *Materials with unusual properties*

The first aramids

After the invention of nylon, chemists began to make sense of the relationship between a polymer's structure and its properties. They were able to predict strengths for particular structures, and research was directed at inventing a 'super fibre'. In the early 1960s, DuPont were looking for a fibre with the 'heat resistance of asbestos and the stiffness of glass'.

The aromatic polyamides seemed promising candidates – the planar aromatic rings should result in rigid polymer chains and, because of the high ratio of carbon to hydrogen, they require relatively large concentrations of oxygen before they burn.

The first polymeric aromatic amide – an **aramid** – was made from 3-aminobenzoic acid. The polymer could be made into fibres and was fire-resistant, but it was not particularly strong. The zig-zag nature of the chains prevented the molecules from aligning themselves properly.

A polymer was needed that had straighter chains, and that could be made from readily available and reasonably cheap starting materials.

This substance turned out to have all the right properties except one. The problem was its insolubility, which made it precipitate out of solution before long polymer chains had been able to grow.

The only suitable solvent seemed to be concentrated sulfuric acid. The company engineers were not impressed! However, the remarkable properties of the 'super fibre' were enough to encourage investment in a plant that uses concentrated sulfuric acid as a solvent. The polymeric material produced was called Kevlar.

Assignment 4

a Draw a small section of the structure of the polymer that would be made from 3-aminobenzoic acid.

b Kevlar is made from two monomers – a diamine and a dicarboxylic acid. Look carefully at Figure 26 and then draw the structural formulae of the two monomers used.

c The intermolecular bonds in Kevlar are disrupted by concentrated sulfuric acid – that is why it dissolves. How do you think the bonds are affected?

d Kevlar fibres are produced by squirting the solution in concentrated sulfuric acid into water. Suggest why the polymer precipitates out when the solution is diluted in this way.

Why is Kevlar so strong?

Kevlar is a fibre that is fire-resistant, extremely strong and flexible. It also has a low density because it is made from light atoms – carbon, hydrogen, oxygen and nitrogen. Weight for weight, Kevlar is about five times stronger than steel! One of its early uses was to replace steel cords in car tyres – Kevlar tyres are lighter and last longer than steel-reinforced tyres.

Kevlar is strong because of the way the rigid, linear molecules are packed together. The chains line up parallel to one another, held together by hydrogen bonds. This leads to sheets of molecules. The sheets then stack together regularly around the fibre axis to give an almost perfectly ordered structure. This is illustrated in Figure 26.

The important thing to realise about Kevlar is that it adopts this crystalline structure because of the way the polymer is processed to produce the fibre. And this is a consequence of the work put in by the DuPont team.

Developing a market for Kevlar

A full-scale commercial plant for the production of Kevlar required an investment of $400 million. It was therefore essential that there would be a market for the product. Although Kevlar has some remarkable properties, it could not simply replace existing materials. Detailed market development had to take place alongside technical development. In other words, new uses had to be found for the new polymer.

Many uses have now been developed in addition to replacing steel in tyres. Kevlar ropes have 20 times the strength of steel ropes of the same weight – they last longer too. A stiffer form of Kevlar is used in aircraft wings, where its strength combined with its low density is important. And it's ideal for making bulletproof vests, and jackets for fencers.

PEEK

The story of PEEK begins in the early 1960s when John Rose, a chemist at University College London, moved to the Plastics Division of ICI. He was put in charge of a team that had a brief to develop new polymers.

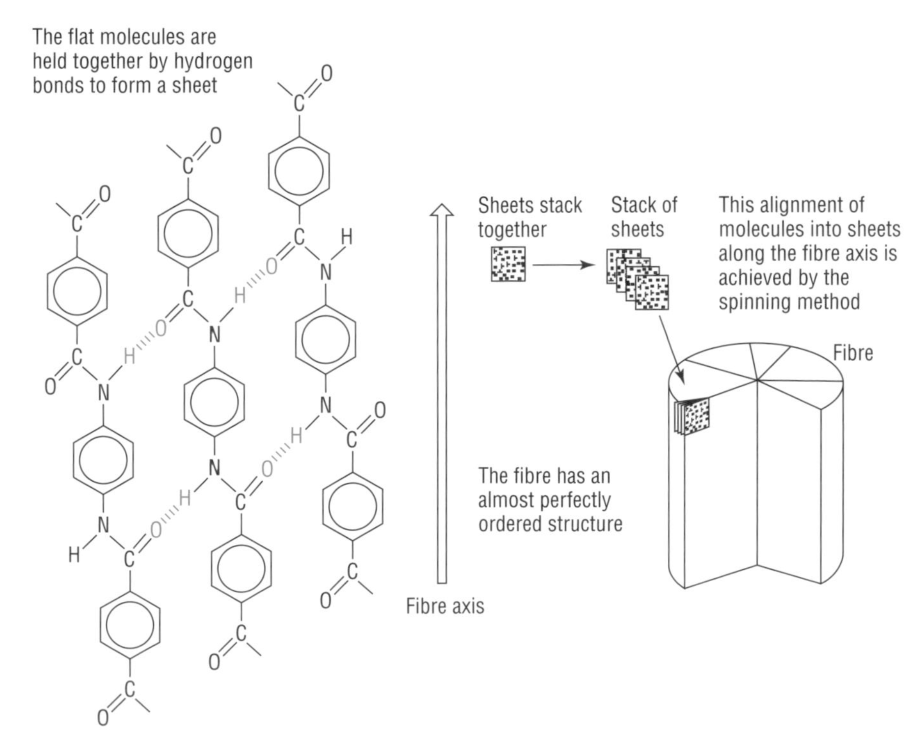

▲ **Figure 26** Crystalline structure of Kevlar.

▲ **Figure 27** Kevlar is used to strengthen the rigging of ocean-going yachts and tennis racquet frames.

Rose decided to investigate high-temperature materials. He knew that these would need to have high melting points and also be resistant to oxidation. For these reasons, he decided that his new polymers would have to be based on aromatic compounds.

The team tried to join aromatic units together in as many ways as possible. They had to develop new types of reaction and solve many other problems along the way. They also had to look for monomers that were reasonably cheap to make.

Out of all this came PEEK – its structure is shown in Figure 28.

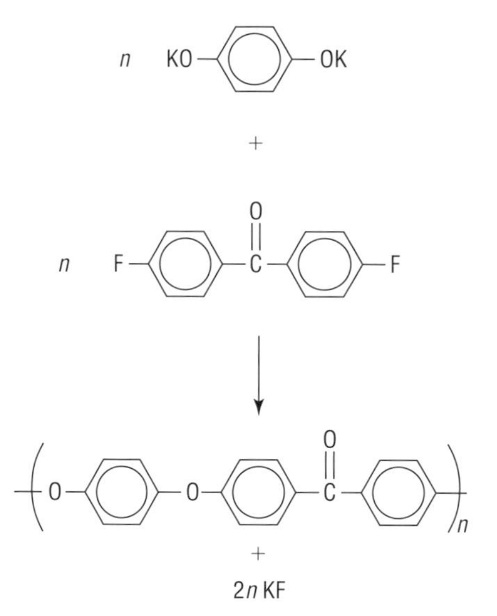

▲ **Figure 28** Equation for the formation of poly(ether-ether-ketone), PEEK.

Although it is expensive, the very wide range of applications of PEEK (from medical implants to aerospace components) made it worth developing.

You can find out more about polymers that have been designed to have specific properties in **Chemical Ideas 5.7**.

A clever idea

Although aeroplane crashes are infrequent, they can be catastrophic when they occur. Even if you survive the impact, you are still in danger from fire. To help reduce this risk, seats, overhead lockers and wall panels can be made from PHA, poly(hydroxyamide). When heated strongly, PHA loses water and rearranges to form another polymer, PBO, which is fire-resistant. No smoke is produced – very important in an aeroplane.

So why not make the seats and lockers from PBO in the first place? The answer is that it is very difficult to fabricate – whereas PHA is easy.

▲ **Figure 29** There are stringent international regulations to ensure air passengers are as safe as possible in the event of fire. The use of PHA is one way chemists are contributing to the solution to this problem.

Mixing it

In this course you have by no means come across all the polymers that are available to manufacturers. But suppose you have a particular application in mind – even a full list would not contain enough substances to allow you to be sure of finding one with just the properties you were seeking. You can bet that at some point you would say, 'What we need is something like X, but that behaves a bit like Y.'

These days, it is much cheaper to modify existing polymer materials than to develop new ones. Well-known polymers are frequently combined to produce new materials that show some of the properties of the individual components. For example, sheets of different polymers can be stacked together to form laminates, or polymers can be mixed to produce composites.

It is also possible to mix things at a molecular level and to make *polymer alloys*, in which polymers are mixed when molten to give a new material with the desired properties, or to make **copolymers**, or to add **plasticisers**. You have already read about these in **Chemical Ideas 5.6** and **5.7**.

Although it is cheaper to modify existing polymers, many new polymer materials have been designed. These include shape memory polymers, some of which were originally made for use in the space industry. Shape memory polymers are examples of copolymers in which there are two main regions per chain – a hard segment and a soft, or switching, segment.

The hard segment has very strong intermolecular bonds that stabilise the permanent shape of the material. The switching unit has a much lower thermal transition temperature, or T_{trans}. At temperatures above T_{trans} the chain segments forming the switching segment are flexible and the material is highly elastic. The hard segments of the chains determine the overall shape of the polymer and this gives the polymer its permanent shape. The flexibility of the switching segments of the chains below T_{trans} is limited and this gives the polymer its temporary shape. T_{trans} can either be a **glass transition temperature (T_g)** or a melting point (T_m). Changing its temperature to values above or below T_{trans} can reshape the polymer.

Earlier in the module you saw a photo of surgeons using a polyester thread for making stitches (Figure 14, page 24). Figure 30 shows how a shape memory polymer might be used for the same purpose. The stitch thread would change shape as it is heated by the warmth from the body, causing it to tighten to just the right level to hold the wound together, without causing damage to the surrounding tissue (Figure 31).

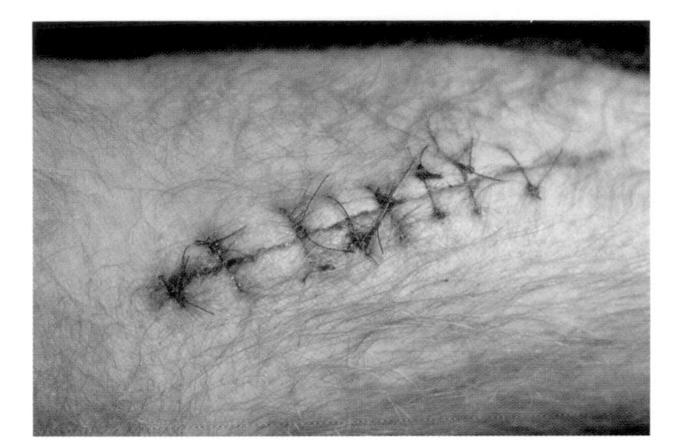

▲ **Figure 31** Stitches made from a shape memory polymer – they tighten themselves as they are warmed by the patient's body.

Activity MR4.1 looks at the effect of temperature on an everyday substance – bubble gum.

Polymer crystallinity is considered in **Activity MR4.2**.

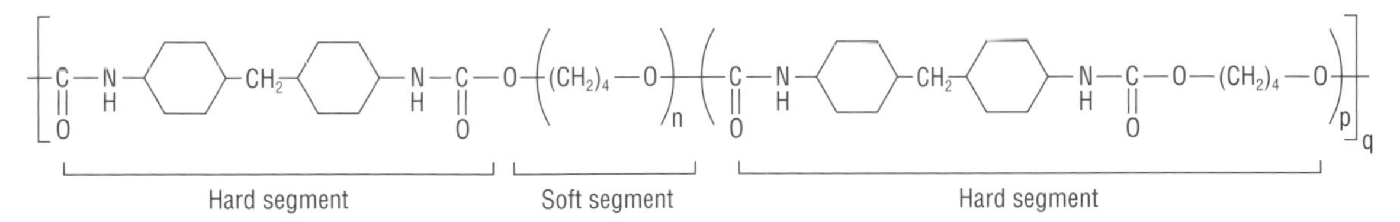

▲ **Figure 30** Structure of part of the chain in a shape memory polymer.

MR5 Summary

In this module you have seen how chemists have learnt to use their knowledge of materials, their bonding and their structure, to help them in producing materials with specific properties. They can do this by modifying existing materials (either chemically or physically) or by designing new materials.

You began by reading about a range of materials that have been produced with specific properties to allow them to be used for particular applications. You went on to read about the discovery of nylon, a polymer designed to imitate the structure of silk. You learned about condensation reactions, and about the properties and reactions of amines and amides. The story then moved on to polyesters, another type of condensation polymer. This allowed you to revisit and extend earlier work on alcohols, carboxylic acids and esters. You met two modern condensation polymers, Kevlar and PEEK, which led you to think about the effect of temperature on polymers.

As in **Polymer Revolution**, the link between structure and bonding and properties is central to the module and this allowed you to apply your knowledge and understanding of intermolecular bonds in new situations.

As well as reading about a range of materials, you considered some environmental and economic issues connected with the disposal and recycling of plastics.

Activity MR5 allows you to check your knowledge and understanding of the important chemistry contained in this module.

TL

THE THREAD OF LIFE

Why a module on 'The Thread of Life'?

This module introduces you to proteins – one of the most versatile classes of chemical, found in all living things. You will learn about the structures of proteins, their importance as enzymes and their synthesis in cells from DNA, the thread of life.

To understand how proteins are formed, you need to know about amino acids and you need to revisit earlier work on carboxylic acids, amines and amides. In that sense the module carries forward your study of organic chemistry.

Explanations of the structures and behaviour of proteins and other macromolecules found in cells are, however, based on physical chemical ideas – in particular, molecular shape and intermolecular bonding. Studying the behaviour of enzymes also provides an opportunity to extend your chemical knowledge – this time about the rates of chemical reactions.

Overview of chemical principles

In this module you will learn more about ideas introduced in earlier modules in this course:

- amines and amides (**Materials Revolution**)
- carboxylic acids (**What's in a Medicine?** and **Materials Revolution**)
- condensation reactions (**Materials Revolution**)
- stereoisomerism (**Polymer Revolution**)
- intermolecular bonds (**Elements from the Sea**, **Polymer Revolution** and **Materials Revolution**)
- the rates of chemical reactions (**The Atmosphere**)
- catalysts (**Developing Fuels** and **The Atmosphere**).

You will also learn new ideas about:

- amino acids
- optical isomerism
- protein structure
- enzymes
- the shapes of molecules and molecular recognition
- rate equations, reaction orders and half-lives
- protein biosynthesis, DNA and RNA.

TL

THE THREAD OF LIFE

TL1 *Man's future is in his genes*

Most of us have heard that genes are important for handing on our hereditary characteristics to future generations, but how do they do it? You may have heard of the Human Genome Project and DNA testing – what are they?

A genome is all of the genetic material contained in the chromosomes of an organism. The Human Genome Project set out in 1993 to map the complete DNA of human beings. With the technological advances that occurred after its inception, the project was completed within 10 years.

DNA carries our genetic code and contains the instructions for making proteins via our genes. Proteins have several vital uses in the body, but one of the most important is the control of cellular functions.

In order to understand a little of what the Human Genome Project involved and how DNA testing works, we need to look first at the structure and function of proteins.

TL2 *What are proteins?*

The name *protein* (meaning 'first thing') was coined by Berzelius in 1838. He had little idea what proteins were, but he recognised their importance because they were so widespread in living things.

Proteins are big molecules – they are *natural polymers* with relative molecular masses up to about 100 000. They play a key role in almost every structure and activity of a living organism (Figure 2). That's why they are regarded as some of the most important constituents of our bodies.

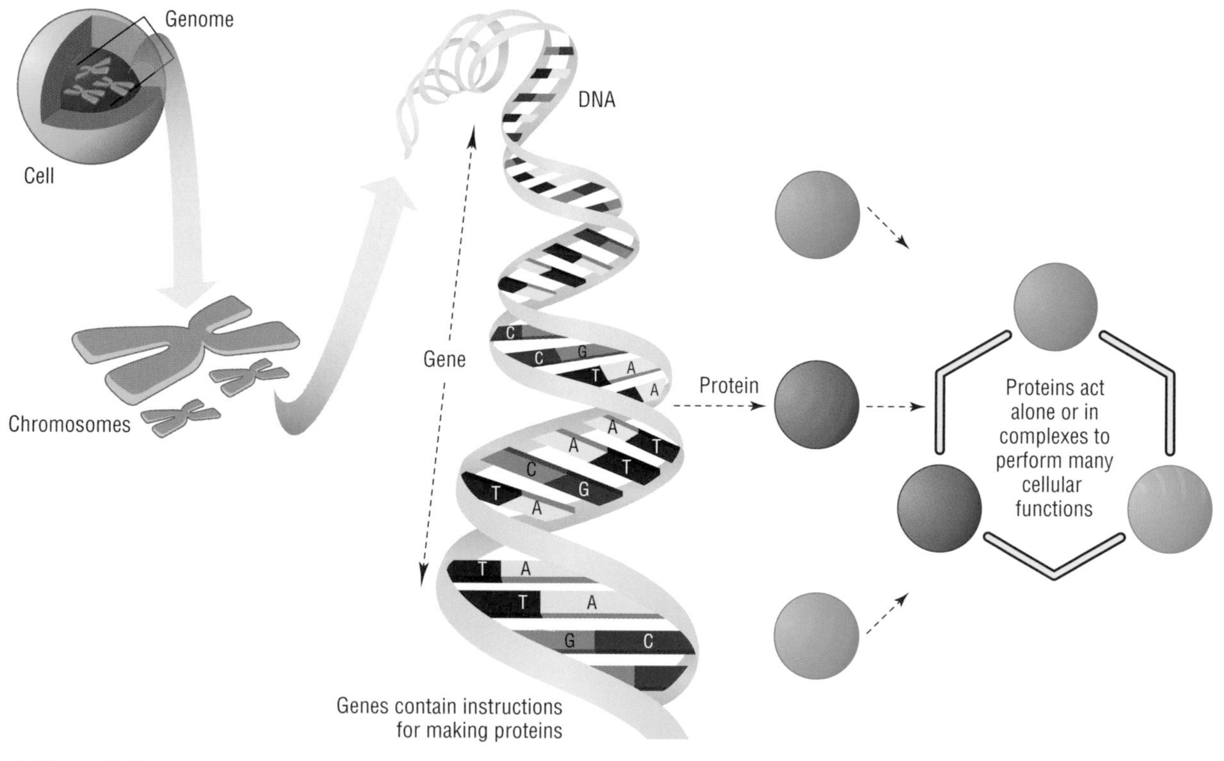

▲ **Figure 1** From genes to proteins.

THE HUMAN GENOME

- Genetic material is stored in chromosomes.
- Each chromosome contains thousands of genes.
- Each gene consists of many *codons* and is responsible for making one protein.
- Each codon consists of three *bases*.
  You will learn about codons and bases later.

The Human Genome Project

The project was completed in 2003. The coordinators were the US Department of Energy and the National Institute of Health. There were major contributions from the Wellcome Trust (UK) and additional contributions from Japan, France, Germany, China and other nations.

The goals of the project were:

- to *identify* all the genes in human DNA, approximately 20 000–25 000
- to *determine* the sequences of the 3 billion chemical base pairs that make up human DNA
- to *store* this information in databases
- to *improve* tools for data analysis
- to *transfer* related technologies to the private sector
- to *address* the ethical, legal and social issues arising from the project.

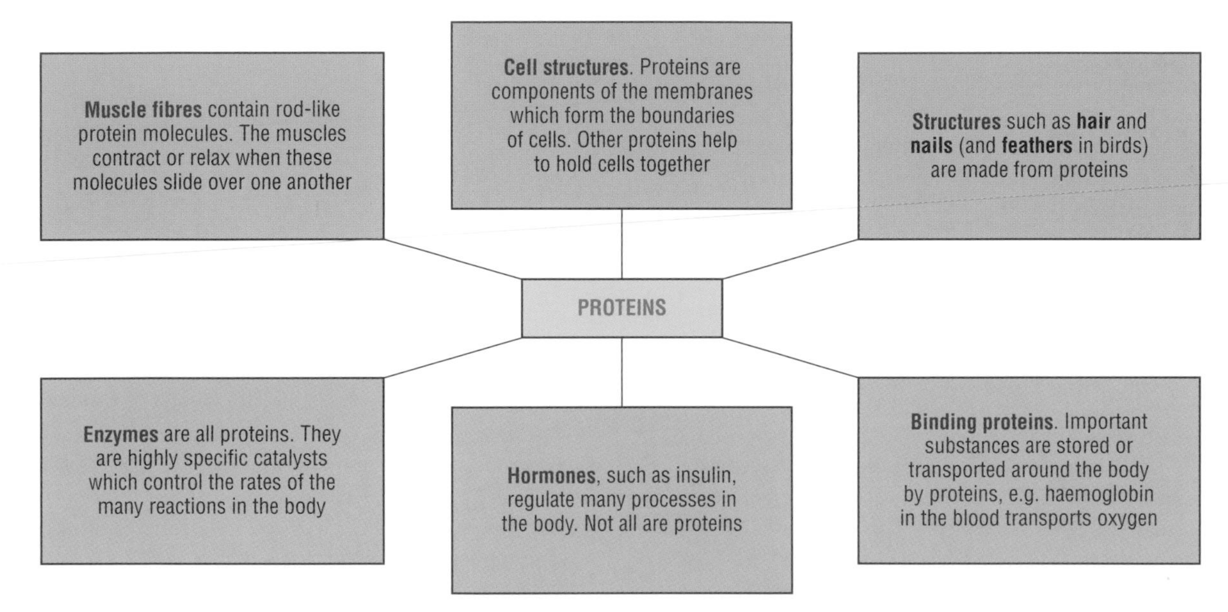

▲ **Figure 2** Proteins perform many functions in our bodies. Fibrous proteins (lilac boxes) are the major structural materials; globular proteins (orange boxes) are involved in maintenance and regulation of processes and include hormones and enzymes.

▲ **Figure 3** Hair and fingernails are both made up from protein.

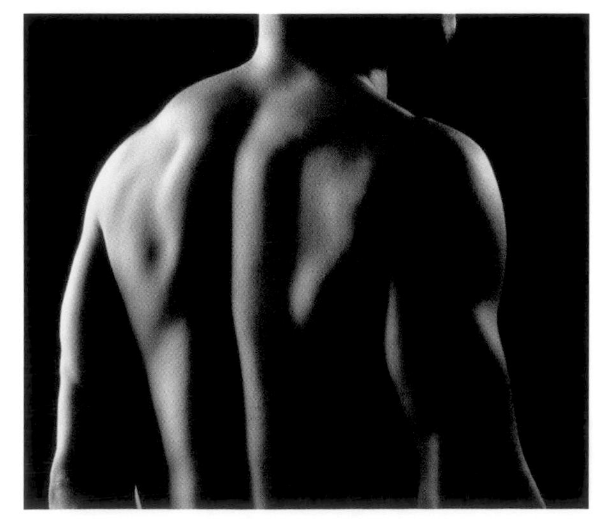

▲ **Figure 4** Muscle is made up from protein.

Amino acids – the building blocks of proteins

Figure 5 illustrates the composition of a molecule of human insulin. Insulin is a hormone and is one of the simplest proteins.

The abbreviations in circles in Figure 5 represent the **α-amino acids** that have combined to form insulin. There are two short chains of these **amino acid** residues – the parts of the original amino acid molecules that are joined together to form the protein. They are joined together by *condensation* reactions. All the proteins in the living world are made from just 20 α-amino acids, all of which have the same general structure as shown in Figure 6.

This is a good time to remind yourself of the chemistry of some important functional groups. You can read about carboxylic acids and their reactions in **Chemical Ideas 13.3** and **13.4**, and about amines and amides in **Chemical Ideas 13.8**.

Activity TL2.1 allows you to investigate the reactions of an amino acid.

You can study the reactions of amino acids in **Chemical Ideas 13.9.**

Each amino acid has a different side chain, labelled R (see Figure 6). Table 1 shows the structures of these R groups and the names of the amino acids, together with their abbreviated symbols.

Proteins need to be replaced continuously in all of us, even in adults who have stopped growing. This is obvious sometimes – like when your hair and nails grow after they have been cut. It is also important that hormones and some enzymes (such as the digestive enzymes) are made when needed, and then destroyed once they have done their job, so that they do not go on producing their effects after the need for them has passed.

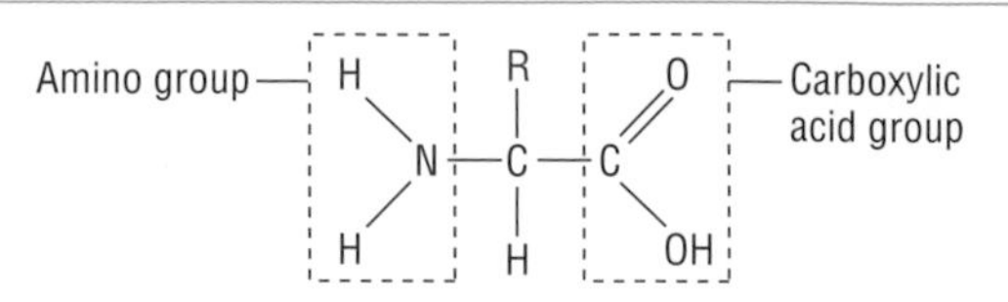

▲ **Figure 6** General structure of an α-amino acid. The amino group ($-NH_2$) and the carboxylic acid group ($-COOH$) are both attached to the same carbon atom, the α-carbon – this is why they are called α-amino acids. The R group is different in different amino acids (see Table 1).

We replace our proteins from the food we eat, but we don't need to eat proteins that are identical to the ones being replaced – we don't need to eat human hair or finger nails to grow our own! In fact, we don't even have to eat animal protein; we can supply our needs by eating plants.

This is because our bodies break proteins down into their constituent 'building blocks' – amino acids. About half of these are classified as *essential amino acids* – our bodies cannot synthesise these and we must take them in from food. The others can be made from carbohydrates and other amino acids in the body.

The amino acids are then reassembled to make our own collection of proteins. For each of us, this collection is unique – most of your proteins will be identical to those found in other humans, but some will be different. Human proteins are also different from the proteins of other animals. However, in some cases very similar proteins are found, not just throughout the animal kingdom but in plants and microorganisms as well – for example, the enzymes that oxidise glucose in cell metabolism.

All these millions of proteins are built up from the same small number of amino acids. What makes each protein different is the *order* in which the amino acids are joined to one another. This is called the **primary structure** of the protein. You have already seen the primary structure of human insulin in Figure 5.

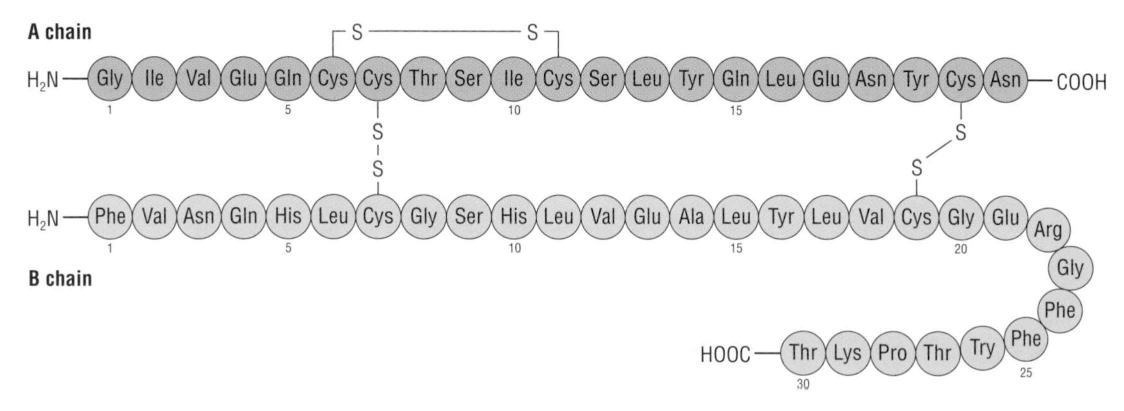

▲ **Figure 5** Human insulin – the two chains are held together by –S–S– links.

Table 1 The 20 amino acids that make up proteins
(for clarity, the whole amino acid has been drawn out in the case of proline).

Amino acid	Abbreviation	R group	Amino acid	Abbreviation	R group
glycine	Gly	$-H$	cysteine	Cys	$-CH_2-SH$
alanine	Ala	$-CH_3$	methionine	Met	$-CH_2-CH_2-S-CH_3$
valine	Val	$-CH(CH_3)CH_3$	aspartic acid	Asp	$-CH_2-C(=O)OH$
leucine	Leu	$-CH_2-CH(CH_3)CH_3$	glutamic acid	Glu	$-CH_2-CH_2-C(=O)OH$
isoleucine	Ile	$-CH(CH_2-CH_3)CH_3$	asparagine	Asn	$-CH_2-C(=O)NH_2$
phenylalanine	Phe	$-CH_2-$(benzene ring)	glutamine	Gln	$-CH_2-CH_2-C(=O)NH_2$
proline	Pro	(ring structure: HN, COOH)	tyrosine	Tyr	$-CH_2-$(benzene ring)$-OH$
tryptophan	Trp	$-CH_2-$(indole ring, N–H)	histidine	His	$-CH_2-$(imidazole ring: HN, N)
serine	Ser	$-CH_2-OH$	lysine	Lys	$-CH_2-CH_2-CH_2-CH_2-NH_2$
threonine	Thr	$-CH(CH_3)OH$	arginine	Arg	$-CH_2-CH_2-CH_2-NH-C(NH_2)=NH$

Assignment 1

The amino acids in Table 1 can be grouped according to key features of their R groups – for example, whether their side chains are polar and will interact strongly with water, or whether they are non-polar and will disrupt water's intermolecular bonding.

Use the abbreviated symbols for the amino acids to answer the questions below.

a List four amino acids, in each case, that you think have:
- **i** non-polar side chains
- **ii** polar side chains
- **iii** ionisable groups on their side chain.

b Look at the structures of leucine and isoleucine. Explain why isoleucine is so named.

c List one amino acid, in each case, in which the R group contains:
- **i** a primary alcohol group
- **ii** a secondary alcohol group
- **iii** a phenol group
- **iv** a carboxylic acid group.

REPRESENTING AMINO ACID SEQUENCES

The dipeptide obtained by condensing the carboxylic acid group of glycine with the amino group of alanine has the structure

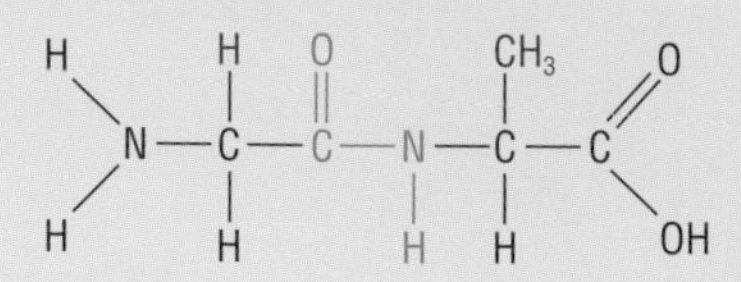

This dipeptide would be abbreviated to *Gly Ala*. The convention of reading peptide groups in the direction with the free NH_2 group on the left is very important if the amino acid sequence is to be read unambiguously.

Assignment 2

a Draw the structure of the dipeptide *Ala Gly*.

b In the tripeptide *Ser Gly Ala*, which amino acid has an unreacted:
- **i** NH_2 group?
- **ii** COOH group?

Making peptides

When amino acids combine to form a protein such as insulin, the carboxylic acid group on one amino acid joins on to the amino group on the next, and a molecule of water is lost from between them. This type of process is called a **condensation reaction**, and the –CONH– (secondary amide) group that links the amino acid residues in the product is called a **peptide link**. Two amino acids joined in this way make a *dipeptide*:

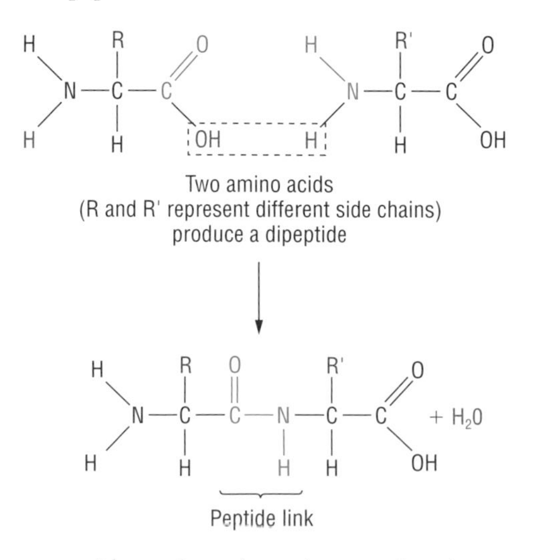

You met amides and condensation reactions in **Materials Revolution**, section **MR2**.

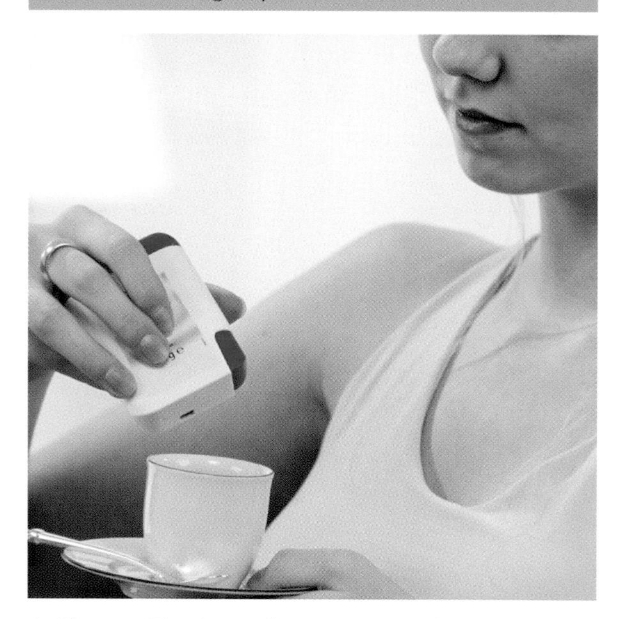

▲ **Figure 7** The dipeptide aspartame is a low-calorie sweetener.

In **Activity TL2.2** you can study peptide structures, using models or ICT. In **Activity TL2.3** you can break down aspartame and identify its amino acid components. In doing this activity, you will carry out *paper chromatography*, which you need to be aware of. You can find out more about paper chromatography in **Chemical Ideas 7.3** and **Chemical Ideas Appendix 1: Experimental Techniques** (Technique 4).

Chemists cannot make amino acids react together directly. They have to make the –COOH group more reactive – for example, by turning it into an acyl chloride. To make proteins, they also need to take into account another property of amino acids.

Amino acid molecules are not flat – they have a three-dimensional shape based on the tetrahedral arrangement of the four bonds around the α-carbon atom. When we look at them properly in this way, all the amino acids in Table 1, with the exception of glycine, exist in *two* isomeric forms known as D or L **optical isomers**. Proteins are built up from only the L isomers.

Chemical Ideas 3.5 tells you about optical isomerism. You can remind yourself about the shapes of molecules by reading **Chemical Ideas 3.2** and you can revise earlier work on structural isomerism in **Chemical Ideas 3.3** and *E/Z* isomerism in **Chemical Ideas 3.4**. You can also remind yourself how to draw different types of structural formulae by looking at **Chemical Ideas 12.1**.

Cells are able to build up proteins directly from L-amino acids. That's why chemists are learning how to use bacterial and yeast cells to make proteins – in many ways it's better than using traditional techniques. You will discover more about how cells make proteins in section **TL5.**

In **Activity TL2.4** you can build models for some amino acids and investigate optical isomerism further.

TL3 *Proteins in 3-D*

Folded chains

Scientists need to know about more than just a primary structure before they understand how a protein works. One thing they now know is that most proteins have a precise shape that arises from the *folding* together of the chains. The action of many proteins is critically dependent on this shape.

As long as different molecules fold to the same shape, they may have similar actions. Chain folding gives proteins their three-dimensional shape – it also places chemical groupings in positions where they can bond most effectively.

The chains in a protein are often folded or twisted in a regular manner as a result of hydrogen bonding. This results in the formation of a *helix* or a *sheet*, which is called the **secondary structure** of a protein (see Figure 8).

The chains may then fold up further. The overall shape of a protein is called its **tertiary structure** and

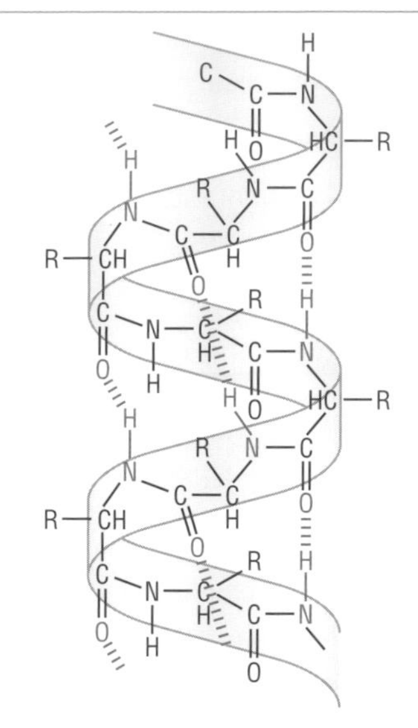

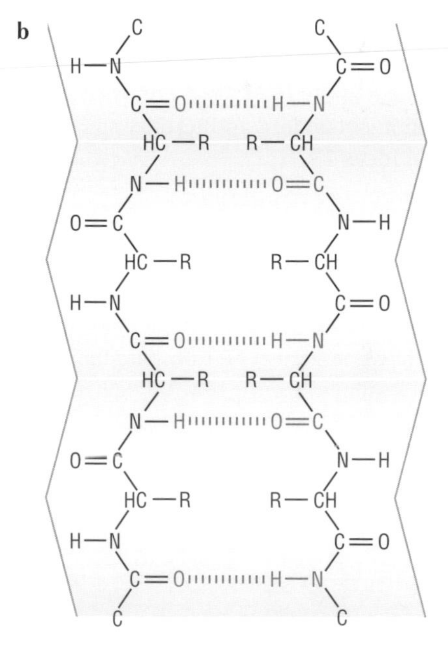

▲ **Figure 8** The secondary structure of a protein involves folding as a result of hydrogen bonding – this figure shows the protein chain folded into a helix **a** and a sheet **b**.

this is stabilised by the bonds listed below – these occur between the R-groups of the amino acids in the chains. Four types of interaction are important in chain folding:

- *instantaneous dipole–induced dipole* bonds between non-polar side chains on amino acids such as phenylalanine and leucine. The centres of protein molecules tend to contain amino acids like these, so that the non-polar groups do not interfere with the hydrogen bonding between the surrounding water molecules.

- *hydrogen* bonds form between the peptide groups that link the chain together in secondary structures. They also form between polar side chains (e.g. $-CH_2OH$ in serine and $-CH_2CONH_2$ in asparagine). If amino acids with polar side chains are situated on the outside of proteins, then hydrogen bonds can also form to water molecules surrounding the protein.
- *ionic* bonds between ionisable side chains, such as $-CH_2COO^-$ in aspartic acid and $-CH_2CH_2CH_2CH_2NH_3^+$ in lysine.
- *covalent* bonds. The –SH groups on neighbouring cysteine residues can be oxidised to form –S–S– links. Look back at Figure 5 on page 36. There are three such links in human insulin – two hold the A and B chains together.

You can revise earlier work on instantaneous dipole–induced dipole bonds and hydrogen bonding by reading **Chemical Ideas 5.3** and **5.4**.

You can read about how these bonds hold a protein in its **secondary** and **tertiary** structure in **Chemical Ideas 13.9**.

'Ribbon diagrams' are often used to represent the tertiary structure of proteins. Ribbon diagrams for the enzyme *ATPase* and the hormone *insulin* are shown in Figures 9 and 10. An electron density map graphic for insulin is shown in Figure 11. Compare the ribbon diagram of insulin in Figure 10 with Figure 5 on page 36, which shows its primary structure.

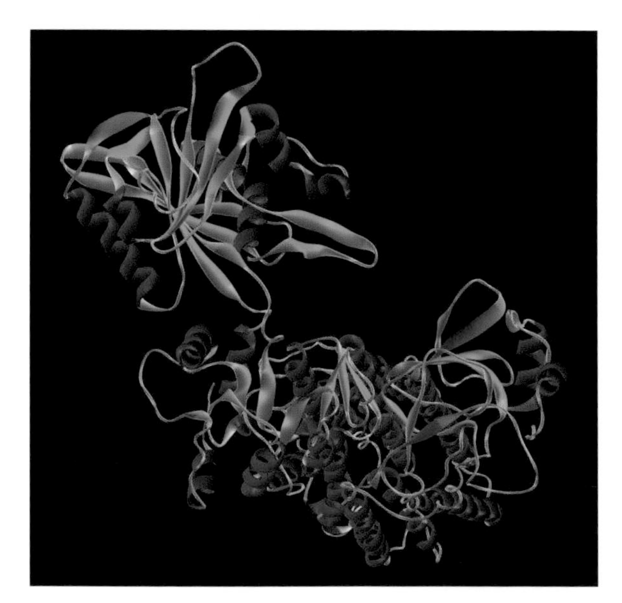

▲ **Figure 9** Tertiary structure of a protein; here the calcium pumping ATPase muscle enzyme is illustrated as a ribbon diagram. The helices (in red) and the pleats (in blue) show up clearly in this representation.

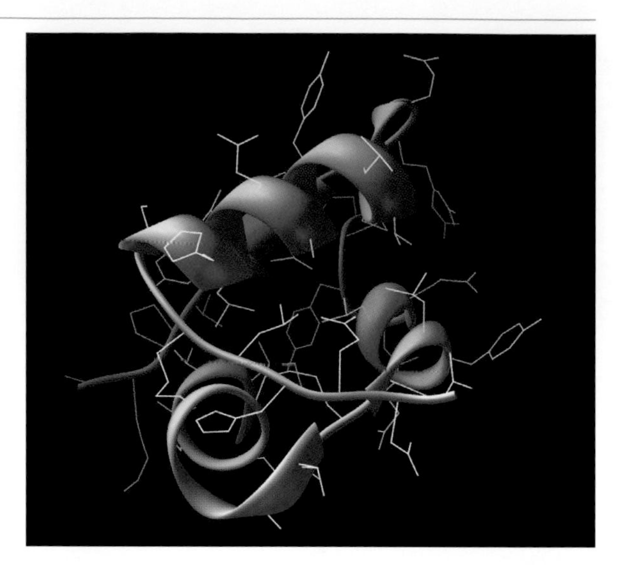

▲ **Figure 10** Ribbon diagram showing two chains in insulin. There are two short helical sections (bottom) in the A chain. The B chain contains a helical section (top, in orange/red) between amino acids 9 (Ser) and 19 (Cys).

▲ **Figure 11** An electron density map graphic of an insulin monomer.

To summarise – there are three levels of protein structure:

- *primary structure* – the order of the amino acid residues
- *secondary structure* – the coiling of parts of the chain into a helix or the formation of a region of sheet
- *tertiary structure* – the folding of the secondary structure.

In **Activity TL3** you can make models to represent protein structures.

WORKING OUT THE STRUCTURE OF INSULIN

Insulin was recognised as a protein in 1928. Its primary structure was revealed in 1955 by Professor Fred Sanger and his team at Cambridge. They broke the protein into shorter bits by hydrolysis, and then used chromatography to identify these bits (similar to the techniques used in **Activity TL2.3**.) By piecing together the bits, they eventually worked out the amino acid sequence. Professor Sanger received a Nobel Prize for his work on the amino acid sequencing of proteins, and a second Nobel Prize for more recent work on sequencing DNA.

▲ **Figure 12** Nobel Prize medal, presented by The Royal Swedish Academy of Sciences. A Latin inscription on the reverse (*Inventas vitam juvat excoluisse per artes*) can be loosely translated as 'And they who bettered life on Earth by new-found mastery'.

The three-dimensional structure of insulin was worked out from X-ray diffraction studies carried out by Professor Dorothy Hodgkin's group at Oxford. Professor Hodgkin received a Nobel Prize in 1964 for earlier work, particularly work on the structure of vitamin B_{12} and penicillin. She is only the third woman, after Marie Curie and her daughter Irène Joliot-Curie, to receive a Nobel Prize in Chemistry.

▲ **Figure 13** Professor Dorothy Hodgkin.

TL4 *Enzymes*

Diabetic people are unable to control the level of glucose in their blood. They can buy some reagent strips from a pharmacy to test for glucose in their urine. (When glucose builds up to a high level in the blood, it is excreted in the urine.) The strips contain an *enzyme* that produces a colour reaction when glucose is present. You can find more detail on the strips in the 'Testing for glucose' green box. Another method involves putting a drop of blood on a detector strip and placing this strip in a hand-held detector that gives a digital readout of the blood glucose level.

The urine test strips illustrate four important points about enzymes. Enzymes are

- *catalysts*
- *highly specific* – for example, the test strips work only with glucose; they give no response with other sugars
- *sensitive to pH* – many work best at a particular pH; they become inactive if the pH becomes too acidic or too alkaline
- *sensitive to temperature* – many enzymes work best at temperatures close to body temperature; most are destroyed above 60–70 °C.

You can use some glucose test strips with different sugar solutions in **Activity TL4.1**. The activity illustrates the important points about enzyme behaviour.

Active sites

Enzymes are so *specific* because they have a precise tertiary structure that exactly matches the structure of the **substrate** – the molecule that is reacting. It's an example of molecular recognition. Figure 15 shows a

TESTING FOR GLUCOSE IN URINE

The fresh reagent strip has a small coloured square at one end that turns a different colour in the presence of glucose. The square is impregnated with four reagents:

- glucose oxidase, an enzyme that catalyses the reaction

  glucose + oxygen → gluconic acid + hydrogen peroxide

- an indicator, sometimes 2-methylphenylamine – this is present in its reduced form (XH_2) which is colourless, but it turns into a coloured form (X) when it is oxidised
- peroxidase, an enzyme that catalyses the oxidation of the indicator by hydrogen peroxide

  hydrogen peroxide + XH_2 (colourless) → water + X (coloured)

- a buffer – a mixture of chemicals that keeps the reagents at a fixed pH during the test.

The manufacturer's instruction sheet recommends storing the test strips below 30 °C, but not in a refrigerator.

▲ **Figure 14** A doctor discusses the results of a girl's test.

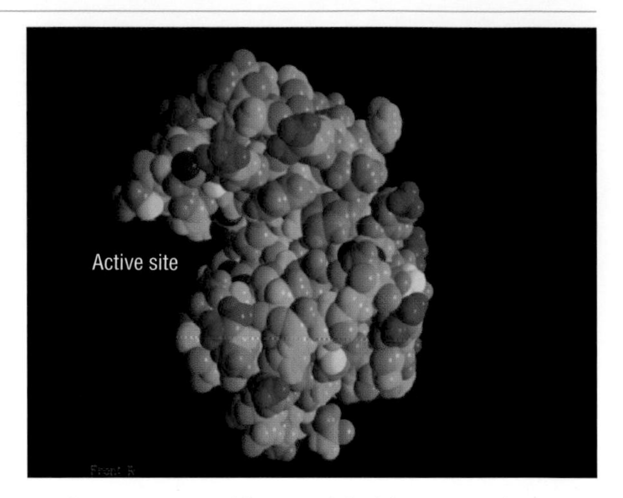

▲ **Figure 15** Space-filling model of the enzyme lysozyme – the cleft that forms the active site is clearly visible.

space-filling model of the enzyme *lysozyme*, which catalyses the breakdown of the cell walls of bacteria and helps to protect us from infection.

There is a cleft in the enzyme surface formed by the way the protein chain folds. The shape of the cleft is tailored for the substrate molecules to fit into. Within the cleft are chemical groups, some of the R groups on the amino acid residues, that bind the substrate and possibly react with it. This region of the enzyme is called its **active site**.

The bonds that bind the substrate to the active site have to be weak so that the binding can be readily reversed when the products need to leave the active site after the reaction. The bonds are usually hydrogen bonds or interactions between ionic groups. The binding may cause other bonds within the substrate to weaken, or it may alter the shape of the substrate so that atoms are brought into contact to help them to react.

After reaction, the product leaves the enzyme, which is then free to start again with another molecule of substrate. The whole process is summarised in Figure 16.

What happens, of course, is more complex than Figure 16 shows. In many cases, scientists believe that the substrate is not quite a perfect fit and must alter its shape to fit into the active site. This means that both the substrate and the active site will be in strained arrangements, and this can help the reaction to occur.

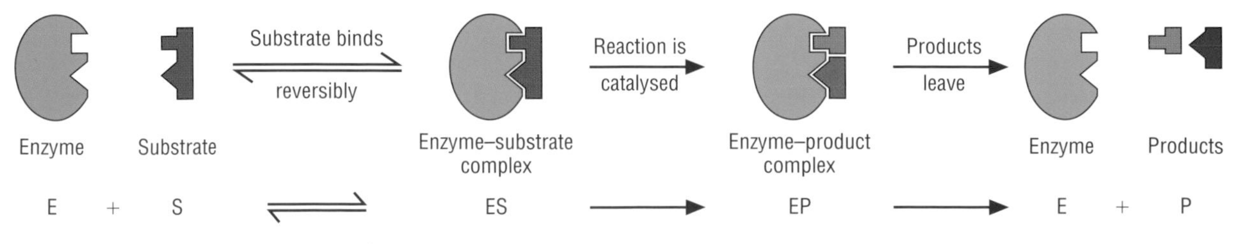

▲ **Figure 16** Model of enzyme catalysis.

Molecules that fit on to the active site but cannot be catalysed are called **inhibitors** (in more advanced work, 'competitive inhibitors'). An example is shown in the green box below.

ENZYME INHIBITION

The enzyme succinate dehydrogenase catalyses one of an important series of reactions in our bodies by which glucose is oxidised to carbon dioxide and water. The reaction in question is:

$$HOOC\text{–}CH_2\text{–}CH_2\text{–}COOH \rightarrow HOOC\text{–}CH\text{=}CH\text{–}COOH + H_2$$

succinic acid
(butanedioic acid)

Propanedioic acid ($HOOC\text{–}CH_2\text{–}COOH$) is found to be an *inhibitor.* It fits into the active site of the enzyme (because of the two carboxylic acid groups), but it cannot lose hydrogen in the same way as butanedioic acid because it does not have two $\text{–}CH_2\text{–}$ groups.

Such inhibitor studies are used by biochemists to study the nature of the active site of enzymes. Here, the two carboxylic acid groups are clearly important for the substrate to bind to the active site.

Assignment 3

Look at the 'Enzyme inhibition' green box.

a Name the product of dehydrogenation of butanedioic acid.

b This product exists as two *E/Z* isomers. Draw the two structures and label them *E* and *Z*.

c Would you expect both the *E* isomer and the *Z* isomer to be formed by the action of the enzyme? Explain your answer.

d Write the structure of pentanedioic acid. Consider whether succinate dehydrogenase would catalyse the dehydrogenation of pentanedioic acid and what the product might be.

Enzymes as catalysts

Reactant molecules in a catalysed reaction have a lower minimum energy to react when they collide than they have in an uncatalysed reaction. In other words, the *activation enthalpy* is lower. This is illustrated in Figure 17 for an enzyme-catalysed reaction. When the activation enthalpy is lower the reaction takes place more quickly.

You were introduced to the idea of activation enthalpy in **Chemical Ideas 10.1** and **10.2** and to catalysts in **Chemical Ideas 10.5** and **10.6**.

Factors that affect the performance of enzymes

Many industrial catalysts consist of inorganic substances that are relatively unaffected by changes in temperature and pH. However, the complex organic structure of enzymes means they are much more sensitive to such conditions.

If an enzyme's active site contains ionisable groups, the enzyme's action will be affected by a change in pH. For example, if there is a $\text{–}COOH$ group that acts by donating an H^+ to the substrate, raising the pH will turn it into $\text{–}COO^-$ and the enzyme will not be able to function correctly.

An enzyme will also become inactive if its shape is destroyed. Figure 18 illustrates how the active site of an enzyme can be made up from the side chains of amino acids in different parts of the protein molecule. They are held close together by the enzyme's tertiary structure.

If the tertiary structure is broken, the enzyme loses its shape and the side chains are no longer close together. The active site is destroyed and the enzyme is said to be *denatured*.

The tertiary structure is held together by weak dipole–dipole bonds and hydrogen bonds. These can be broken easily by raising the temperature, which

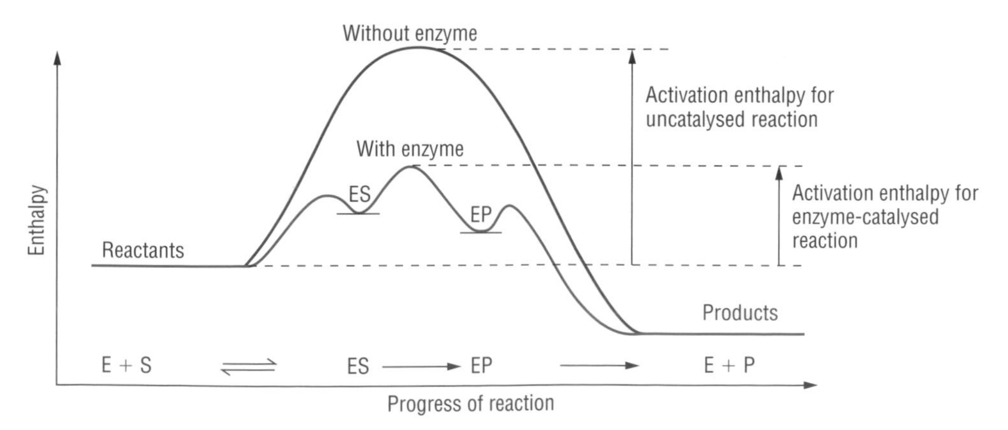

▲ **Figure 17** How an enzyme provides a route of lower activation enthalpy.

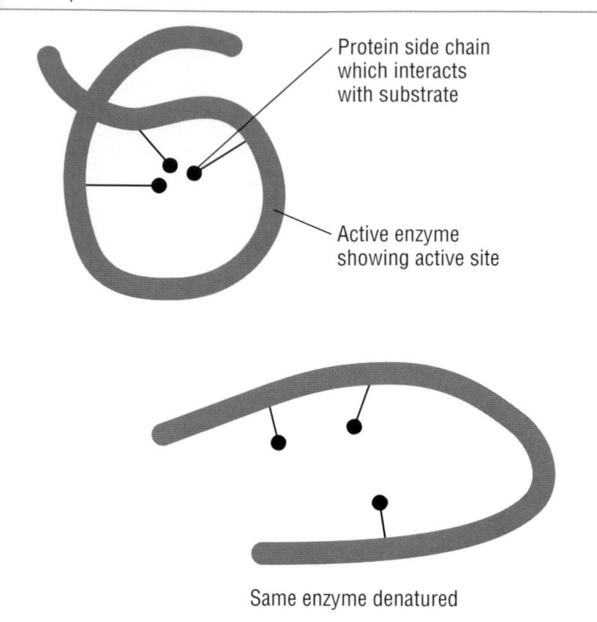

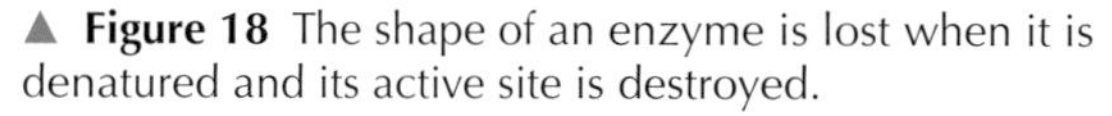

▲ **Figure 18** The shape of an enzyme is lost when it is denatured and its active site is destroyed.

▲ **Figure 19** The proteins in egg white are denatured when an egg is cooked.

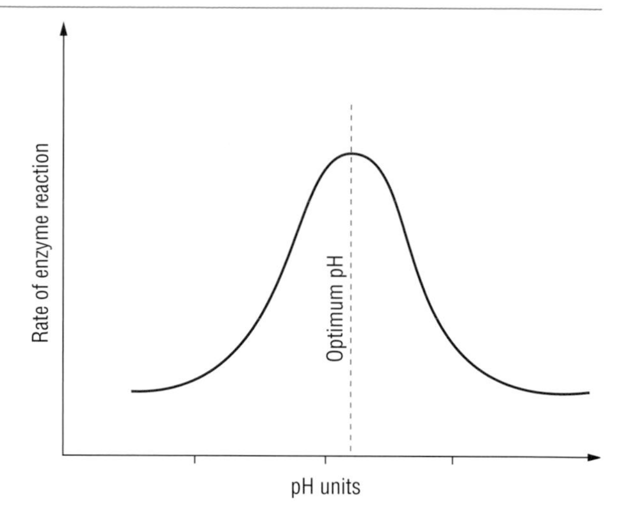

▲ **Figure 20** Graph showing how the rate of a typical enzyme reaction varies with pH.

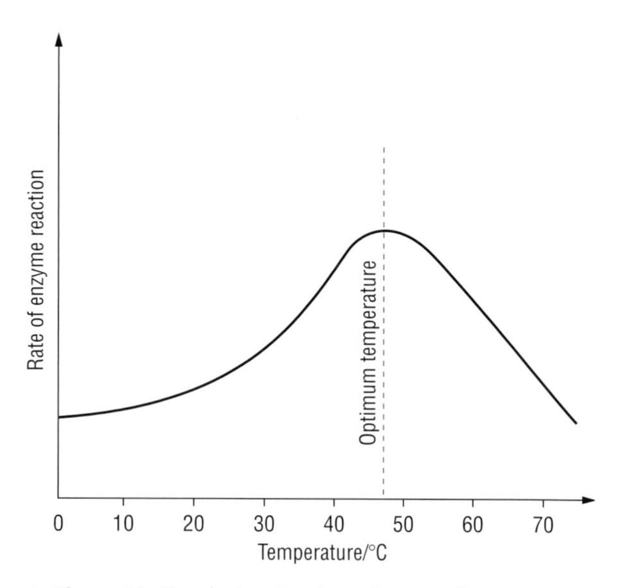

▲ **Figure 21** Graph showing how the rate of an enzyme reaction varies with temperature.

causes them to vibrate more vigorously and weaken or break. Ionic bonds holding the tertiary structure together can be broken by changing the pH. Enzymes are *sensitive* to small changes in *temperature* or *pH*.

Graphs of enzyme activity against pH and temperature show the effect of change in these factors on the rate of enzyme reaction.

Figure 20 shows a graph of enzyme activity plotted against pH. You can see that an enzyme has an *optimum* pH at which it works best. This is usually around pH 7, but don't forget that stomach bacteria, for example, work at much lower pH values. Even small differences from the optimum can reduce the activity, because the ionisation of $-COOH$ and $-NH_2$ groups is changed by these small differences.

Figure 21, on the other hand, shows enzyme activity plotted against temperature. You can see that the enzyme activity rises with temperature at first, as for any other reaction. This is because more molecules have enough energy to collide so that their combined energy is greater than the activation enthalpy. However, at higher temperatures the enzyme is denatured and its activity falls. Enzyme systems have evolved to operate at close to their optimum. For example, most human enzymes have an optimum temperature between 35 and 40 °C – body temperature is 37 °C.

Rates of enzyme (and other) reactions

The study of the rates at which reactions occur is often referred to as **chemical kinetics** (from the Greek word 'kinesis' meaning *movement*). In **Chemical Ideas 10.3** you will learn about **rate equations, rate constants** and **orders of reaction**. You will study experimental methods of investigating orders of reaction, including the use of *half-lives*. Then you will come back and apply these ideas to *enzyme kinetics*.

You can learn more about the effect of concentration on the rates at which chemical reactions take place in **Chemical Ideas 10.3**.

In **Activities TL4.2** and **TL4.3** you can study the effect of concentration on reaction rate.

Enzymes are usually present only in small amounts and their relative molecular masses are very large, so their molar concentrations are very low. If the substrate concentration is high enough, all the enzyme active sites will have substrate molecules bound to them. If the substrate concentration is now increased further, no more enzyme–substrate complexes can be formed, and the rate at which substrate molecules pass through the reaction pathway and change into products remains constant. In this situation the reaction rate does not depend on the substrate concentration – the reaction is *zero order* with respect to substrate. The **rate-determining step** is EP → E + P (Figure 16, page 42).

When the substrate concentration is low enough, not all the enzyme active sites will have a substrate molecule bound to them. The overall reaction rate will now depend on how frequently enzymes encounter substrates, which will depend on how much substrate there is – twice as much substrate, twice as many encounters. So the reaction is now *first order* with respect to the substrate. The **rate-determining step** is E + S → ES (Figure 16). The graph in Figure 22 indicates the variation of the kinetics of an enzyme reaction with respect to substrate concentration.

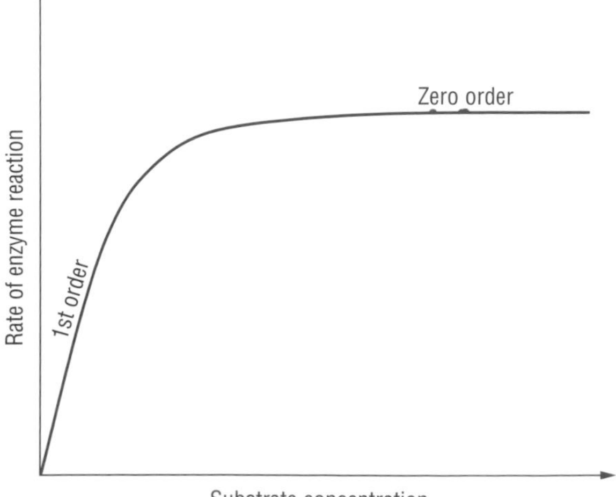

▲ **Figure 22** Graph showing how the rate of a typical enzyme reaction varies with substrate concentration.

You can read more about rate-determining steps in **Chemical Ideas 10.3** and enzyme kinetics in **Chemical Ideas 10.4**.

You can investigate enzyme kinetics further in **Activity TL4.4**, which uses *urease* as an example.

You can consolidate your understanding of *chemical kinetics* in **Activity TL4.5**, which revises the types of graph you meet when studying rates of reaction.

Enzymes at work

The use of an enzyme in a diabetes testing strip to detect glucose is just one of many applications of enzymes. Many medical diagnostic kits are based on the use of enzymes.

However, medical applications use only small quantities of enzymes. Very much larger amounts are used in the food industry and in the manufacture of washing powders. Almost all of these are *hydrolases* – enzymes that hydrolyse fats, proteins or carbohydrates.

In the food processing industry, two of the major uses of enzymes are:

- producing glucose syrup (used as a sweetener in food products) by breaking down starch with enzymes like α-amylase
- making cheese using rennet enzymes – these break down the milk protein casein, and cause the separation of the curds (solid) from the whey (liquid).

▲ **Figure 23** Enzymes are important in the production of bread, cheese and wine.

Other uses include baking, brewing and processing fruit in order to make fruit juices.

Many biological washing powders contain one or more enzymes to assist in the removal of stains. The enzyme is usually a *protease* to hydrolyse proteins in blood and food, but a *lipase* may also be added to break down fats. More recently, *cellulases* have been added too – these break down the tiny surface fibres that give older clothes a fluffy, dull look. Protein engineering is being used to make enzymes that are more stable in hot washes, or to create a wider range of active enzymes that will do their job at lower temperatures.

Enzymes are finding applications in wider areas of waste treatment. For example, an enzyme is being used to destroy cyanide ions, which are left over after gold extraction or after the production of some polymers. Enzymes can also be used to help break up oil spillages. They also find their way into fashion where they are used, for example, to give the 'stonewashed' appearance in jeans.

Enzymes are often used because they are 'greener' than other catalysts. They allow reactions to occur at lower temperatures, conserving energy, and they often reduce the number of steps needed to make a product. This is important because it increases the *atom economy* of the process.

You can read more about atom economy and enzymes in **Chemical Ideas 15.9**.

TL5 *DNA for life*

From our study of proteins, we can see that they are very important molecules in the body. Enzymes are particularly important because they control the reactions that occur in our cells. We need now to consider how cells make proteins.

As a chemist you would need three things before you could synthesise a protein in the laboratory:

- a set of instructions for the protein – in other words, something that told you its primary structure
- supplies of the pure amino acids ready for you to use in the appropriate steps
- a way of making the amino and carboxylic acid groups react with one another more easily.

If we look at how a cell makes its proteins, we can see a close parallel with what a chemist would do.

The instructions specifying the primary structure of a protein are carried by molecules of **deoxyribonucleic acid (DNA)**. You can read about the development of our understanding of the structure of DNA through a series of scientific models in the 'Development of models of DNA' purple box on page 48. Crick and Watson finally came up with the structure and their paper (amazingly short for such a fantastic discovery) was released in 1953. Included in the paper was the 'throw-away' sentence: 'It has not escaped our notice that the specific pairing we have postulated immediately suggests a possible copying mechanism for the genetic material'.

The 'building blocks' for DNA are shown on the left-hand side of Figure 24. These are also shown in the **Data Sheets** that you are allowed to take into examinations. These 'building blocks' are deoxyribose (a sugar), phosphate groups ($H_2PO_4^-$) and four different bases. These bases are adenine (A), thymine (T), cytosine (C) and guanine (G).

Figure 24 shows that a single strand of DNA consists of **sugar** (**deoxyribose**) molecules and **phosphate** groups condensed together to form a long chain of alternating sugar–phosphate groups. This is often referred to as the sugar–phosphate backbone. One of four **bases** (represented by A, T, C and G) is condensed to a deoxyribose unit in the sugar–phosphate backbone. Each sugar/phosphate/base group is called a *nucleotide*. (The order of bases shown in Figure 24 is for illustration only – the sequence is different in different DNA molecules.)

DNA consists of two strands twisted in a helical form with the sugar–phosphate backbone around the outside, held together by hydrogen bonding involving the internal base pairs. It is the bases that form the code for protein synthesis. Crick and Watson were the first to show that DNA consisted of two strands, arranged in a **double helix.** Note that DNA is a 'nucleic acid', *not* a protein. Although many proteins have helical secondary structures, these are *not* double helices!

Assignment 4

This assignment concerns the 'Development of models of DNA' purple box on page 48.

a Comment on Franklin's statements in 1951 that DNA was helical and that the phosphate groups were on the outside. How do her proposals link to the currently accepted model of DNA?

b Draw a diagram of a hydrated magnesium ion (see 'Franklin's response to the 3-chain model'). Why would hydrated ions not be suitable for holding the sugar–phosphate chains together?

c Draw a molecule of unionised phosphoric acid and suggest which groups were involved in Pauling's hydrogen bonds.

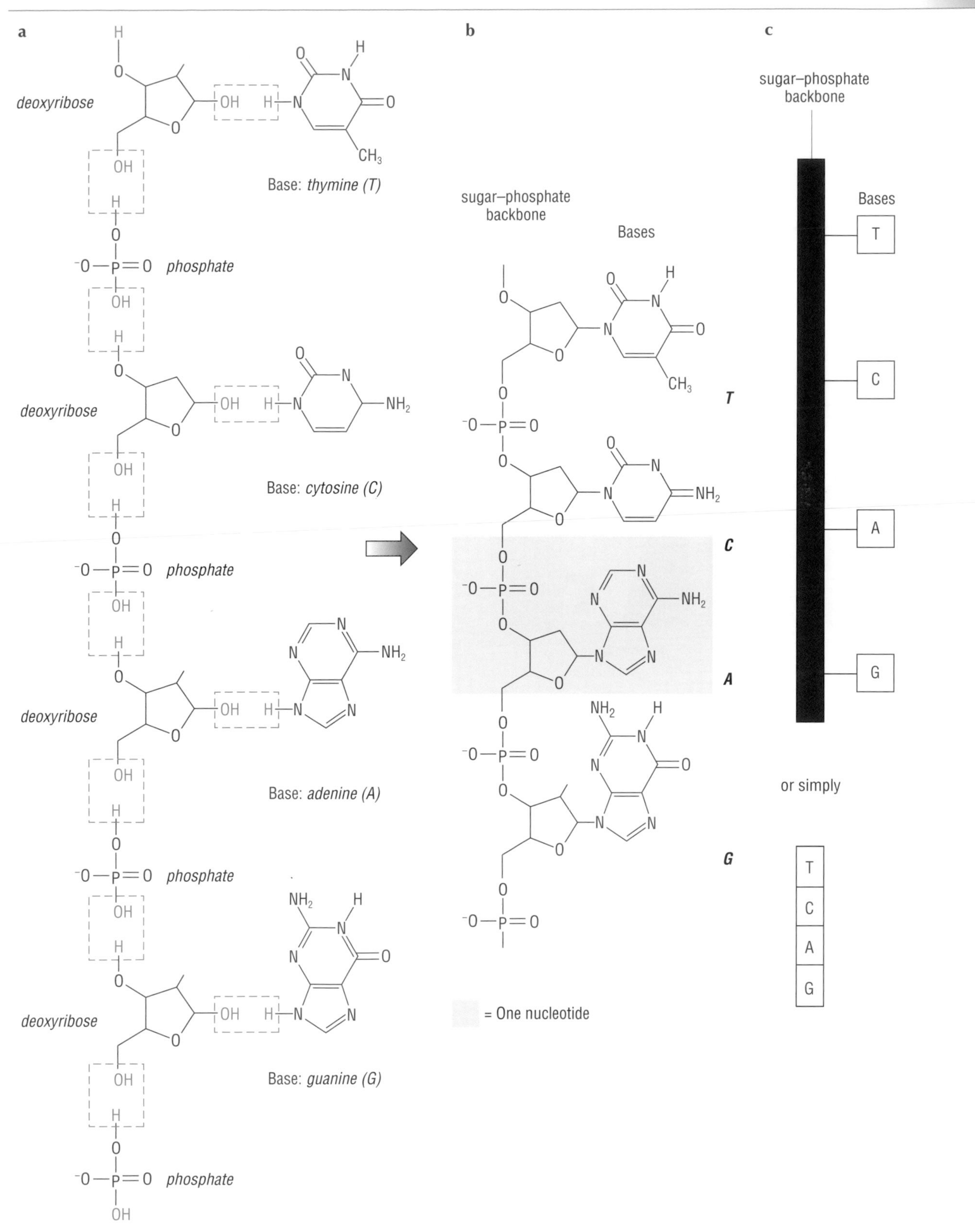

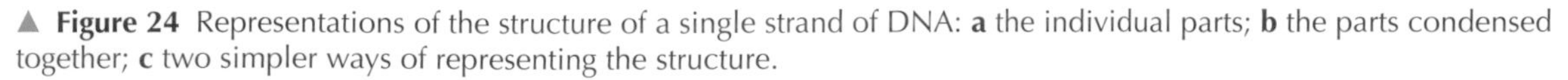
▲ **Figure 24** Representations of the structure of a single strand of DNA: **a** the individual parts; **b** the parts condensed together; **c** two simpler ways of representing the structure.

THE DEVELOPMENT OF MODELS OF DNA

Rosalind Franklin's early ideas (November 1951)

Rosalind Franklin was a brilliant young X-ray crystallographer who worked at Kings College, London. She was at first assistant and then research partner to Maurice Wilkins, a respected physicist who had decide to tackle the structure of DNA using X-ray diffraction. Franklin noted that the diffraction pattern of DNA suggested that the DNA chains were in 'closely packed' helical form, with the phosphate groups near the outside.

▲ **Figure 25** Rosalind Franklin.

▲ **Figure 26** Maurice Wilkins.

Crick and Watson's 3-chain model (1951/1952)

Francis Crick was an English physicist who had his sights set on finding the DNA structure and the almost certain Nobel Prize that would be won for its discovery.

James Watson was an American biologist who came to Cambridge in 1951 to pursue his belief that X-ray diffraction was the clue to understanding macromolecules. He joined a group working on protein structure, but his thoughts, too, kept returning to DNA.

▲ **Figure 27** Francis Crick (right) and James Watson with their final DNA model in 1953.

They decided at first that the sugar–phosphate backbone was in the centre of the molecule, and that there were three chains twisted together, held by doubly-charged cations such as Mg^{2+}.

Franklin's response to the Crick–Watson 3-chain model (1952)

Wilkins and Franklin were invited to Cambridge to be shown the 3-chain model. Franklin disagreed with the initial suggestions of Crick and Watson, saying that Mg^{2+} ions could not be responsible for holding the chains together – something that Watson had to admit was quite possibly true.

Pauling's 3-helix model (1952/1953)

Linus Pauling was a successful and established chemist working at the California Institute of Science and Technology (Cal Tech). He had recently been co-discoverer of the helical structure of proteins. He proposed a 3-helix model of phosphate groups in the middle, with sugars around them and bases on the outside.

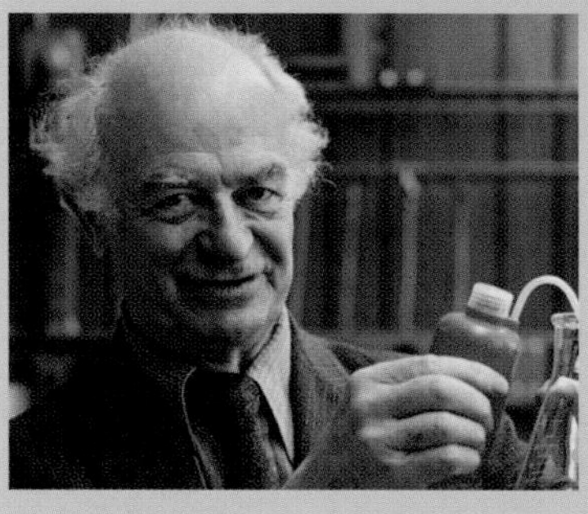

▲ **Figure 28** Linus Pauling.

Watson's response to Pauling (1953)

Watson noticed that the phosphate groups in the centre were not capable of being ionised because the hydrogen bonds that Pauling had postulated as holding his structure together depended on their not being so. It therefore followed that the proposed structure was incorrect.

Crick and Watson's crucial paper (1953)

Crick and Watson finally proposed the structure for DNA that is accepted today (with the now familiar sugar–phosphate double helix surrounding horizontally positioned base pairs holding the chains together) in a paper first published in the journal *Nature* in 1953. For this, they were awarded the 1962 Nobel Prize in Physiology or Medicine, in conjunction with Wilkins. Rosalind Franklin had sadly died by this time.

You can see from this account that over time various models were devised to account for the structure of DNA. Scientists worked to refine preceding models until the currently accepted version was produced.

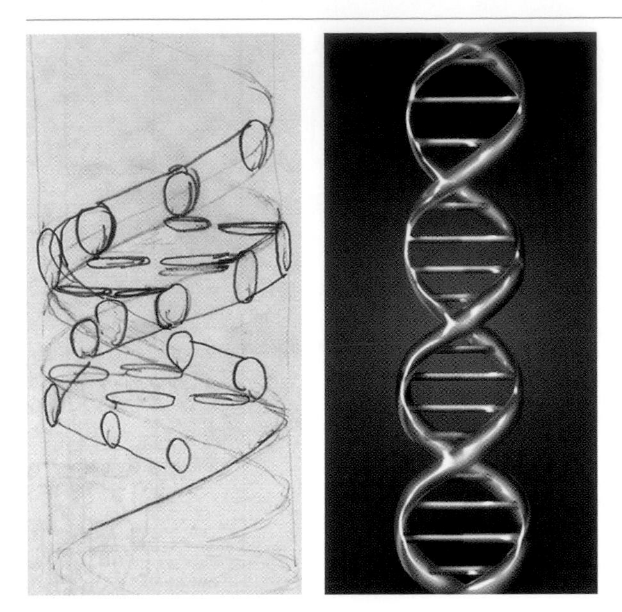

▲ **Figure 29** Francis Crick's original sketch of the structure of DNA, made in 1953 (left). Compare this with the modern 3D representation of DNA as we know it today (right).

Assignment 5

a In the formation of a strand of DNA, what type of reaction is responsible for the linking of:

i the deoxyribose and phosphate?

ii the deoxyribose and base?

b The skeletal formula of deoxyribose is shown in the illustration of DNA given in Figure 24. Draw a full structural formula for deoxyribose.

Base pairing

Crucial to Crick and Watson's double-helix model was the understanding that pairs of bases in DNA can form hydrogen bonds together – adenine (A) with thymine (T), and cytosine (C) with guanine (G). Figures 30 and 31 illustrate this in outline and Figure 32 shows the chemistry of the base pairing.

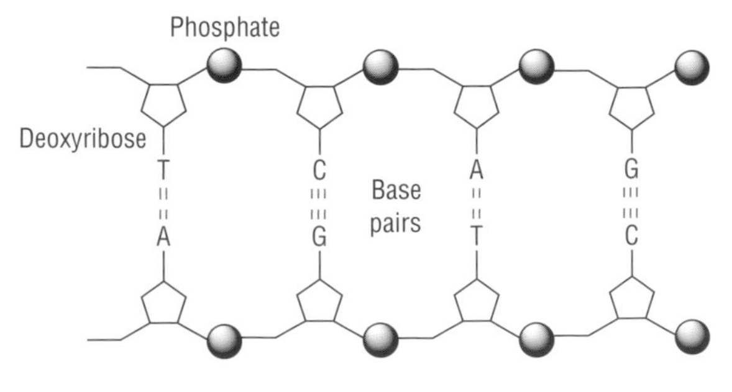

▲ **Figure 30** Two DNA strands held together by hydrogen bonds between pairs of bases.

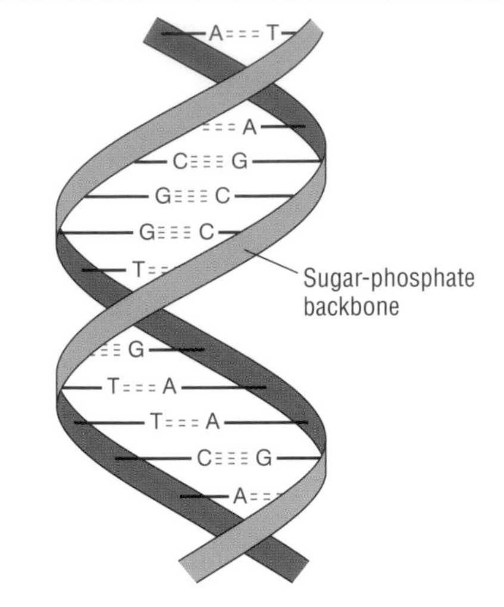

▲ **Figure 31** Illustration of a portion of a DNA molecule showing the double helix – you can think of the sugar–phosphate backbone as the sides of a ladder with the rungs formed by pairs of bases.

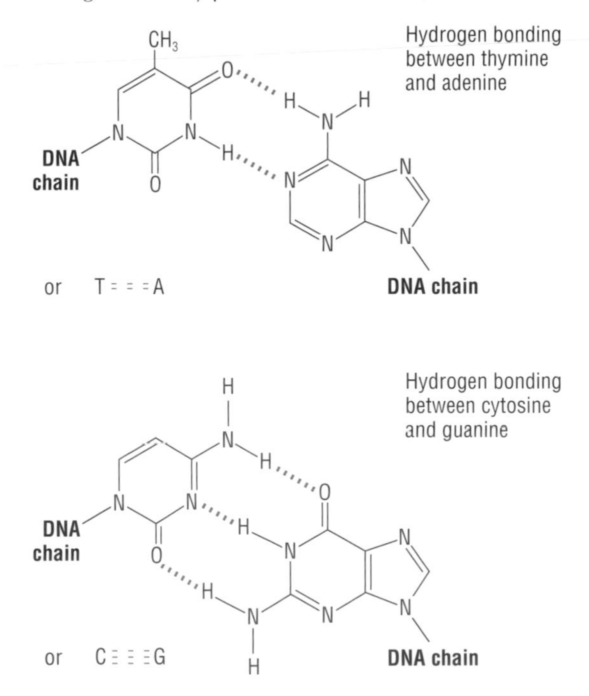

▲ **Figure 32** Base pairing on DNA – note how the molecular shapes of T and A (top) allow them to fit together to form two hydrogen bonds between them, while the molecular shapes of G and C allow them to fit together to form three hydrogen bonds between them.

In **Activity TL5** you can make a model of the DNA double helix.

Chemists have found that the base pairing in DNA enables it to do two things:

- One strand, with the aid of enzymes, can synthesise a complementary copy of itself (i.e. identical to the other strand with which it is normally paired).

For example, TCGAT on the original strand will appear as AGCTA on the complementary replicated strand. This **replication** of DNA happens prior to cell division, so the same information can be found in each new cell.

- One strand can make a complementary copy of another nucleic acid, called messenger RNA, mRNA. It does this by a process known as *transcription*. mRNA is used in the process of protein synthesis.

Transcription

Using enzymes, a section of the DNA relating to one protein unzips – this strand of DNA acts as a template for copying. RNA nucleotides present in the cell nucleus are joined together using enzymes. The order in which the RNA nucleotides join together is determined by the order of bases on the DNA template. When one section of DNA has been 'copied' the double helix then re-forms (zips back up) and the mRNA produced is able to pass out of the cell nucleus. What happens next is described later in this section.

mRNA differs from DNA in the following ways:

- it has ribose as the sugar, not deoxyribose – hence it is called *ribo*nucelic acid
- it has the base uracil (U), rather than thymine (T) – these differ by a methyl group ($-CH_3$) at a place that does not affect base pairing, so U pairs with A, just as T pairs with A in DNA
- it exists only as a single strand and does not pair up.

The similarities and differences between RNA and DNA components can be seen in Table 2.

Table 2 Comparison of DNA and RNA.

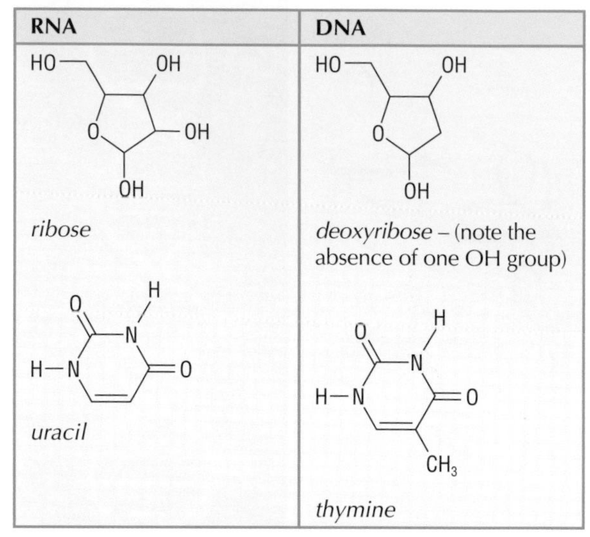

RNA	DNA
ribose	*deoxyribose* – (note the absence of one OH group)
uracil	*thymine*

mRNA carries a code in its sequence of bases that corresponds to particular amino acids – the code is a *triplet code*, with three bases coding for each amino acid. The group of three bases is called a *codon*. Table 3 shows the codes – for example, GUG codes for valine and AAG codes for lysine. Because there are 64 arrangements of three bases and only 20 amino acids, the code is said to be 'degenerate' – there is more than one code for each amino acid. The 'start' code is AUG (that always puts a methionine at the start of the protein sequence – sometimes this is hydrolysed off later). There are three 'stop' codons – UAA, UGA and UAG – that end the synthesis of a chain of amino acids.

Table 3 Triplet base codes (codons) for each amino acid used in mRNA.

First base	Second base				Third base
	U	C	A	G	
U	UUU Phe	UCU Ser	UAU Tyr	UGU Cys	U
	UUC Phe	UCC Ser	UAC Tyr	UGC Cys	C
	UUA Leu	UCA Ser	UAA Stop	UGA Stop	A
	UUG Leu	UCG Ser	UAG Stop	UGG Trp	G
C	CUU Leu	CCU Pro	CAU His	CGU Arg	U
	CUC Leu	CCC Pro	CAC His	CGC Arg	C
	CUA Leu	CCA Pro	CAA Gln	CGA Arg	A
	CUG Leu	CCG Pro	CAG Gln	CGG Arg	G
A	AUU Ile	ACU Thr	AAU Asn	AGU Ser	U
	AUC Ile	ACC Thr	AAC Asn	AGC Ser	C
	AUA Ile	ACA Thr	AAA Lys	AGA Arg	A
	AUG Met	ACG Thr	AAG Lys	AGG Arg	G
G	GUU Val	GCU Ala	GAU Asp	GGU Gly	U
	GUC Val	GCC Ala	GAC Asp	GGC Gly	C
	GUA Val	GCA Ala	GAA Glu	GGA Gly	A
	GUG Val	GCG Ala	GAG Glu	GGG Gly	G

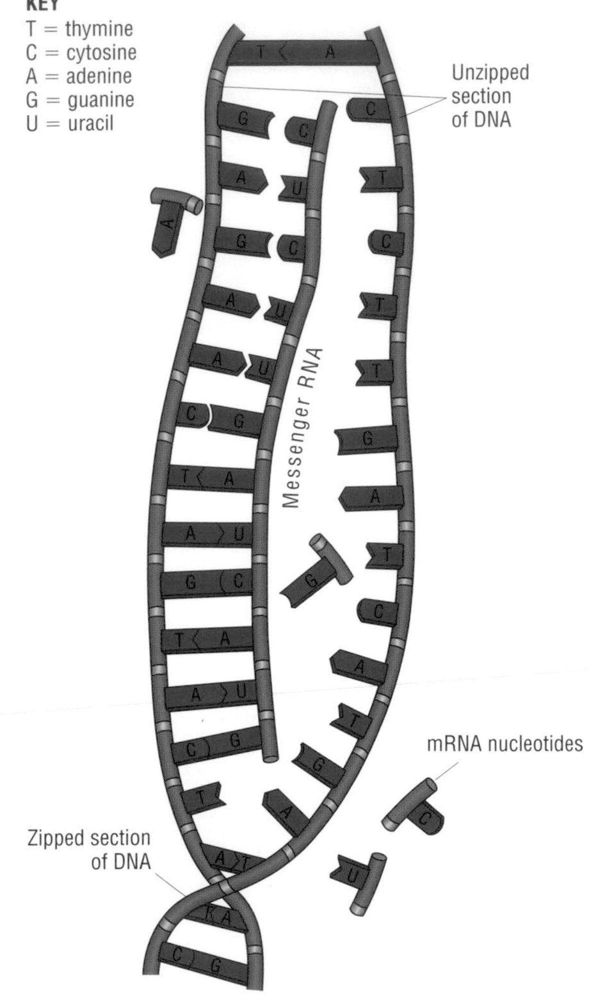

▲ **Figure 33** Diagram representing transcription – note how the mRNA nucleotide bases pair with the bases on the unzipped section of the DNA.

Assignment 6

Table 3 shows that only the first *two* bases of the RNA codon are important for some amino acids.

a The identity of the third base does not matter. Make a list of these amino acids.

b For other amino acids, it is important that all *three* bases are correct. Make a list of these amino acids.

Translation

Protein synthesis takes place in structures called *ribosomes* within a cell. mRNA passes here from the nucleus after being synthesised by transcription (see earlier). During protein synthesis, amino acids need to be joined together in the correct primary structure – mRNA provides a template through which this can happen. Amino acids are taken to the mRNA joined to small lengths of RNA called transfer RNA (tRNA) (Figure 34).

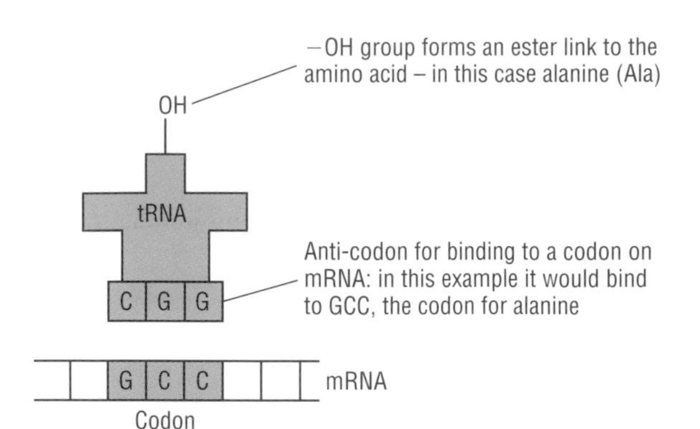

▲ **Figure 34** Schematic representation of a tRNA molecule showing the three bases that form the anti-codon.

The code on the tRNA that fits to the mRNA is called an *anti-codon*, as it is the complementary sequence of bases to the codon. Thus GCC on the mRNA codes for alanine, so one of the tRNAs that attaches to alanine has the anti-codon CGG (Figure 34).

Figure 35 summarises the roles of mRNA and tRNA in protein synthesis. Figure 36 shows how the codons on the mRNA are read as they are in contact with the ribosome and 'translated' into a protein chain.

You can think of the ribosome as rolling along the mRNA chain. tRNA molecules, each carrying an amino acid, feed into the front of the ribosome and the protein chain grows from the back. Part of this diagram is reproduced on the **Data Sheets** that you are allowed to take into examinations.

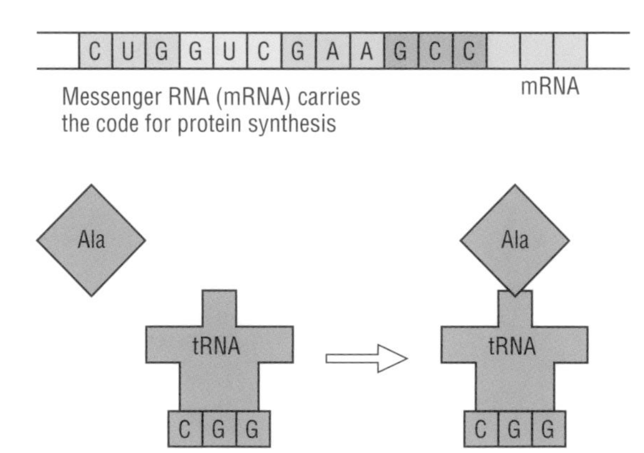

▲ **Figure 35** Messenger RNA and transfer RNA.

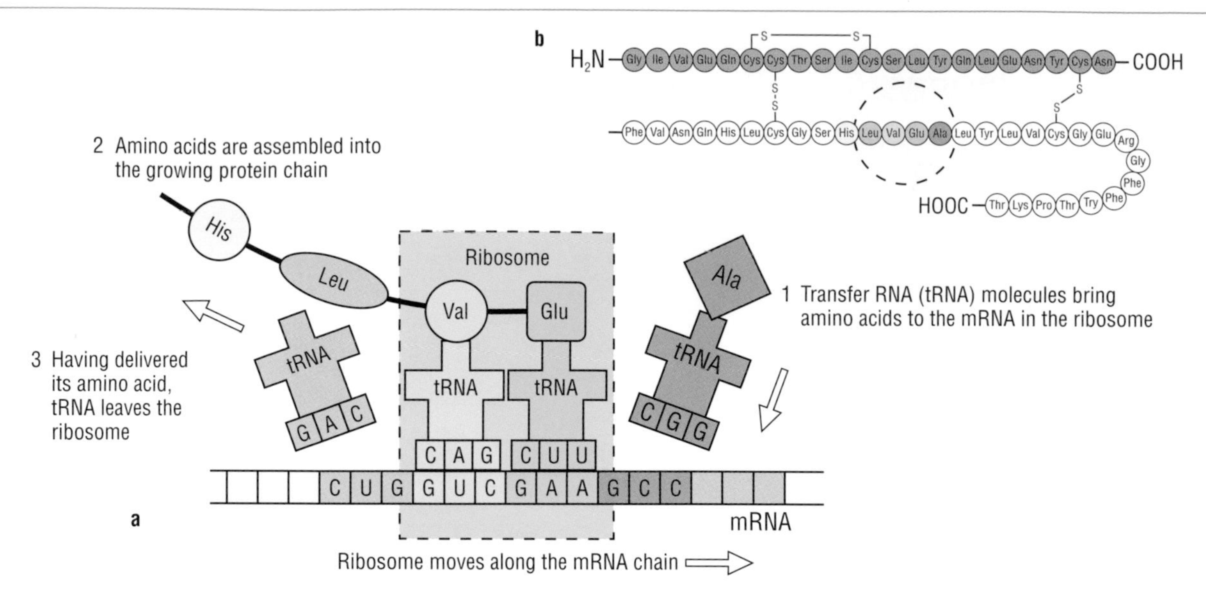

▲ **Figure 36** Protein synthesis and the reading of codons on mRNA: **a** section of insulin under construction; **b** complete insulin molecule.

Assignment 7

a Use the codons from Table 3 to predict the peptides obtained if RNA molecules with the following patterns of bases were used:

i AAAAAA

ii CGCGCGCGC

iii UACCUAACU

b Predict the anti-codons for these amino acids:

i Trp

ii Asp.

In summary:

- DNA codes for RNA (transcription)
- RNA codes for proteins (translation).

Another important contrast between DNA and RNA is that each DNA molecule contains the information for the production of many different mRNA molecules, but, in general, each mRNA molecule is a set of instructions for just one protein. A DNA segment responsible for a particular protein is called a *gene*. Protein synthesis is summarised in Figure 37.

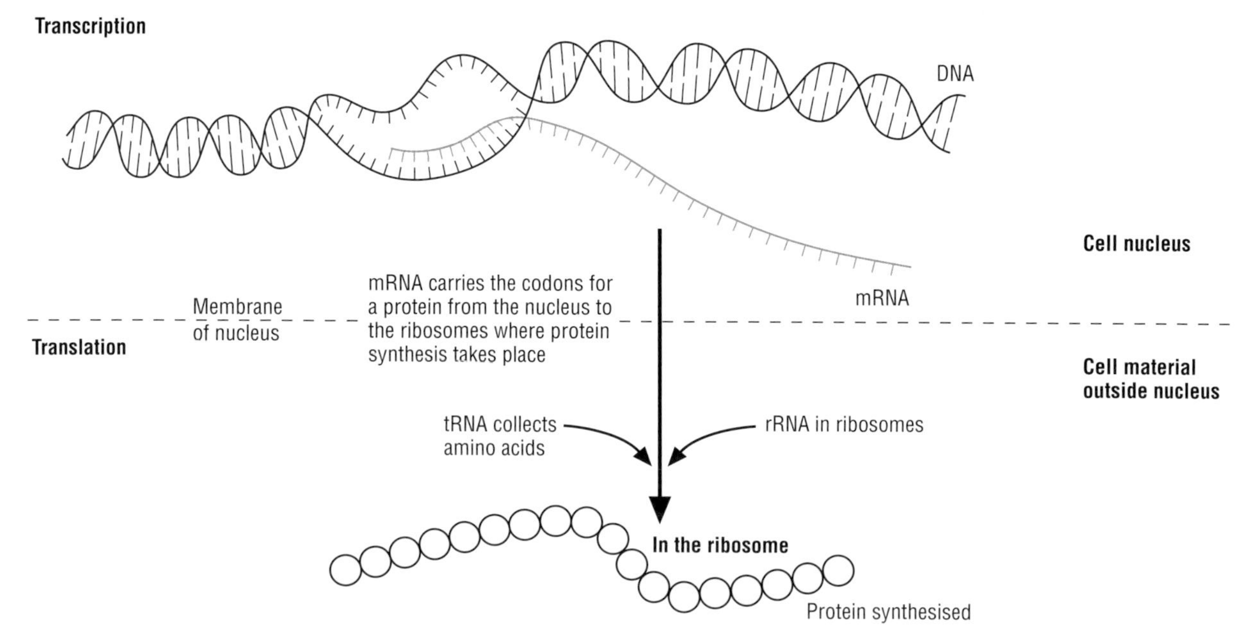

▲ **Figure 37** Summary of protein synthesis in higher organisms.

Assignment 8

One strand of DNA in a cell nucleus carries the sequence of bases CAGT.

a Write down the corresponding sequence of bases on the mRNA strand that copies from it.

b Write down the corresponding sequence of bases on the other DNA strand in the double helix.

Every cell in your body contains a full set of genes in its nucleus and so has a DNA molecule that carries, among others, the gene for insulin production. But the gene is only 'switched on' in the special pancreatic cells that make insulin.

In addition to numerous genes, DNA molecules contain stop codons (base combinations that begin or end RNA production) as well as regions of 'junk DNA' that appear to have no function – perhaps relics of predecessor organisms, or with a function yet to be discovered.

TL6 *Making use of DNA*

DNA finger-printing

DNA finger-printing is based on the fact that no two people (apart from genetically identical twins) share the same DNA sequence. It is now used regularly to help to investigate very serious crimes such as murder or rape.

A trace of blood, semen or skin can be used. A solution of an enzyme is added to the sample – this cuts the DNA at particular sites into a specific pattern of fragments. Fragments from genes are not used in this process, because they do not vary much between individuals. However, areas of 'junk' DNA are used. These nearly always differ (apart from within families, where there are similarities) especially if at least four DNA sites are sampled.

The resulting solution is applied to a gel, which is then subjected to an electric field. The DNA fragments are electrically charged owing to the negative charges carried by the phosphate groups and so the different-sized DNA fragments move at different speeds through the gel towards the positive electrode. The process is known as *gel electrophoresis*.

To see the pattern of a DNA profile test, radioactive tracers are added that bind to DNA fragments. The plate on which the gel is spread is then exposed to a photographic film. A series of bands are seen that are compared with the DNA sample from the suspect.

DNA profiling is also used in paternity disputes, to prove who is the father of a child (Figure 39), in medical analysis of genetic relationships between different people and populations, and in the identification of body remains (Figure 40).

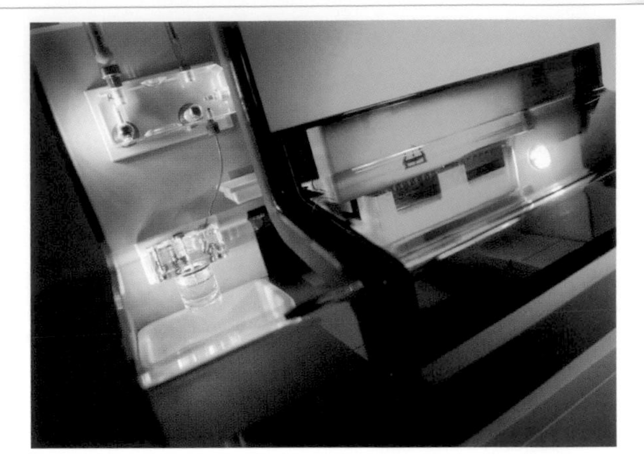

▲ **Figure 38** A DNA sequencing machine – it can sequence DNA fragments to identify genes and their make-up.

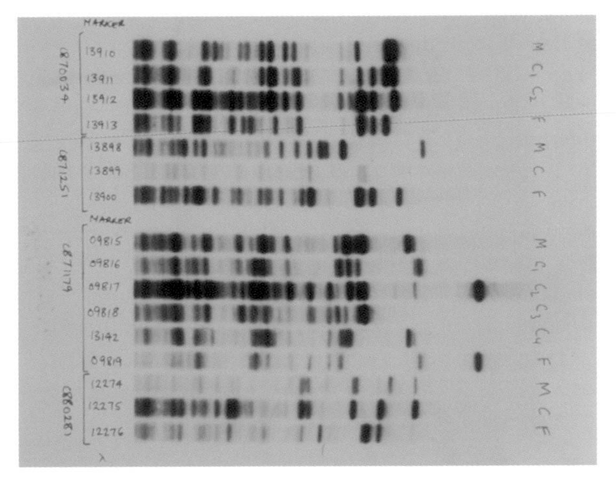

▲ **Figure 39** DNA fingerprinting is used to analyse family relationships. Here we see an X-ray of bands of DNA produced by electrophoresis – the bands are marked M for mother, C for child and F for father (see the right-hand side of the image).

▲ **Figure 40** The identity of the bones thought to be those of the last Tsar of Russia was confirmed by comparing DNA from the bones with DNA from some of the Tsar's living relatives. Tsar Nicholas II was killed in 1918 during the Russian Revolution.

Ethical issues connected with using and storing data from human DNA

One of the main controversies surrounding the use of DNA data is holding them in a national database. In England, anyone arrested on suspicion of a recordable offence must submit a DNA sample to the database, which is then kept on permanent record. The law is different in Scotland and most individuals are removed from the database if they are acquitted.

Should the DNA data of an innocent person be kept? Those conscious of civil liberties point to possibilities of future Governments being less benign than at present as a possible concern. Others argue that having a national database of DNA for the entire population would assist in eliminating innocent suspects from police enquiries and help to assist in prosecuting guilty parties.

Another controversy concerns individual DNA testing. If an individual's genes are analysed, the likelihood of him or her contracting specific genetically linked illnesses could be estimated. On the plus side this could lead to targeted screening and treatment from an early age in order to reduce the impact of the disease. On the minus side, might an individual identified as being at greater risk of suffering from a specific disease in the future be asked to pay higher medical or life insurance? Should they be told that they are likely to contract certain diseases – even if there is no cure for the disease? These are dilemmas to be faced in the future, but they are approaching quickly.

A summary of the main concerns are:

Ownership:

- Who should have access to personal genetic information?
- Who owns and controls genetic information?
- Who owns genes and other pieces of DNA?

Genetic testing of unborn children, children and adults:

- Should parents have the right to have their children tested?
- Should testing be performed for genetic diseases for which there is no cure?
- Should an individual always be given his/her personal genetic information?
- Should insurance companies make use of the data to fix premiums?

Assignment 9

a Consider the 'ownership' of personal genetic data.

i Give some examples of people who, in your opinion, should have access to this data – include your reasons.

ii Give some examples of people who, in your opinion, should *not* have access – include your reasons.

b Choose one or more of the questions under the heading 'Genetic testing of unborn children, children and adults' (left-hand column). For the one(s) you have chosen:

i Give arguments for saying 'yes'

ii Give arguments for saying 'no'

iii Give your opinion, weighing these arguments, of whether the answer should be 'yes' or 'no'.

WHOSE IS THE HUMAN GENOME?

DNA was collected from a large number of donors (using sperm from men and blood from women). Samples of this were used at random for sequencing different regions of the genome, so no one knew whether their sample had been used or what for. Many small regions of DNA that vary among individuals (called polymorphisms) were identified during the project. Most of these appear to make no difference at all to human characteristics. A minority of polymorphisms contribute to the diversity of humanity – for example, eye colour. A much smaller minority affect an individual's susceptibility to disease and response to medical treatments.

Bearing all of this in mind, who do *you* think owns the human genome?

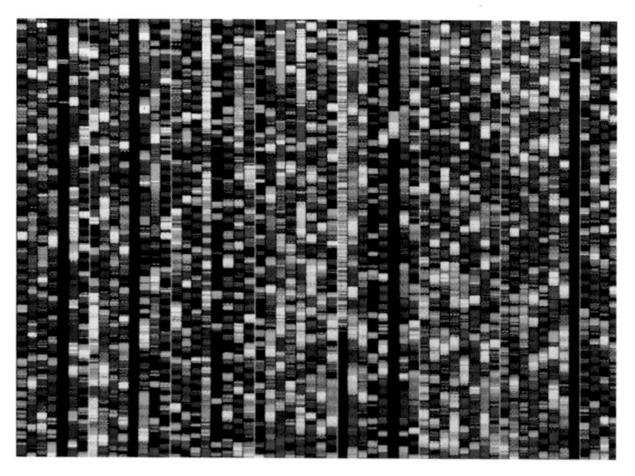

▲ **Figure 41** Computer representation of part of the human genome – each of the coloured bands represents one base.

TL7 *Summary*

This module has introduced you to some of the things chemists have found out about proteins and their building blocks, amino acids. Amino acids (except glycine) are chiral molecules and this introduced you to optical isomerism, a second type of stereoisomerism. Studying amino acids led you to revisit earlier work on carboxylic acids, amines and amides.

Proteins play a wide variety of roles in our bodies – they are molecules of almost infinite diversity, despite all being made by the same type of condensation reaction, which results in the formation of a secondary amide group.

This diversity can be explained in terms of:

- the sequencing of the different structural units of the 20 naturally occurring α-amino acids (primary structure)
- the bonds between these units that give proteins their precise three-dimensional shapes (secondary and tertiary structures).

You then considered one class of proteins – enzymes – which control the rates at which many of the chemical reactions in our bodies take place. This led to the need to develop a more detailed understanding of catalysis, and of how reaction rates are affected by the concentrations of chemicals. It led also to an understanding of the uses of enzymes in industry, particularly their contribution to 'green chemistry'.

The module went on to consider how proteins are synthesised in cells using the DNA that is our 'the thread of life'. The ethical issues of using and storing data from human DNA analysis were also discussed.

Activity TL7 will help you to check your knowledge and understanding of this module.

THE STEEL STORY

Why a module on 'The Steel Story'?

This module tells the story of the production of steel, with emphasis on the redox reactions involved and the huge scale of the process. You will learn about the large variety of steels possible and discover how the composition of a steel is related to the job it has to do.

The story of steel continues with the problems of corrosion. Rusting is introduced as an electrochemical process and various methods of rust prevention are considered. You will use **standard electrode potentials** to explain observations and make predictions about redox reactions. The importance of recycling steel is then discussed, together with some of the problems that must be overcome.

Looking at the composition of steel leads into a detailed study of the properties of iron and other d-block transition metals. These are the structural metals used by engineers to make the things we need in everyday life. The metals and their compounds are also of great importance as catalysts, both in industry and in biological systems. The unique chemistry of transition metals is closely related to their electronic structure – variable oxidation state, catalytic activity, complex formation and the formation of coloured compounds.

Overview of chemical principles

In this module you will learn more about ideas introduced in earlier modules in this course:

- metallic bonding (**Elements of Life**)
- electron energy levels in atoms (**Elements of Life** and **Elements from the Sea**)
- atomic absorption and emission spectra (**Elements of Life**)
- redox reactions and oxidation numbers (**Elements from the Sea**)
- chemical equilibrium (**The Atmosphere** and **Agriculture and Industry**)
- catalysis (**Developing Fuels, The Atmosphere** and **The Thread of Life**).

You will also learn new ideas about:

- the nature and production of steel
- how the properties and uses of steel are related to its composition
- redox reactions and electrochemical cells
- predicting the direction a redox reaction can take
- the properties of the d-block elements
- why compounds are coloured
- complex formation.

SS

THE STEEL STORY

SS1 *What is steel?*

There is no one material called steel – just as there are many plastics, so there are thousands of different steels. Steel is the general name given to a large family of **alloys** of iron with carbon and/or a variety of other elements.

The composition of a steel is determined by the job it has to do. Table 1 shows the wide range of uses of different steels and illustrates the importance of steel in our everyday lives.

Even small differences in the composition of steel have a dramatic effect on its properties. This is particularly true in the case of carbon. Iron containing 4% carbon is extremely brittle and has a limited range of uses. With 0.1% carbon, however, it is easily drawn into wire form and is ideal for making staples or paper-clips. With 1% carbon, steel is stronger without being too brittle – this alloy was used to make the huge cables supporting the Humber Bridge.

ALLOYS

Steel is an example of an alloy – an alloy is a mixture of a metal with one or more other elements. The components are mixed together while molten and allowed to cool to form a uniform solid. The presence of other elements in a metal changes its properties and can often increase its strength. Some examples of different alloys and their uses are shown in Table 2.

As a guide, alloy composition is often expressed as a percentage by mass of the individual elements in the alloy. However, the composition of alloys is not fixed and they cannot be represented by a specific formula in the same way as compounds can be.

▲ **Figure 1** The Humber Bridge – the cables are made of steel.

Table 1 Some of the many ways in which steels affect our lives.

Living with steel	Living with steel
 **Keeping you entertained** From the steel strings of the piano or guitar to the triangle and the steel drum, and on to the lighting that illuminates the top musical or band you are watching – all rely on steel.	 **Material of history** Steel has been used for centuries. This magnificent elephant armour, made in India in the late sixteenth century, recently entered the *Guinness Book of Records* as the largest animal armour in the world!
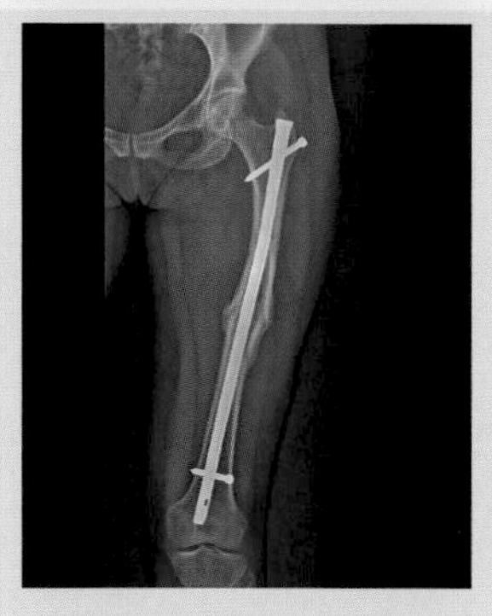 **A caring alloy** Steel protects us and looks after us in so many ways. Apart from sheltering us, consider the role of steel in the metal body protecting us in the car, in gates, door locks, keys, fire extinguisher casings, sprinkler systems and the surgical steel used by surgeons as apparatus and to help repair broken limbs. The list goes on!	 **Green steel** Apart from its use in conventional energy supplies, steel helps to make clean, renewable energy available. It is used in the construction of wind turbines, photovoltaic solar panels and fuel cells. It is also the most widely recycled material in the world – more steel is recycled than all other recyclable materials combined.
 **Movers and shakers** Cars, bicycles, trains, ships, submarines, planes and spacecraft, not to mention the transport infrastructure including bridges, tunnels and roads – all use the strength and versatility of steel to transport us and our goods.	
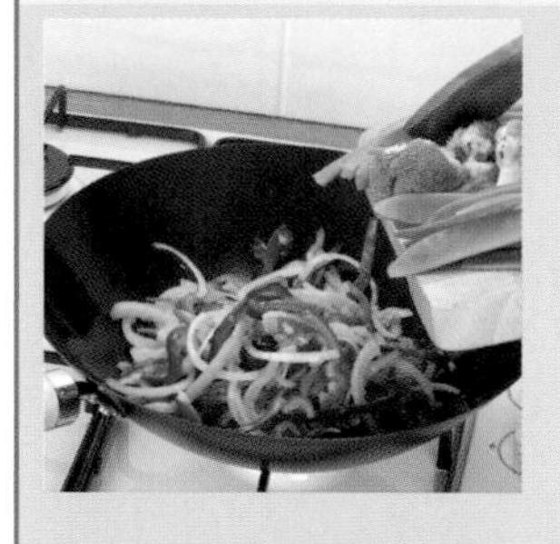 **Nourishing you** The taste sensations of contemporary cuisine cannot be enjoyed without the key part played by steel, either in the kitchen or the stainless steel cutlery on the table – or even in the agricultural machinery used in growing the food we eat. And what about all those 'tin' (steel, really!) cans?	
 **Material of the future** Humans have made innovative use of steel for shelter, and will continue to do so. 17 000 tonnes of steel were used in the construction of Heathrow Airport's Terminal 5.	

In **Activity SS1.1** you can use a microscale method to find out for yourself how much of one element, copper, there is in a sample of the alloy brass.

In **Activity SS1.2** you use colorimetry to determine how much of the element manganese there is in a familiar item, such as a nail.

Chemical Ideas 1.1, **1.2** and **1.3** may be useful in helping with the calculation needed in **Activities SS1.1** and **SS1.2**.

You can find out about the principles of colorimetry in **Chemical Ideas Appendix 1: Experimental Techniques** (Technique 12). This technique measures the intensity of colour formed following a redox reaction.

You can read about why some compounds are coloured in **Chemical Ideas 6.7**.

You can revise ideas about redox reactions in **Chemical Ideas 9.1** and naming inorganic compounds in **Chemical Ideas 3.1**.

Table 2 Some common alloys and some of their uses.

Alloy	Typical composition	Special properties	Typical uses
Steel	**Fe** (99%), C (1%)	Stronger than iron, and more corrosion resistant	Building materials
Duralumin	Al (95%), **Cu** (4%), **Mn** (0.5%), Mg (0.5%)	Increased hardness and tensile strength, light weight	Aircraft parts
Bronze	**Cu** (85%), Sn (15%)	Stronger than copper, corrosion-resistant	Weapons and tools, coinage, decorations
Stainless steel	**Fe** (85%), **Cr** (14%), **Ni** (1%)	Corrosion-resistant	Cutlery, domestic appliances, furnace parts, valves, nuclear reactors
Pewter	Sn (85%), **Cu** (7%), Bi (6%), Sb (2%)	Stronger than tin, but still easy to etch and engrave	Domestic utensils, jewellery
Nichrome	**Cr** (80%), **Ni** (20%)	High resistivity and high melting point	Flame loops, toaster elements
Cupronickel	**Cu** (75%), **Ni** (25%)	Attractive appearance for coins (looks like silver), very ductile	Coinage
Solder	**Pb** (67%), **Sn** (33%)	Harder than lead, with a much lower melting point	To hold other metals together
Brass	**Cu** (60%), **Zn** (40%)	More easily shaped by stamping and machining than bronze	Jewellery, doorknobs, ornaments
Dental amalgam	Sn (44%), Hg (33%), **Ag** (22%)	Resistant to corrosion from the acidic products excreted by mouth bacteria, malleable until set	Fillings
White gold	**Au** (75–85%), **Ni** (8–10%), Zn (2–9%)	Attractive and easily worked, corrosion resistant	High-quality jewellery
Nitinol	**Ni** (55–56%), **Ti** (44–45%)	Shape memory metal	Medical implants, shower safety devices and spectacle frames

Note: the elements shown in bold are transition elements (see section **SS5**).

SS2 *From iron to steel*

Nearly all new steel, whatever its final composition, is made from the same starting material – this is impure iron from a blast furnace, produced as a molten liquid.

You should have studied the extraction of various metals in your earlier studies. The green box on page 60 reminds you of some of the general principles behind metal extraction and purification.

Not all elements are useful when it comes to making steel. Even small amounts of phosphorus, sulfur or dissolved gases (such as oxygen, nitrogen and hydrogen) can lead to poor-quality steel. Brittle steel is a problem in many applications – where it is used in an oil rig or oil pipeline in the North Sea, for example, it could lead to serious consequences. The proportions of elements in a steel need to be carefully monitored during production.

Changing the composition of steel is not the only method used by steelmakers to adjust its properties. The final steel can be subjected to varying degrees of heating and cooling (heat treatment) and worked it by rolling or hammering (work treatment). Both processes modify the metal structure and affect its properties.

Steel is a highly versatile material because both its composition and structure can be adjusted to tailor its properties exactly to the uses in mind. In this module, you will be studying the composition of steel and how this is controlled during manufacture.

Every batch of steel is destined for a particular customer, who specifies the requirements for its composition and any necessary treatments according to its eventual use. This detailed recipe is called the *specification* for that batch of steel.

The impure iron (from the blast furnace) that is the starting material for steelmaking has many other elements dissolved in it. The most important of these are carbon, silicon, manganese, phosphorus and sulfur. Table 3 (page 60) shows the composition of a typical sample of blast furnace iron.

EXTRACTION OF METALS – GENERAL PRINCIPLES

Metal ores are naturally occurring chemical compounds (or elements) that are mined on an economic basis in order to process them to obtain the metal in elemental form.

After removing the surrounding worthless material ('gangue'), the metal is obtained by the process of reduction. This can be either chemical or electrolytic and the choice of method depends on various factors – the relative reactivity of the metal being one of the most important.

Very reactive metals

Highly reactive metals, such as the alkali and alkaline earth metals of Groups 1 and 2, along with aluminium, are usually obtained by electrolysis of a molten compound of the element. For example, sodium metal is obtained by electrolysis of molten sodium chloride. The electrode reactions are:

at the cathode: $Na^+ + e^- \rightarrow Na$

at the anode: $Cl^- \rightarrow \frac{1}{2}Cl_2 + e^-$

Aluminium is obtained by electrolysis of molten aluminium oxide. The electrode reactions are:

at cathode: $Al^{3+} + 3e^- \rightarrow Al$

at the anode: $O^{2-} \rightarrow \frac{1}{2}O_2 + 2e^-$

Less reactive metals

Less reactive metals such as iron, manganese, zinc, tin and lead can often be obtained from their oxides by reduction using carbon (or carbon monoxide). For example:

$$SnO_2 + C \rightarrow Sn + CO_2$$
$$Fe_3O_4 + 4CO \rightarrow 3Fe + 4CO_2$$

The latter reaction occurs in the blast furnace.

Important metals

Certain important metals are extracted by reduction using a more reactive metal. This is often the case when carbon reduction is not feasible, either because the temperature needed to melt the metal compound is very high and/or the carbon reacts with the metal at these temperatures to form a carbide. For example:

$$TiCl_4 + 2Mg \rightarrow Ti + MgCl_2$$
$$Cr_2O_3 + 2Al \rightarrow 2Cr + Al_2O_3$$

Very unreactive metals

Finally, some very unreactive metals are found naturally in their elemental form – gold and silver, for example – and only need extraction from the material they are found in.

Table 3 Analysis of typical blast furnace iron.

Element	Fe	C	Si	Mn	P	S
% by mass	94.0	4.42	0.66	0.41	0.085	0.027

Metal of this composition is very brittle. To turn it into steel, the carbon content must be lowered, most of the phosphorus and sulfur removed and other elements added before the material is allowed to solidify.

All of this is achieved in the basic oxygen steelmaking (BOS) process, in which batches of about 300 tonnes of high-quality steel are made in just 40 minutes. Making steel on a large scale with the right composition is a highly skilled business involving sophisticated technology. The chemistry is spectacular!

Assignment 1

The percentage of carbon by mass in the iron may seem rather low and insignificant (see Table 3). If we look at it in a different way its importance becomes clearer.

Use the information in Table 3 to calculate the relative amounts *in moles* of iron and carbon in blast furnace iron. Neglecting the contributions from the other elements, work out the *percentage of moles* of iron and carbon.

Approximately how many atoms of carbon are there in every 100 atoms of product from a blast furnace?

Removing unwanted elements

Many of the reactions involved in removing unwanted elements from steel are *redox reactions*.

In **Activity SS2** you carry out a redox titration to determine the amount of iron in a sample of the herb, thyme.

You can revise ideas about redox reactions in **Chemical Ideas 9.1**.

You might also want to revise calculations involving concentrations of solutions – you can do this by looking at **Chemical Ideas 1.5**.

The flow diagram in Figure 2 follows the progress of the chemistry involved in the BOS process as impure blast furnace iron is converted to steel – it shows chemical inputs on the left-hand side and outputs on the right. Process flow diagrams (PFDs) are used extensively in industry to provide a clear overview of the stages involved in an industrial process. Each stage is clearly marked and referred to in the following sections using the labels on the diagram in Figure 2.

Removing sulfur

Stage 1

Scrap steel and typically about 300 tonnes of molten iron from a blast furnace are poured into a huge container called a *ladle*. Sulfur is the first element to be removed. This is done in a separate reduction process before the main steelmaking reactions take place. Several hundred kilograms of powdered magnesium are injected through a vertical tube, called a *lance*, into the molten iron in the ladle. In a violent exothermic reaction, the sulfur is reduced to magnesium sulfide, which floats to the surface and is raked off:

$$Mg + S \rightarrow MgS$$

You may find **Chemical Ideas 5.1** will remind you how to write ionic equations.

▲ **Figure 3** This photo showing the charging of the converter gives you an idea of the huge scale of the BOS process.

Assignment 2

a Write an ionic equation to show what happens to sulfur during the reaction with magnesium, and explain why this process is called reduction.

b Draw a dot–cross diagram to show the bonding in magnesium sulfide.

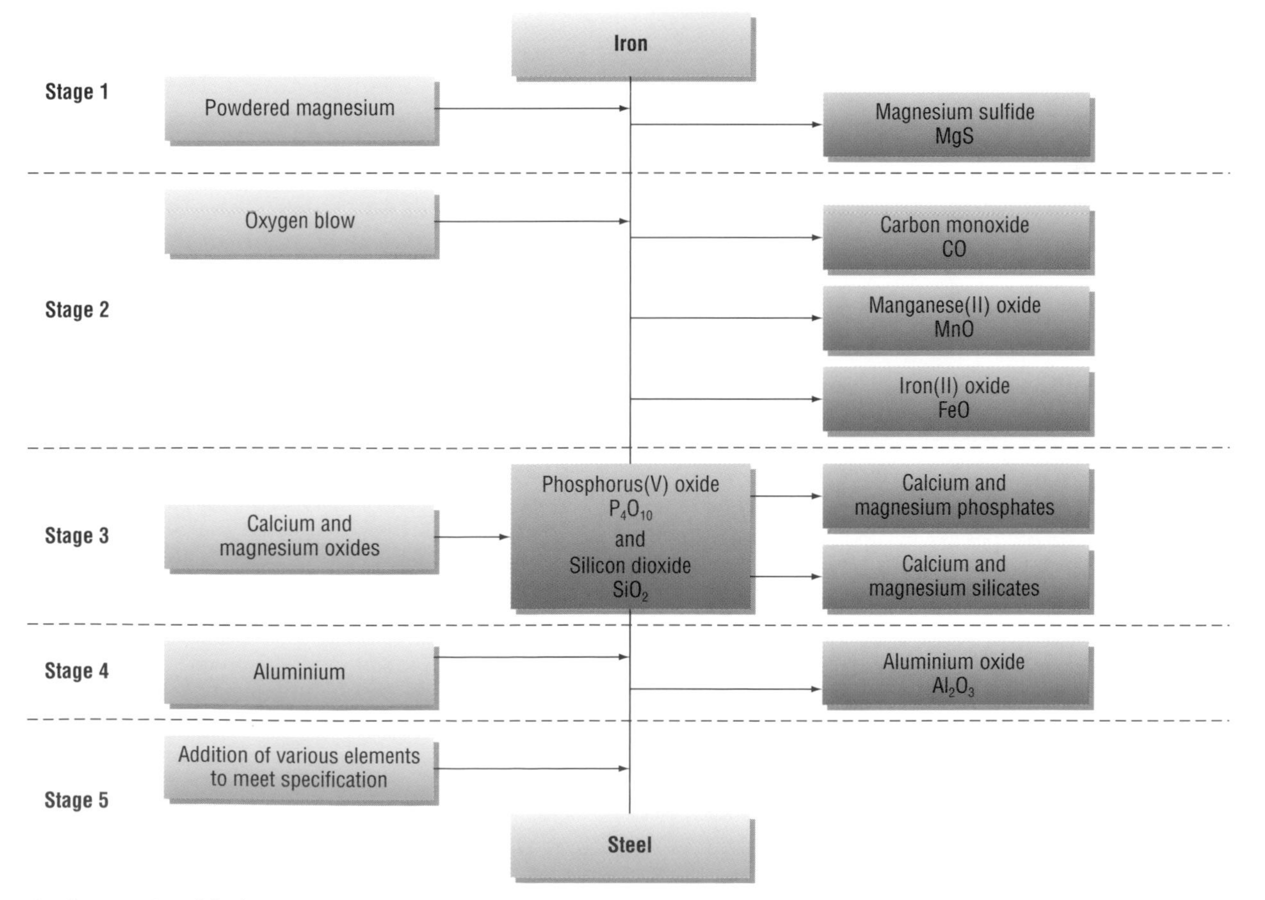

▲ **Figure 2** Simplified process flow diagram for the BOS steelmaking process.

Removing other elements
The huge ladle brings the molten desulfurised iron to the steelmaking vessel or *converter*, which already contains some scrap steel. Now is the time for some very violent chemistry!

Stage 2
Carbon, phosphorus and other elements are removed by direct oxidation with gaseous oxygen (the O in BOS).

$$C + \tfrac{1}{2}O_2 \rightarrow CO$$
$$Si + O_2 \rightarrow SiO_2$$
$$Mn + \tfrac{1}{2}O_2 \rightarrow MnO$$
$$4P + 5O_2 \rightarrow P_4O_{10}$$

Also, unavoidably, $Fe + \tfrac{1}{2}O_2 \rightarrow FeO$

In preparation for the oxygen blow, stage 2 in the process, the converter turns into a vertical position and a water-cooled lance gradually inches its way down close to the surface of the iron. A supersonic blast of oxygen under pressure forces its way into the vessel and creates a seething foam of molten metal and gas that is blasted up the walls of the converter (Figure 4). Over the next 20 minutes or so, most of the impurities of carbon, silicon, manganese and phosphorus, as well as some of the iron, are oxidised.

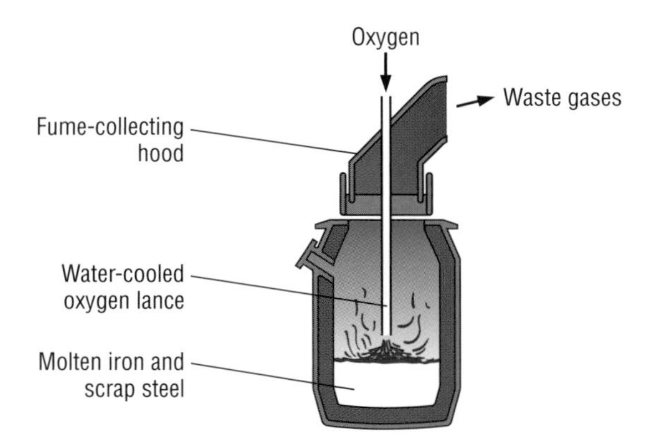

▲ **Figure 4** During the oxygen blow most of the impurities are oxidised and a slag forms on the surface of the molten iron.

Look again at the equations at the start of this section. Carbon monoxide escapes as a gas and is collected via a hood over the vessel. The other oxides remain in the converter and must be separated from the molten metal.

Stage 3
The oxides of phosphorus and silicon are acidic and will react with bases to form salts. So, soon after the oxygen blow has started, a mixture of calcium oxide and magnesium oxide (made by heating limestone and dolomite) is added to the converter.

These are *basic* oxides (the B in BOS) and react with the *acidic* oxides to form a molten 'slag' that floats to the surface. The oxides of manganese and iron also collect in the slag.

The slag can be separated from the molten metal because the two have different densities and form two layers. The slag forms the upper layer and the molten metal the lower layer. You may have used the same principle to separate two immiscible liquids in the laboratory using a separating funnel.

Assignment 3

a Suggest how the carbon monoxide that is collected may be used elsewhere in the plant.
b Apart from economic reasons, why is it not a good idea to release the carbon monoxide into the air?
c When calcium oxide reacts with the acidic oxides SiO_2 and P_4O_{10}, the products are *salts*. Write down possible names and formulae for the two salts formed.
d Explain why CaO is considered to be a basic oxide and P_4O_{10} an acidic oxide.

Keeping track

The process is closely monitored in the control room, where a computer models the conditions in the converter. Two minutes before the predicted end of the oxygen blow, an automatic sampling device, called a *sublance*, descends into the converter. It measures the temperature and carbon content, and removes a sample of metal for analysis.

The computer uses the up-to-date information on temperature and carbon content to predict exactly how much more oxygen is needed to reach the target composition and temperature. The sample withdrawn is rushed to the analytical laboratory to determine the percentages of the elements in the steel.

MEASURING THE COMPOSITION

Analytical chemists quickly measure the composition of steel using *atomic emission spectroscopy*. This involves making the steel sample into an electrode for an electric arc, so that each element present emits a characteristic line spectrum. The intensities of the lines are proportional to the concentration of atoms of each element. At this stage, the analysis involves only a few elements, but the method can be used to monitor up to 20 elements.

You can revise earlier work on atomic emission spectroscopy in **Chemical Ideas 6.1**.

During the oxygen blow (stage 2), the elements are oxidised in a sequence illustrated in Figure 5. When most of the impurities have been removed, some of the iron is also oxidised – this is unavoidable, but is kept to a minimum.

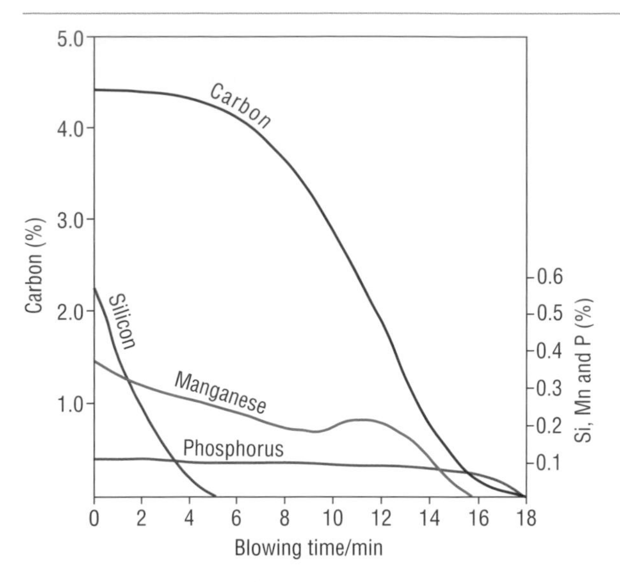

▲ **Figure 5** Removal of elements during steelmaking (note that %C and %Si, Mn and P are plotted on different scales – the rise in %Mn after 9 minutes is because the scrap steel used had a relatively high manganese content).

Controlling the temperature

Careful control of temperature is a vital part of making steel. At the end of the oxygen blow, a temperature of about 1700–1740 °C must be reached – this is the *target tapping temperature*. Higher temperatures waste expensive energy and can cause severe damage to the converter linings. Lower temperatures can be even more costly if the metal solidifies before it is supposed to.

Remember, there is no external heating. The oxidation reactions are all highly exothermic and generate a tremendous amount of energy. Some of this energy is, of course, lost to the surroundings. The rest is absorbed in raising the temperature of the converter contents. Some energy is absorbed by scrap steel (which was added to the converter at the start) as it heats up and melts, maintaining a 'heat balance'. The scrap therefore neatly serves both as a coolant and as a source of recycled steel.

Assignment 4

During the oxygen blow, the elements in the converter compete for the oxygen. The order of their removal as oxides depends on their affinity for oxygen at the high temperature involved, and on the amount of each element present.

Use Figure 5 to answer these questions.

a Which element is the first to be removed?

b Which element is the last to be removed?

c Is this what you would have expected from what you know about the reactivity of these elements with oxygen at lower temperatures?

d At the high temperatures in the converter, sulfur has a similar affinity for oxygen to that of iron. Suggest why it is better to remove most of the sulfur *before* the oxygen blow in a separate process.

At the end of the blow

At the end of the blow, the converter is rotated to pour off the molten steel through a hole near the top into a ladle (*tapping*), and then it is tilted in the opposite direction to remove the slag (*slagging*). Figure 6 illustrates the various positions the converter has to move to during the whole BOS process.

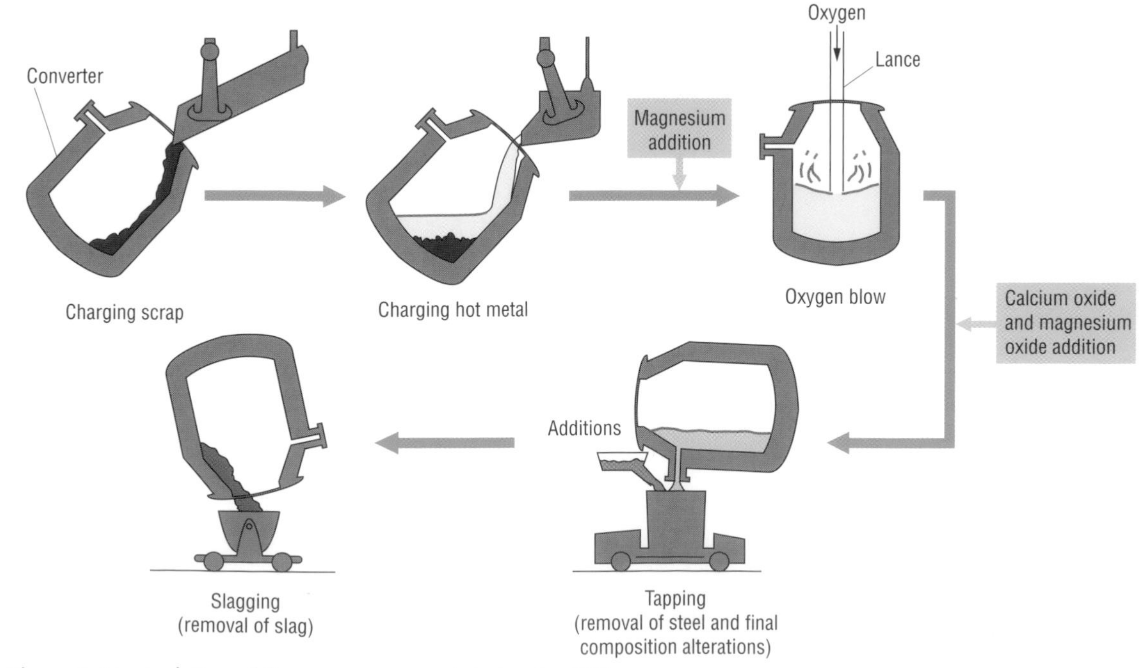

▲ **Figure 6** Manipulation of the converter vessel during the BOS process.

Stage 4

Lumps of aluminium are thrown into the ladle as it is being filled with the molten metal. The aluminium combines with excess oxygen that dissolved in the metal during the oxygen blow. Aluminium oxide forms and floats to the surface.

Meeting the specification

Stage 5

By now the elements that are not wanted – and also some that are – have been removed from the iron. The elements that are needed have to be put back. The computer model predicts the exact quantity of each substance to achieve the *specification* for the particular batch of steel.

Assignment 5

Steelmaking is a batch process. It is found to be most cost-effective in producing batches of around 300 tonnes. However, it is also possible to make steel by a continuous process.

a Describe the main features of a batch and of a continuous process.

b Draw a table to compare the advantages and disadvantages of batch and continuous processes.

Reading **Chemical Ideas 15.1** will help you with this assignment.

Carbon, manganese and silicon, which were removed in the blow, are often added again at this stage. Chromium and more aluminium are other common additions. The addition of cobalt results in a very hard steel used in cutting tools (and maybe even in robots – see Figure 7).

More exotic elements such as niobium, molybdenum and tungsten may also be added. Argon is blown into the liquid steel, using a lance, to stir the mixture and make sure that the composition and temperature are uniform.

Several processes are now available for refining the metal while in the ladle. The *ladle arc furnace*, for example, allows steel in the ladles to be reheated electrically so that it is possible to control the temperature precisely. The ladle arc furnace is particularly important in producing high-carbon/low-phosphorus steels that are used to make steel rods.

The temperature is closely monitored throughout the addition processes and samples are taken frequently for analysis. The computer constantly updates its predictions with this information.

A final 'trimming' adjusts the concentrations of elements in the steel to the required values, and the molten steel is now ready for casting. This is usually a continuous process, carried out by casting into very long strands of solid metal, or it may be poured into moulds in which it solidifies into ingots.

▲ **Figure 7** An android robot called Cog, used to study how humans learn through interactions with others.

Can steelmaking be improved?

World production of steel continues to rise even though plastics and other synthetic materials are taking over the role of steel in some areas. There is a tremendous and growing demand for even higher-quality specialist steels with uniform and consistent compositions.

The BOS process is kept constantly under review and new technology is introduced to produce batches of steel to these very precise specifications as efficiently and economically as possible. For example, recent advances in the continuous casting process have resulted in large energy savings. It may be possible to direct roll the hot steel from the caster – another way of saving energy.

The process is continuously monitored by computers and there are quality measurements at all stages of the process. Computers alert plant engineers when there are problems and the programs give advice on possible solutions – or perhaps even correct the problem.

THE ELECTRIC ARC FURNACE

An electric arc furnace uses old scrap steel that is melted by the heat generated when a spark is produced between carbon electrodes. Lime is added and the impurities are removed as a slag.

By carefully selecting the scrap and making necessary additions, relatively small batches of steel are made to meet given specifications.

As we become more concerned about the environment and recycle more and more steel, larger and more efficient *electric arc furnaces* are taking over a greater share of the production.

On a much smaller scale

Steelmaking has traditionally been a large-scale process. If only small quantities of a particular steel are required, it is uneconomic to use a large-scale operation and the *mini-mill* route is becoming used routinely.

This involves an electric arc furnace, using scrap steel, where as few as 75 tonnes can be produced. Adjacent to the furnace is a casting machine to process the product.

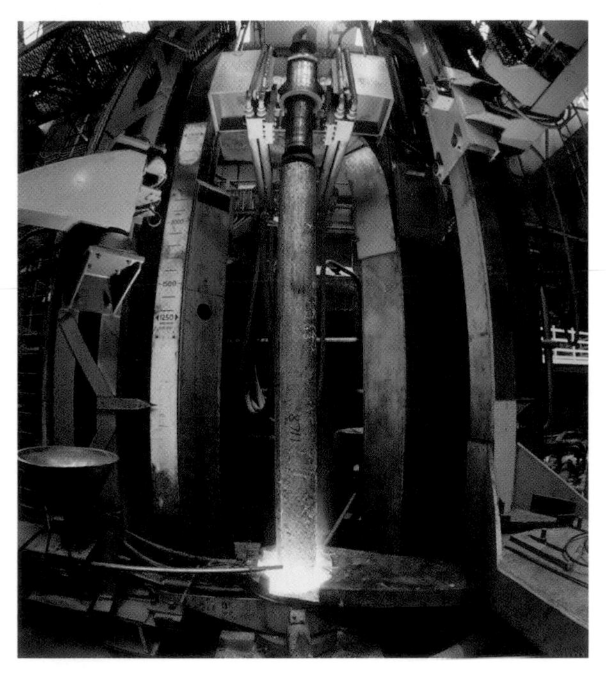

▲ **Figure 8** An electric arc furnace.

SS3 *Steel for a purpose*

Stainless steel

Many of the steel items you use at home, particularly those that come into regular contact with water – such as the kitchen sink, cutlery and the drum of your washing machine – are made of stainless steel, which needs no extra protection.

Stainless steel was developed in 1913 by a Sheffield chemist called Harry Brearley. He was investigating the rapid wear of rifle barrels and decided to try a steel containing a high level of chromium to see if this would prolong their life.

Routine analysis of steel at that time involved dissolving it in acid, and here Brearley met an unexpected difficulty. His high-chromium steel would not dissolve. He also noticed that samples of it left lying around the laboratory for some time stayed shiny.

Brearley immediately realised that he had found a steel that would make excellent cutlery. It would not need to be dried carefully after washing and it would not need frequent polishing.

He did encounter some prejudice about the idea. One of the foremost cutlers in Sheffield thought the idea 'contrary to nature', while another is said to have remarked that 'rustlessness is not so great a virtue in cutlery, which of necessity must be cleaned after each using'! We now take it for granted that our stainless steel knives and forks stay shiny and are not attacked by the acids in food.

▲ **Figure 9** Brearley and some early stainless steel cutlery.

A return to nature

The addition of high levels of chromium to produce stainless steels is one way of stopping a perfectly natural and spontaneous process – the tendency of elemental iron to return to a compound state.

Many metals, including iron, occur in the Earth's crust as oxides. This is because the change from a metal to its oxide is an energetically favourable process – in other words, the oxide is more stable than the metal. Indeed, to reverse the process and extract the metal from its ore requires a great deal of energy. Just think about the high temperatures needed in the blast furnace. No wonder then that iron tends to re-form its oxide – in other words, it *rusts*. Rusting is a common name for the *corrosion* of iron.

Rusting is a redox reaction. To understand this section you will need to find out about electrode potentials and electrochemical cells – you can do this by reading **Chemical Ideas 9.2**.

In **Chemical Ideas 9.3** you can see how electrode potentials can be used to predict the direction a redox reaction can take.

Activities SS3.1 and **SS3.2** will help you understand these ideas.

The rusting of steel is a familiar problem for many car owners – the iron simply returns to its oxide. Cars rust because the steel they are made from reacts with oxygen and water in the atmosphere. When iron or steel rusts a hydrated form of iron(III) oxide with variable composition ($Fe_2O_3 \cdot xH_2O$) is produced. This oxide is permeable to air and water and does not form a protective layer on the metal surface – so the metal continues to corrode under the layers of rust.

STAINLESS STEEL

Contrast the rusting of mild steel with stainless steel. Stainless steel does not rust because it forms a surface layer of chromium(III) oxide (Cr_2O_3). Unlike rust, this oxide is not hydrated and adheres closely to the metal surface. The oxide layer is invisible to the naked eye, being only a few nanometres thick, and allows the natural brightness of the metal to show through. It is impervious to air and water and so protects the metal beneath it. Furthermore, if you scratch the surface film it quickly reforms and restores the protection.

Iron and steel will rust whenever they are in contact with moist air, but the rate of rusting is greatly influenced by other factors, such as impurities in the iron, the presence of acids or other electrolytes in the solution in contact with the iron, and the availability of dissolved oxygen in this solution.

What happens during rusting?

Rusting is an electrochemical process. Electrochemical cells are set up in the metal surface, where different areas act as sites of oxidation and reduction. Two half-reactions involved in rusting are

$$Fe^{2+}(aq) + 2e^- \rightarrow Fe(s) \quad E^{\ominus} = -0.44\,V$$
$$\tfrac{1}{2}O_2(g) + H_2O(l) + 2e^- \rightarrow 2OH^-(aq) \quad E^{\ominus} = +0.40\,V$$

The reduction of oxygen to hydroxide ions occurs at the more positive potential, and so electrons flow to this half-cell from the iron half-cell in which iron is oxidised to iron(II) ions. Figure 10 shows what happens when a drop of water is left in contact with iron or steel.

The concentration of dissolved oxygen in the water drop determines which regions of the metal surface are sites of reduction and which are sites of oxidation.

At the edges of the drop, where the concentration of dissolved oxygen in the water is higher, oxygen is reduced to hydroxide ions:

$$\tfrac{1}{2}O_2(g) + H_2O(l) + 2e^- \rightarrow 2OH^-(aq)$$

The electrons needed to reduce the oxygen come from the oxidation of iron at the centre of the water drop, where the concentration of dissolved oxygen is low. The $Fe^{2+}(aq)$ ions pass into solution:

$$Fe(s) \rightarrow Fe^{2+}(aq) + 2e^-$$

The electrons released flow in the metal surface to the edges of the drop.

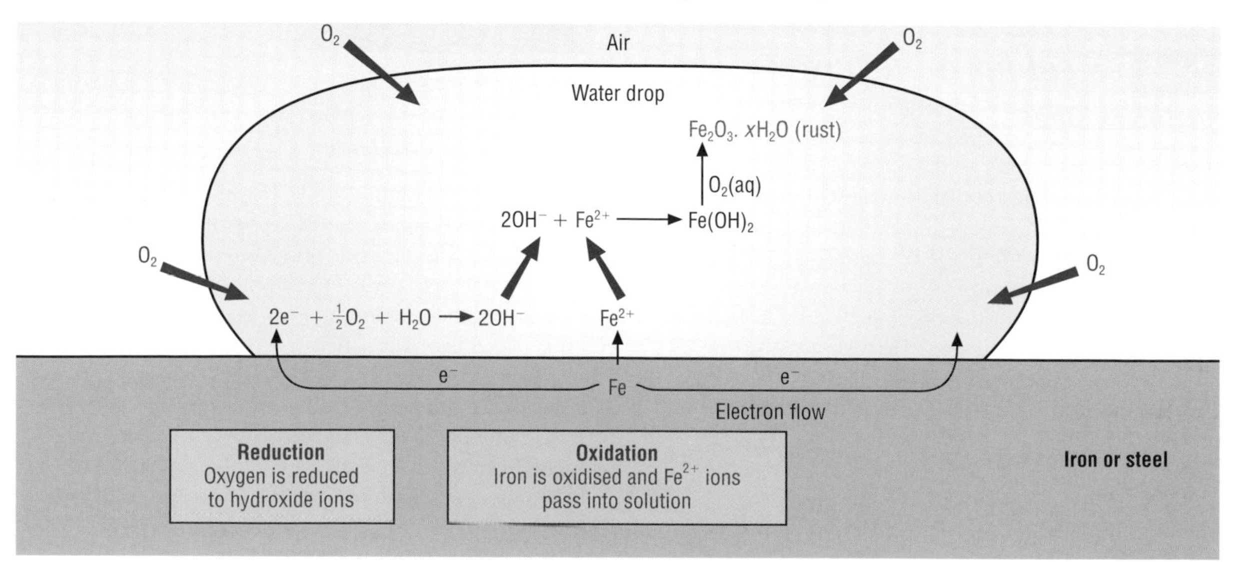

▲ **Figure 10** Rusting is an electrochemical process.

This explains why corrosion is always greatest at the centre of a water drop or under a layer of paint – these are the regions where the oxygen supply is limited. 'Pits' are formed here where the iron has dissolved away.

Rust forms in a series of secondary processes within the solution, as Fe^{2+} and OH^- ions diffuse away from the metal surface. It does not form as a protective layer in contact with the iron surface.

$$Fe^{2+}(aq) + 2OH^-(aq) \rightarrow Fe(OH)_2(s)$$

$$Fe(OH)_2(s) \xrightarrow{O_2(aq)} Fe_2O_3{\cdot}xH_2O(s)$$

Some ionic impurities, such as sodium chloride from salt spray near the sea, promote rusting by increasing the conductivity of water.

Other ionic compounds can interfere with the electrochemical reactions involved and actually inhibit rusting. This might happen if the positive ions form an insoluble hydroxide with the $OH^-(aq)$ ions produced in the oxygen half-reaction, or if the negative ions form an insoluble iron (II) compound. Sodium chloride does neither of these.

The pH of the solution is also important – rusting is accelerated under acidic conditions but inhibited under alkaline conditions.

Activity SS3.3 is an activity that will help to support your knowledge of electrochemical cells.

Keeping nature at bay

The simplest way of protecting steel against rust is to provide a barrier between the metal and the atmosphere. The barrier may be oil, grease or a coat of paint.

Barriers made from organic polymers are increasingly used. The steel is coated with a plastic film – a colourful and flexible answer to the rusting problem. A quick look around at home will provide many examples – sink drainers, refrigerator and dishwasher shelves and even car bodies (see Figure 13).

Sometimes iron is covered with a thin layer of another metal. Many car manufacturers make their car bodies from *galvanised* steel – this has a protective coating of zinc. While the galvanised surface is undamaged, the zinc layer is protected from corrosion by a firmly adherent layer of zinc oxide.

Even if the coating is scratched the protection is still maintained, because the zinc corrodes in preference to the iron – the zinc is being used as a *sacrificial metal*. One of the earliest examples of using a sacrificial metal was suggested by Humphry Davy in 1824 to protect the metal sheathing on sailing ships from corrosion.

Great improvements in protecting cars from rusting have been made over the last few years (see Figure 13). Indeed, one manufacturer offers a 12-year anti-corrosion

▲ **Figure 11** The 'Angel of the North' sculpture near the A1 road near Gateshead is made from weather-resistant high-strength COR-TEN steel. The steel contains 0.25–0.40% copper and weathers to a rich brown colour. Unlike most steels, in this 'weathering steel' (more correctly called 'atmospheric corrosion-resistant steel') where the oxide matrix on the surface forms a protective layer, preventing further rusting.

▲ **Figure 12** Early advertising signs were saved from the damaging effects of weather and pollution by a corrosion-resistant coating of vitreous enamel. The signs often corroded at the edges, where the protective enamel coating became damaged.

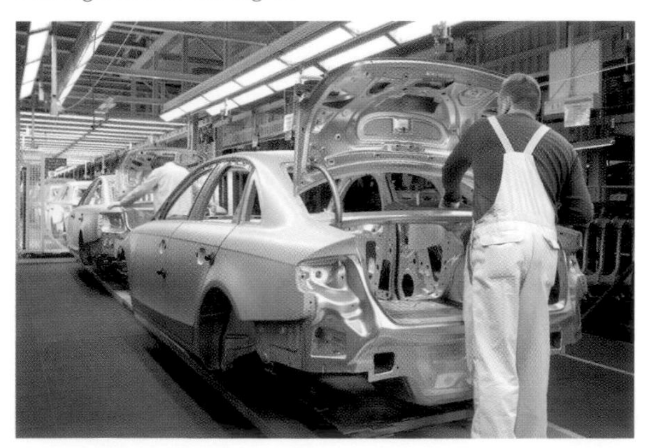

▲ **Figure 13** Steel car bodies are dipped into an aqueous dispersion of tiny particles of polymer. The coating is then heat treated to form a film on the steel that will protect it from rusting.

warranty on all new car models. Even so, if steel is used in car bodies then the rusting problem is only postponed, not eliminated.

Assignment 6

Use the **standard electrode potentials** given below to explain why zinc and magnesium, but not tin, can be used as sacrificial metals to protect steel.

Half-reaction	$E^{\ominus}$/V
$Mg^{2+}(aq) + 2e^- \rightarrow Mg(s)$	–2.36
$Zn^{2+}(aq) + 2e^- \rightarrow Zn(s)$	–0.76
$Fe^{2+}(aq) + 2e^- \rightarrow Fe(s)$	–0.44
$Sn^{2+}(aq) + 2e^- \rightarrow Sn(s)$	–0.14

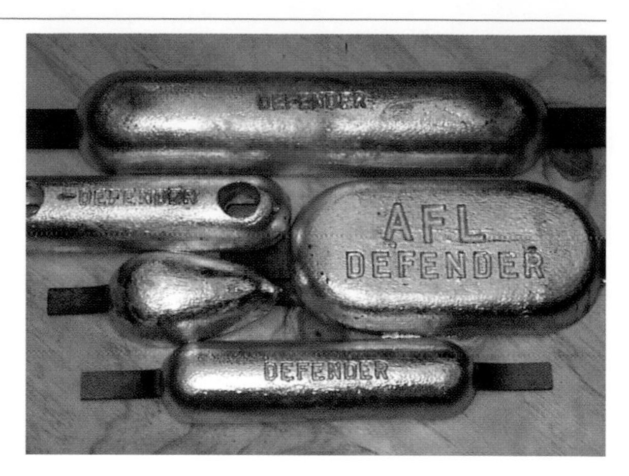

▲ **Figure 15** Zinc blocks of the type that are attached to the steel hulls of ships for sacrificial protection.

In the sea, conditions are far from standard, but the *order* of electrode potentials for these reactions is not changed.

Today blocks of zinc are used to protect North Sea oil rigs and the hulls of ships by transferring corrosion from a valuable complex steel structure to a readily replaceable metal lump (see Figures 14 and 15) .

You can see the reason for using zinc by comparing the standard electrode potentials of the iron and zinc half-cells. Any metal with a more negative value than iron could be used as a sacrificial metal. Tin cannot be used as a sacrificial metal to protect steel, but as long ago as 1812 iron coated with tin was used to make containers – 'tin' cans – to preserve cooked food.

Activity SS3.4 focuses on predicting the feasibility of redox reactions.

Referring to **Chemical Ideas 11.5** (the section on the stability of oxidation states) and **Chemical Ideas 9.3** will help in your understanding of this activity.

A prize-winning invention

Napoleon remarked at the beginning of the nineteenth century that 'an army marches on its stomach'. His armies were widely separated, fighting campaigns in Russia and Spain, and were severely limited by long supply lines from France and the lack of fresh provisions.

Napoleon offered a prize to anyone who could suggest a solution to his food supply problem. In 1812 he awarded 12 000 francs to a French confectioner, Nicolas Apert, for inventing a method of preserving cooked food by sealing it in an airtight glass jar while it was still hot.

Meat, vegetables and fruit could be kept palatable for long periods of time. Later that year an Englishman, Peter Durand, adapted Apert's method by using a tin-plated iron canister instead of a glass bottle – with obvious advantages. So the tin canister, or 'tin can', was born.

In 1824 Captain Sir Edward Parry set out on his third voyage in search of the North-West Passage from the Atlantic to the Pacific Ocean, taking with him a good supply of canned food. His aim was to find an Arctic

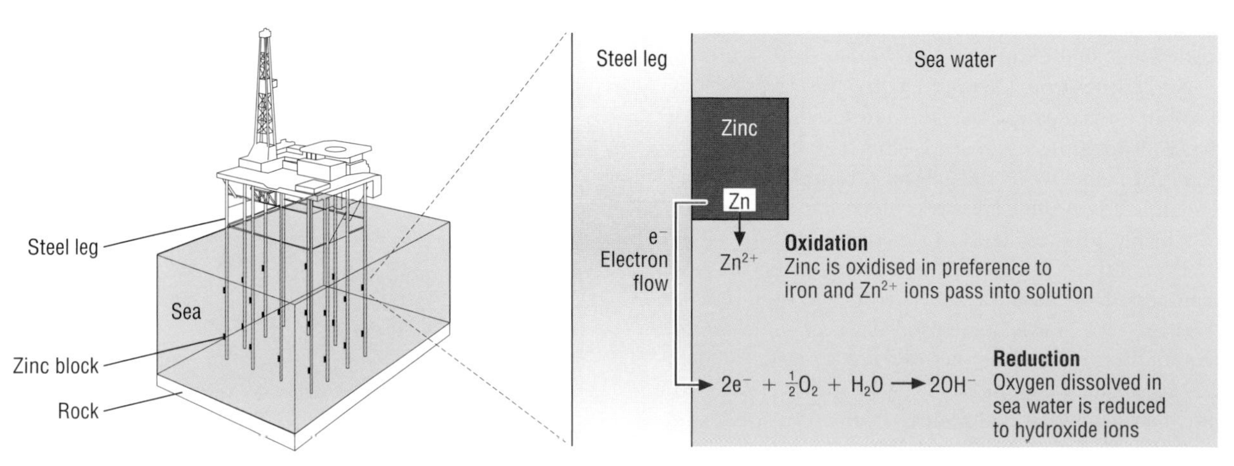

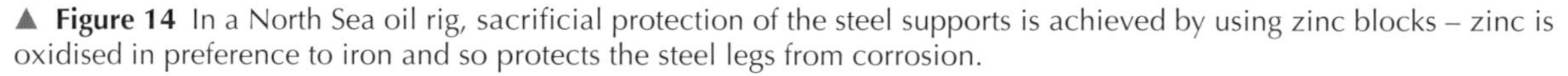

▲ **Figure 14** In a North Sea oil rig, sacrificial protection of the steel supports is achieved by using zinc blocks – zinc is oxidised in preference to iron and so protects the steel legs from corrosion.

route to India and the Far East. One of his ships became ice-bound and he was forced to abandon it together with a large quantity of stores.

Some of the cans were recovered by Captain Ross in a similar expedition four years later (Figure 16). In 1939 one of these cans was opened – after more than a century, the roast veal and gravy inside were still wholesome.

▲ **Figure 16** The famous can of veal taken by Parry on his 1824 expedition from the Atlantic to the Arctic Ocean in search of the North-West Passage.

Assignment 7

a When preserving food, why is it important that the jar or can is airtight and sealed while the food is still hot?

b Peter Durand used iron plates dipped into molten tin and soldered together to make a container. What was the purpose of the tin?

c What would happen if a tin can were scratched on the outside so that the iron showed through? Refer to the standard electrode potentials given in Assignment 6.

d Suggest reasons why it is not a good idea to use zinc instead of tin to coat the can.

SS4 *Recycling steel*

About 45% of the world's steel production is from recycled steel. This means that over 200 million tonnes of iron are recovered each year, with an energy-saving equivalent to 160 million tonnes of coal or 100 million tonnes of oil – about 40% of the UK's annual energy consumption. The magnetic properties of steel make it easy to separate from other materials, and cleaning can usually be done by simple incineration to remove labels and paint.

You saw earlier in the module that recycled scrap is an integral part of the BOS process, where it acts as a coolant. Scrap steel makes up about 18% of every cast of 'new' metal. In the electric arc process for making steel, *only* scrap is used.

Much of the scrap used in steelmaking comes from the steelworks itself – waste from previous batches, miscasts, etc. – or from industries that make the steel products. The composition of this type of scrap is well known. In this respect, the steel industry consumes its own waste.

Scrap from discarded products, such as cars and washing machines, must be carefully graded and selected. Steelmakers need to have a good idea of the content of the scrap metal to avoid adding unwanted elements to the steel. Some of the elements present improve the properties of the steel. For example, many mild steels now contain low concentrations of transition metals such as nickel and chromium from the added scrap. Other elements such as copper can cause problems (by altering softness) if incorporated into the 'new' steel.

▲ **Figure 17** An electric arc furnace (shown in Figure 8, page 65) can be used to recycle scrap steel like this.

Recycling used 'tin' cans

Recycling used 'tin' cans involves removing the tin coating from the steel. This has been done for a very long time using the waste metal from the tin plating works. Only since the 1980s have attempts been made to extract used 'tin' cans from household waste and recycle them on a large scale.

Shredding the cans and removing unwanted paper and residual food is a vital part of the preparation before detinning. One easy way of cleaning the cans is to burn off the unwanted material – unfortunately, the tin diffuses into the steel during burning and makes it less useful. Mechanical shredding devices now shred

▲ **Figure 18** In the UK, we use 13 billion steel cans a year – with every household using approximately 600 cans in that time. The cans weigh 40% less than those produced 30 years ago and the tin coating is less than 6×10^{-3} mm thick, an enormous saving in mineral resources, energy and waste. 70% of all steel packaging is recycled.

and clean the cans and the steel fragments are picked out magnetically. About 98% of the unwanted material can be removed in this way.

The cleaned and shredded tin cans are treated with a hot solution of sodium hydroxide in the presence of an oxidising agent. The tin dissolves as a compound of tin(IV), as shown in the half-equation below:

$$Sn(s) + 6OH^-(aq) \rightarrow [Sn(OH)_6]^{2-}(aq) + 4e^-$$

stannate(IV) ion

The steel left behind is rinsed and pressed into bales for transport to a steel plant. The tin can be recovered by electrolysis.

SS5 *Partners in steel*

There may be many different elements present in a steel in addition to iron. A few, such as carbon and silicon, are non-metals but most are metals. Look back at Table 2 in this module – it lists some of the metals commonly used to make alloys, including steel. Many of the elements present in steel, including iron, are **d-block elements**.

Understanding the chemistry of d-block elements

To understand how steel behaves when exposed to weathering and what can be done to prevent corrosion, or to understand how fruit juices can affect the inside of a food can, you need to know more about the chemistry of d-block elements.

These elements are sometimes called **transition metals** because they show a transition in properties between the reactive s-block metals and the less reactive metals on the left-hand side of the p-block. Their chemistry is very characteristic and is a direct result of their electronic structure.

You can revise the positions of the main blocks in the Periodic Table in **Chemical Ideas 11.1**.

You can also recall formulae of ions and the link between ion charge and group in the Periodic Table in **Chemical Ideas 3.1**.

Chemical Ideas 11.5 tells you about the properties of d-block elements.

You can remind yourself about energy levels in atoms and how electrons are arranged in these energy levels by reading **Chemical Ideas 2.4**.

Typical chemical properties of transition metals include:

- the formation of compounds in a variety of oxidation states
- catalytic activity of the elements and their compounds
- a strong tendency to form complexes
- the formation of coloured compounds.

The elements on the edges of the d-block, such as scandium and zinc, do not show many of these properties and are not usually classed as transition metals.

Variable oxidation states

Metals like sodium or magnesium have just one oxidation state in all of their compounds, but transition

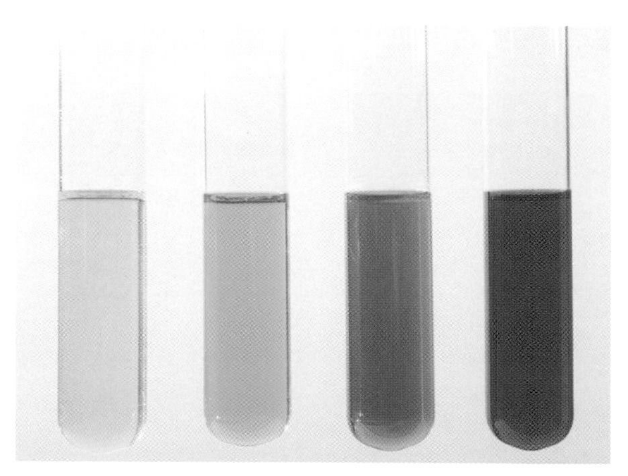

▲ **Figure 19** Oxidation states of vanadium – from left to right, the test tubes contain solutions of vanadium in oxidation states +5, +4, +3 and +2.

metals form compounds in a range of oxidation states, many with beautiful and characteristic colours (Figure 19). **Activity SS3.4** has already introduced you to these ideas.

Complex chemistry

You may have met the blue **complexes** of copper(II) ions with water and ammonia **ligands** in your earlier chemistry studies during this course. **Complex**

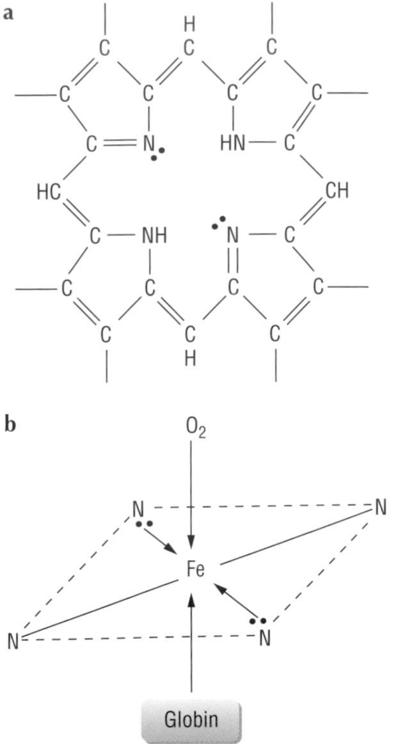

▲ **Figure 20 a** The porphyrin ring system; **b** haemoglobin bound to an oxygen molecule to form oxyhaemoglobin.

formation is a further characteristic property of transition metals – typically, a central metal atom or ion is surrounded by six electron-donating ligands, although complexes with two and four ligands are also common.

Iron forms the red complex haemoglobin, responsible for transporting oxygen in the blood (see Figure 20). The oxygen molecule is relatively loosely attached to the iron. It is carried around the body in the bloodstream and released from the complex when needed.

Complexes even occur inside cans of fruit and this may cause a few problems! (See Figure 21.)

You can read more about the chemistry of complexes in **Chemical Ideas 11.6**.

In **Activity SS5.1** you will explore the properties of some transition metal ions.

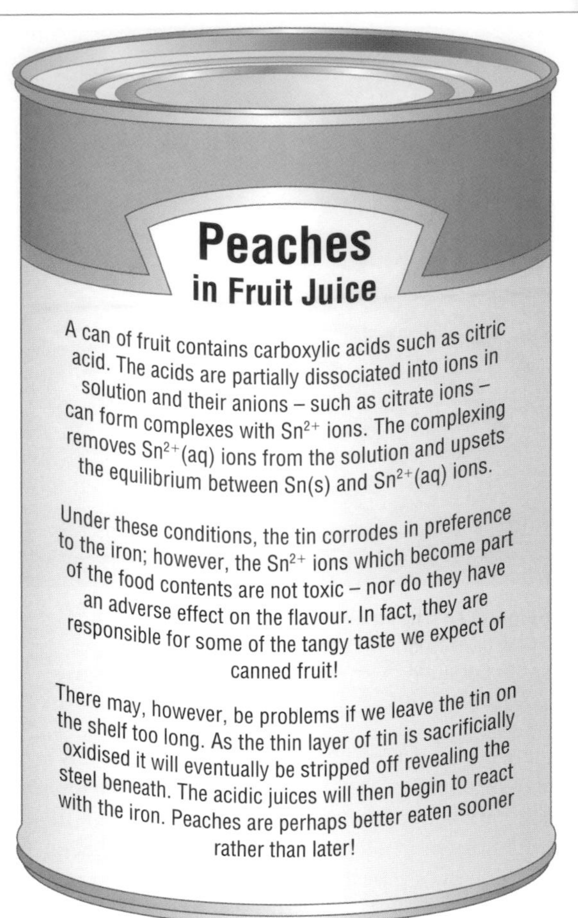

▲ **Figure 21** The formation of complexes inside a can of peaches – a tasty reaction?

Table 4 The colours of many gemstones are due to the presence of traces of d-block metal ions.

Gemstone	Colour	Ion present
blue sapphire	blue	V^{3+} or Co^{2+}
emerald	green	Cr^{3+}
topaz	yellow	Fe^{3+}
turquoise	blue–green	Cu^{2+}
amethyst	purple	Mn^{3+}
ruby	red	Cr^{3+}

Coloured compounds

The colour of a transition metal compound depends on the oxidation state of the metal ion, the nature of the ligands surrounding it and the spatial arrangement of these ligands. A wide variety of colours is observed (Table 4).

Activity SS5.2 is a card-matching activity to help you learn the formulae, names and colours of transition metal compounds.

Catalysis

Another characteristic property of transition metals and their ions is that many can act as catalysts. For example:

- iron is the catalyst in the manufacture of ammonia in the Haber process
- nickel is used in the hydrogenation of vegetable oils to make margarine
- a platinum–rhodium mixture (sometimes with other transition metals) is the catalyst in the catalytic converters in car exhaust systems.

▲ **Figure 22** The Rose Window at York Minster – glass is coloured green with either chromium(III) or iron(II); cobalt(II) gives blue glass and copper(II) gives a blue–green colour.

Catalysts also play a vital role in biological systems enabling complex reactions to occur quickly in dilute aqueous solution at moderate temperature and pH – the conditions in a living cell. Many transition metal ions are required by living things in minute but definite quantities. Cobalt, copper, manganese, molybdenum and vanadium are all ultra-trace elements essential for the catalytic activity of various enzymes.

Assignment 8

Make a list of some of the catalysts you have met already, together with the reactions they catalyse.

How many are transition metals or compounds of transition metals?

You have already read about catalysis in **Chemical Ideas 10.5** and **10.6**.

In **Activity SS5.3** you will investigate the catalytic activity of cobalt(II) ions.

d-Block metals can be expensive

The most abundant d-block element in the Earth's crust is iron, which is relatively cheap. However, some of the other d-block elements can be expensive – the cost of the steel rises if these are used in steelmaking.

For example, stainless steel contains a minimum of 12% chromium (by mass) and usually nickel as well to make it corrosion-resistant. A typical stainless steel might contain 74% iron, 18% chromium and 8% nickel. It can be as much as five or six times more expensive than mild steel and so is used selectively.

The price of a metal depends on many factors other than its abundance in the Earth's crust. These include the cost of mining the ore, the ease of extraction of the metal from the ore, the demand for the metal, transport costs and political factors in the countries involved.

Some elements are said to be *strategically critical*, because one or two countries have a monopoly over their supply. For example, the Republic of South Africa holds more than 70% of the world's known reserves of chromium, with Zimbabwe having over half of the remainder, whereas most of the world's deposits of molybdenum are found in the US and Canada.

SS6 *Summary*

In this module you looked at the different methods for extracting metals from their ores, depending on their reactivity, and followed the story of steel production from the iron ore fed into the blast furnace to a finished steel product. This is not one story but many. Each starts with iron from the blast furnace but leads to one of a multitude of different steels, each tailored to the job it has to do. Steel production has to be a batch process and must be closely monitored. Recent advances have been aimed at producing higher-quality steels to exact specifications, quickly and reliably – both on a huge scale and in small batches.

Many of the reactions involved in the BOS process are redox reactions and this allowed you to revise earlier work in this area. You also saw how atomic emission spectroscopy is used to check the composition of a steel during the process.

The story of steel next led to the problems of corrosion. Here you found out about the mechanism of rusting as an electrochemical process and what steps can be taken to slow it down. This led to a more detailed study of electrode potentials and electrochemical cells. You learned how electrode potentials can be used to predict the direction of a redox reaction.

Next, you considered the importance of recycling steel and some of the problems that must be overcome, for example, when recycling 'tin' cans.

Looking at the composition of steel and other alloys led you into a more detailed study of iron and other d-block metals. These metals form compounds in a variety of oxidation states and you used this idea to explain the coloured compounds and catalytic properties of d-block elements. You were able to use electrode potentials to predict the relative stability of these variable oxidation states. You also learned about ligands and complex formation.

Activity SS6 will help you to summarise what you have learned in this module.

AI

AGRICULTURE AND INDUSTRY

Why a module on 'Agriculture and Industry'?

Growing crops for food has been a major human activity for a long time. The rapidly increasing world population means that the need to provide enough food without destroying our environment is one of the biggest challenges facing us.

The biological, chemical and physical processes occurring in soil are highly complex, and involve a number of interrelated reactions. This module looks at the processes that occur as plants grow and decay. It then looks at ways in which a knowledge of these processes can be used to optimise crop yields and to ensure our food supply, in the context of both organic and non-organic (conventional) farming.

Finally, chemical and organic methods of pest control are studied. In particular, the module covers the development of pesticides that do not persist in the environment, and also explores the applications of some herbicides.

In considering these aspects of agriculture you will apply knowledge and understanding of chemical principles developed earlier in the course. These include ideas about chemical bonding and shapes of molecules, rates of reaction and redox chemistry. You will also revise some of the issues that need to be considered in industrial processes, such as the manufacture of an agrochemical. You will also learn about the redox chemistry of nitrogen, equilibrium reactions and the industrial manufacture of ammonia, as well as gaining a general overview of bonding, structure and properties.

Overview of chemical principles

In this module you will learn more about ideas introduced in earlier modules in this course:

- the relationship between properties of substances and their structure and bonding (**The Atmosphere**)
- redox reactions (**Elements from the Sea**)
- electronic structure – sub-shells and orbitals (**Elements from the Sea**)
- chemical bonding (**Elements of Life**)
- the shapes of molecules (**Elements of Life**)
- rates of reaction (**The Atmosphere**, **Developing Fuels** and **The Thread of Life**)
- the raw materials used in an industrial process (**Elements from the Sea**)
- the costs and efficiency of an industrial process (**Elements from the Sea**)
- health and safety issues involved in running a chemical plant (**Elements from the Sea**)
- atom economy and industrial processes (**Elements from the Sea**).

You will also learn new ideas about:

- the relationship between the properties of substances and their structure and bonding
- the redox chemistry of nitrogen
- chemical equilibria
- the selection of optimum conditions for the industrial manufacture of ammonia
- the ways in which chemists are involved in food production.

AI

AGRICULTURE AND INDUSTRY

AI1 *What do we want from agriculture?*

The world population is growing. As Figure 1 shows, a rapid increase in human numbers began in about 1850 and, while the rate of increase has slowed down, numbers are set to continue to increase throughout the twenty-first century. The world population doubled from 3 billion in 1961 to 6 billion in 1999, and the United States Census Bureau forecast that the world population will reach 9 billion by 2042 – an increase of 50% in 43 years. Table 1 shows estimates of when each billion milestone was or will be met.

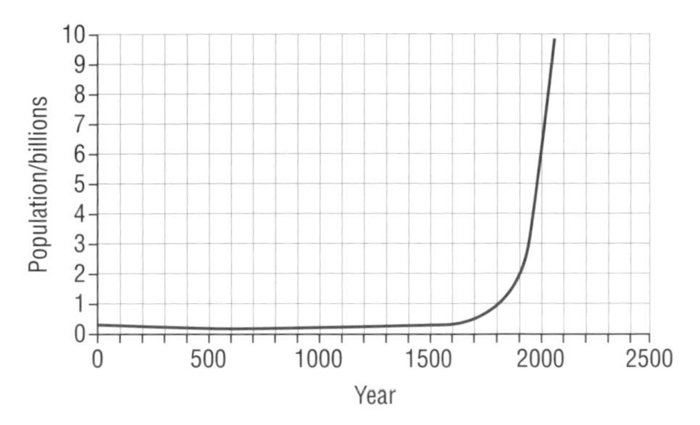

▲ **Figure 1** World population increase – with projected estimates.

> **Assignment 1**
>
> Look at the world population milestones in Table 1.
>
> **a** Calculate the time taken in years for the world population to increase by each billion.
>
> **b** Describe the trend in the *rate* of increase in world population from 1804 to 2042.

This rapid increase in human population poses a huge challenge to agriculture – our planet cannot produce an *unlimited* supply of food for its human population. The issue of increasing birth rate needs to be tackled, but the challenge to agriculture remains. The ultimate goal is to feed everyone adequately, without harming the environment. This involves producing enough food of the right kind, in the right place and at the right time.

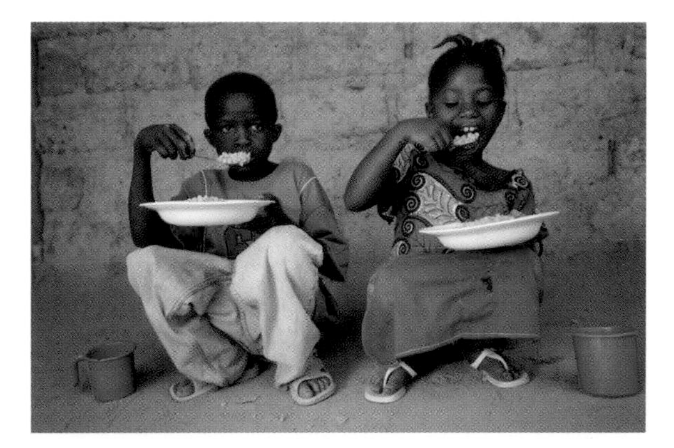

▲ **Figure 2** The ultimate goal is to feed everyone adequately, without harming the environment.

How can we increase food production *without* encroaching on the world's remaining forests and wildernesses? It can be done by making the most efficient use of *existing* agricultural land – particularly by improving crop varieties and planting techniques, and making sensible use of added plant nutrients (in manure and fertilisers) and pesticides.

▲ **Figure 3** Harvesting wheat in Montana, US.

Table 1 World population milestones (United States World Census Bureau).

Population	1 billion	2 billion	3 billion	4 billion	5 billion	6 billion	7 billion	8 billion	9 billion
Year	1804	1927	1961	1974	1987	1999	2011	2024	2042

Mistakes have been made. Growing one crop again and again on the same soil can destroy its fertility by depleting specific nutrients. Agricultural technology that is appropriate in one place cannot just be exported to other regions with different soil types and growing conditions.

To avoid repeating past mistakes, we need to use knowledge and understanding, gained both from scientific research and from the experience of farmers, to develop *sustainable* systems of agriculture – that is, agriculture that can go on indefinitely without degrading the environment.

What do plants need for growth?

As they grow, plants take carbon dioxide from the air into their leaves, and water and nutrients from the soil via their roots. The most important nutrient elements essential for plant growth are shown in Table 2.

Table 2 Essential nutrient elements and their sources – other minor nutrients (sodium, fluorine, iodine, silicon, strontium and barium) are not needed by all plants.

Elements used in relatively large amounts	Elements used in relatively small amounts	
carbon	nitrogen	iron
hydrogen	phosphorus	manganese
oxygen	potassium	boron
	calcium	molybdenum
	magnesium	copper
	sulfur	zinc
		cobalt
		chlorine
mostly from carbon dioxide and water	from soil	

Figure 4 shows the importance of soil as the source of nutrients for plants. When we harvest crops we disturb the natural nutrient cycles by removing large quantities of plants before the natural processes of decay take place – these decay processes would return nutrients to the soil. The nutrients removed when plants are harvested need to be replenished before a crop with similar nutrient needs can be grown in the same soil.

In this module you will look at the production of crops for food – but do remember that arable agriculture also involves the cultivation of plants such as cotton (grown for fibre) and oil seed rape (grown to make biofuels) and forests producing timber.

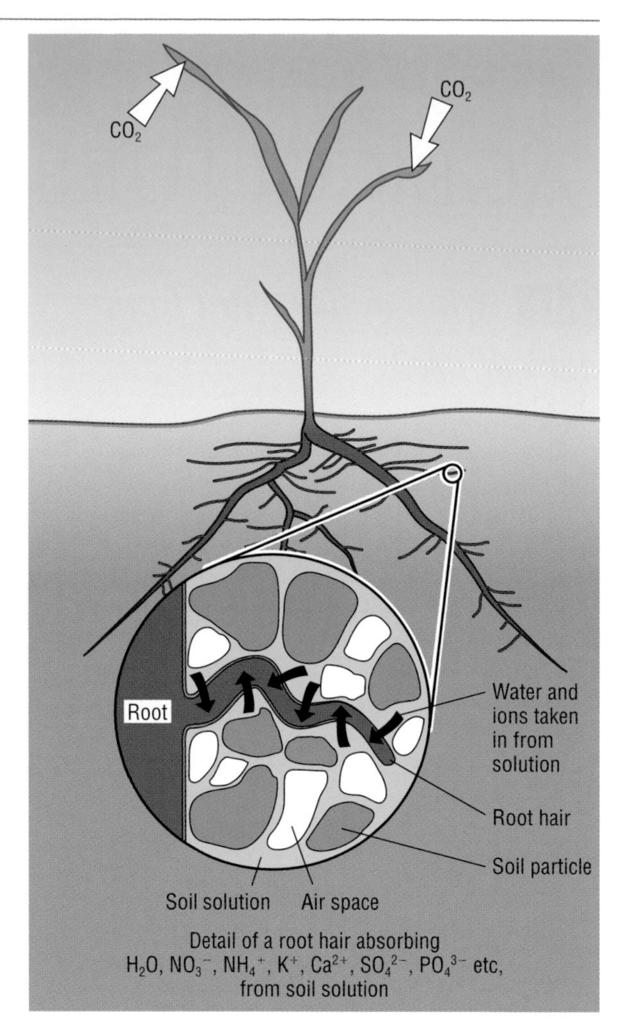

▲ **Figure 4** Sources of plant nutrients.

AI2 *The organic revolution*

The UK organic food market has boomed in recent years, growing by 25% annually on average. At the start of the twenty-first century many people are choosing to buy organic produce for a number of reasons, including concern for the environment. Eating organic produce is one way to reduce consumption of pesticide and herbicide residues. Consumers may also choose to buy organic food because they believe it to be safer and more nutritious than conventionally farmed food. In 2008, the Food Standards Agency stated that the balance of scientific evidence does *not* support this view. Some scientific papers *do* reach this conclusion (see the work cited below). Other scientific papers find no difference between organic and conventionally farmed foods. In order to reach a robust conclusion it is necessary to evaluate evidence from a range of sources.

In October 2007, the results of a £12 million, four-year investigation led by Newcastle University into organic and conventional farming were published, claiming evidence that organic foods do have the greater nutritional value. Fruit and vegetables were grown on adjacent organic and conventional (non-organic) sites across Europe, including a 725-acre farm attached to the university. Some of the early findings from the study included claims that potatoes, kiwi fruit and carrots contained more vitamin C than their non-organic counterparts, and that organic lettuce,

▲ **Figure 5** The organic food market is booming in the early part of the twenty-first century.

▲ **Figure 6** In 2008, the Food Standards Agency stated that there were scientific studies both supporting and refuting the idea that organically farmed food is more nutritious than conventionally farmed food.

spinach and cabbage contained higher levels of minerals than their non-organic equivalents.

So how is organic farming different from traditional farming? Firstly, it is important to understand that the word 'organic' in the context of farming has a very different meaning from the one you have encountered in studying chemistry on this course. The precise definition of the word 'organic' in these two contexts is given in the purple box (above right).

Organic farming adopts a holistic approach to food production. Key aspects of organic farming and food include:

- restricted use of artificial fertilisers and pesticides
- emphasis on soil health, maintaining this through crop rotation and the application of manure and compost.

WHAT IS ORGANIC?

Organic chemistry – the study of the compounds of carbon, with the exception of CO_2, CO and the carbonates, which are traditionally included in inorganic chemistry.

Organic farming – the process of farming using fertilisers and pesticides of only plant or only animal origin. For a product to be classified as organic, farmers must adhere to a strict set of standards during its production.

Organic farmers must adhere to a strict set of standards that define what they can and can't do. They are severely restricted in their use of chemical fertilisers and pesticides. (See sections **AI3** and **AI4** for more details.)

Soil health

How can crops be grown again and again on the same soil without decreasing its fertility? The fertility of a soil depends on many complex interactions between the biological, chemical and physical processes occurring.

To really understand how organic farmers can keep soil supplied with the nutrients essential for plant growth (outlined in section **AI1**), you first need to understand a lot more about soil structure and nutrient recycling.

What is soil?

The Earth's crust makes up just 1.5% of the volume of the planet (Figure 7). In this module we are concerned with the thin layer, typically 1–2 m thick, on the top of the crust – this is *soil*.

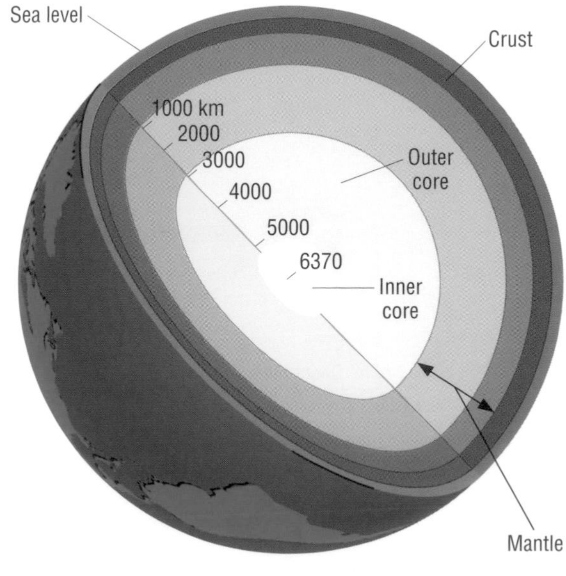

▲ **Figure 7** Layers of the Earth.

Soil is a mixture of weathered rock fragments and organic matter. Together they make a porous fabric that can hold both air and water. The approximate proportions by volume for a typical moist soil are shown in Figure 8.

The quantities of air and water in a soil are variable, but the ratio is important in determining how well the soil can support plant growth. The minerals play a vital part in making nutrient ions available to plants

Soil minerals are produced by *weathering* the rocks that make up the Earth's crust. The fragments of rock and minerals vary enormously in size, from coarse stones and gravel to fine clay particles.

The coarse fragments are relatively inactive chemically. It is the very fine clay fraction that is most active, because clay minerals can bind positive ions to their surfaces. In this way, essential ions are held in the soil.

Weathering

Weathering involves the action of wind, rain, frost and sunlight – and nothing on the surface of the Earth's crust escapes it. Weathering breaks up the rocks. It changes their physical and sometimes their chemical composition. It carries away soluble materials – and some of the solid fragments as well. Biological processes also contribute to weathering – for example, roots can work their way into crevices in rocks and break them apart.

But weathering is creative as well as destructive – it makes a soil out of the uppermost layers of weathered rock. Figure 9 summarises the main pathways of weathering in the moderately acidic conditions of humid, temperate climates. In different climates the rates of the reactions alter, altering the composition of the soils.

Soil organic matter

The organic matter in soil is made up of plant debris, animal remains and excreta – and the products formed by decomposition of all these things. Soil organic matter forms a store from which the next generation of plants will get their nutrients. A variety of organisms living in the soil make use of the debris deposited. During the processes of decomposition, elements in organic compounds are converted into inorganic ions such as ammonium, nitrate(V), sulfate(VI) and phosphate ions – this process is called mineralisation.

Some of the carbon in decomposing organic matter is released into the atmosphere as carbon dioxide. A range of new organic molecules is synthesised to make **humus**. There are many macromolecules in

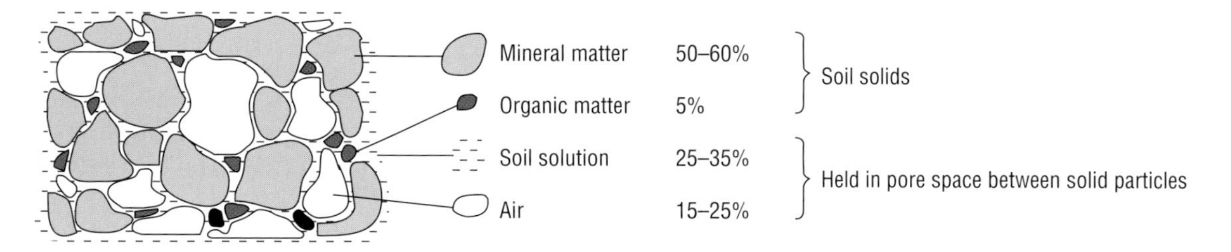

▲ **Figure 8** The four major components of soils – showing the approximate proportions by volume for a typical moist soil.

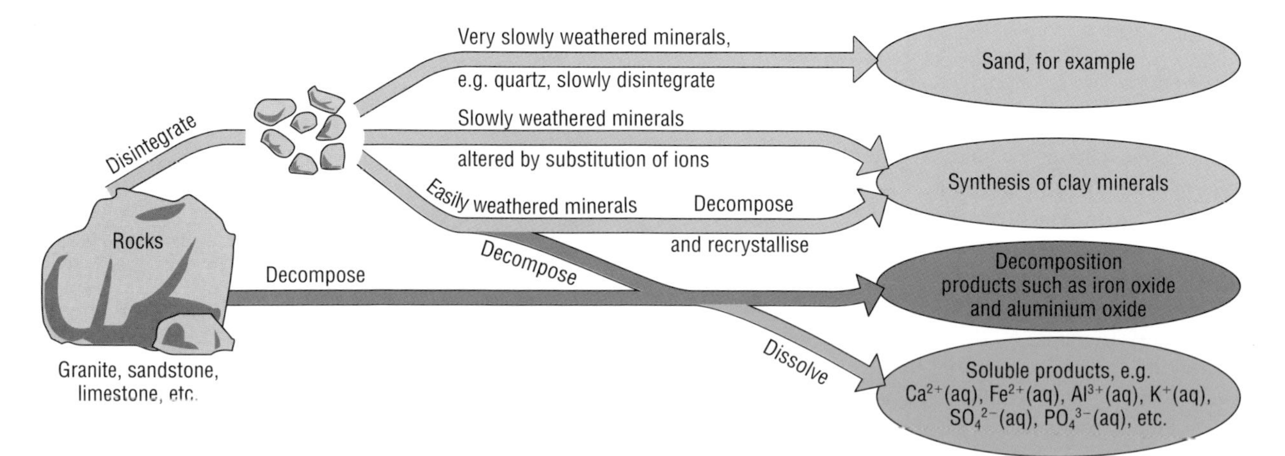

▲ **Figure 9** Processes of weathering in the moderately acidic conditions common in humid temperate climates.

humus, with relative molecular masses up to 500 000. Major components of humus are esters of carboxylic acids, carboxylic acid derivatives of benzene, and phenolic compounds. The following two groups are therefore common:

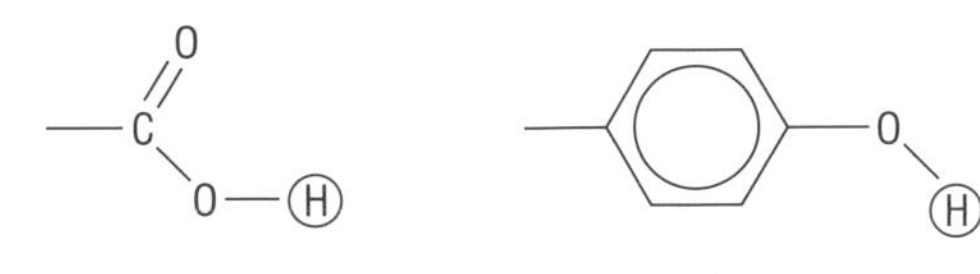

Both groups can lose H^+ ions (those ringed in the diagram above). If they do lose these ions, negatively charged groups are left on the molecules in humus. These are able to form ionic bonds with metal cations, so that humus can hold a variety of nutrient ions in a similar way to clays.

Keeping soil fertile

Nutrient cycling

Apart from carbon dioxide from the air, plants get all their nutrients from the soil (see Figure 4 in section **AI1**).

These nutrients are drawn from an inorganic store in the soil, and an organic store partly on top of and partly in the soil. Elements are cycled continuously between living systems, the organic store and the inorganic store. The general routes are shown in Figure 10.

The organic store is replenished by organic manure, animal excretions and by the death and decay of living organisms. Microorganisms act on organic matter and convert it into humus. They also convert it into inorganic ions (mineralisation) producing ammonium, nitrate(III), nitrate(V), phosphate and sulfate(VI) ions.

Weathering of soil and rock minerals releases more ions into the inorganic store. Nutrients can be lost by being leached out of the top layers of soil by rainwater, and nitrogen can be lost by conversion into gases such as NH_3, N_2 and N_2O – these disperse into the atmosphere. Figure 11 shows typical values for the total quantities of some elements in different soils.

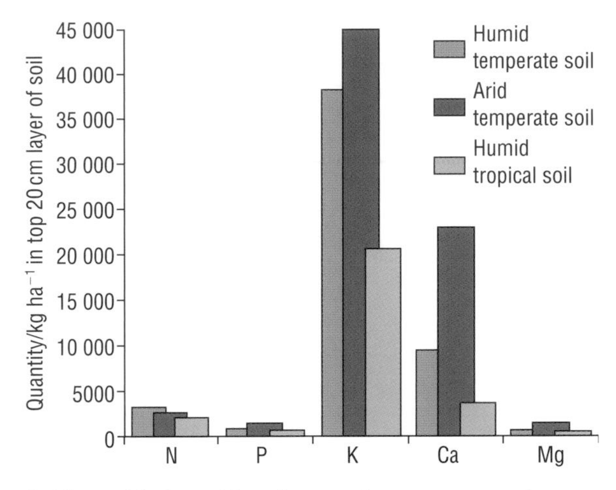

▲ **Figure 11** Quantities of some elements present in different soils.

> **HECTARES**
>
> Land areas are conveniently measured in **hectares** (ha):
>
> $$1\ \text{ha} = 1 \times 10^4\ \text{m}^2\ (= 2.47\ \text{acres})$$
>
> A football pitch is about 0.5 hectares. Fields range from about 6 hectares to about 60 hectares.

Only a small fraction of each element in the soil is actually available to plants in the form of ions that can be absorbed through their roots. The nitrogen content of soil can be measured in the form of nitrate(V), nitrate(III) and ammonium ions extracted from a known mass of soil. Analytical methods are also used to determine quantities of potassium and phosphate ions. The extracted quantities are used to classify a soil as low, medium or high in nutrients present in forms that plants can use. Figures for three major nutrients are given in Table 3.

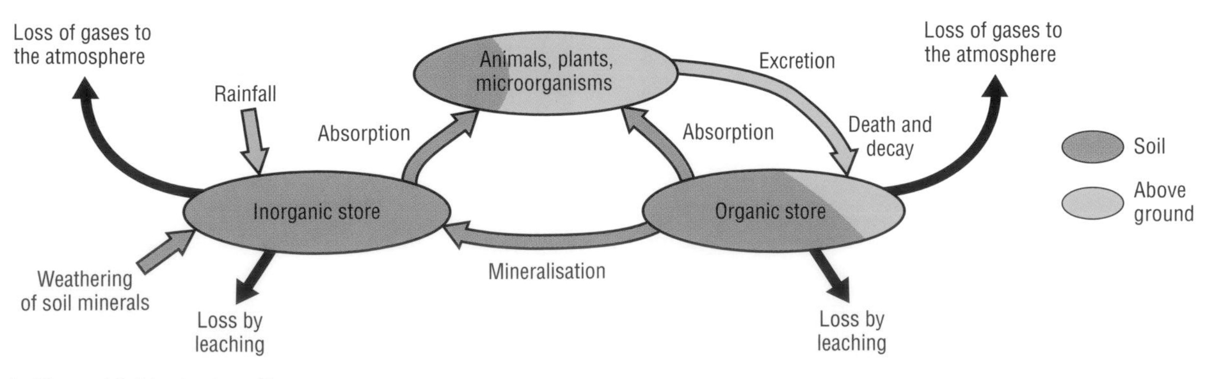

▲ **Figure 10** Nutrient cycling.

Table 3 Nutrients in soils in forms that plants can use.

Soil range	Nutrient extractable in forms that plants can use/kg ha^{-1} in top 20 cm layer of soil		
	Nitrogen	Potassium	Phosphorus
low	40	0–150	0–23
medium	90	150–600	23–63
high	140	over 600	over 63

Assignment 2

Referring to Table 3 and Figure 11, in a humid temperate soil rated high in nutrients, approximately what percentage of:

a the total potassium content
b the total nitrogen content is in a form that plants can use?

Healthy crop growth depends on the ability of the soil to supply nutrients – this makes the rates at which nutrients are interconverted in the cycles very important. The rate of supply of ions to plant roots has to be rapid enough during the peak growing period to meet the demands of the crops. Crop yields are reduced if there is a shortage of even one nutrient (see Figure 12). Figure 13 shows one of the effects of a shortage of potassium ions.

In the next section, you will look at one nutrient cycle in more detail.

Chemical Ideas 11.3 gives a summary of the chemistry of nitrogen and other elements in Group 5.

It may help you to revise earlier work on redox reactions and oxidation states in **Chemical Ideas 9.1**.

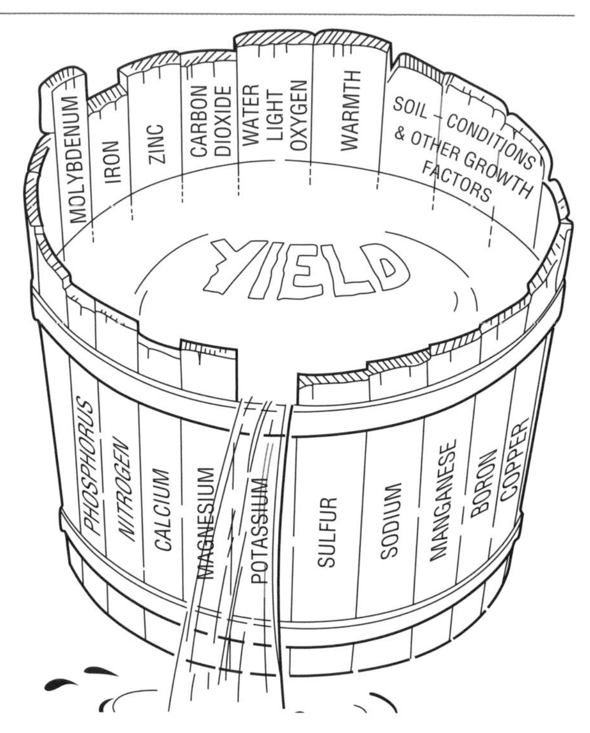

▲ **Figure 12** The German chemist Justus von Liebig used this barrel to illustrate to his students that a deficiency of any single plant nutrient is enough to limit growth – here potassium is the limiting nutrient.

The nitrogen cycle

Almost all the nitrogen in soil is present in complex organic compounds and so is not readily available to plants. Various processes convert gaseous nitrogen and organic nitrogen compounds into the soluble ammonium and nitrate(V) ions that plants *can* use. The main processes in the nitrogen cycle are listed below – refer to the diagram of the nitrogen cycle in Figure 14 as you read through.

▲ **Figure 13** Leaves from **a** healthy plant; **b** plant with potassium deficiency.

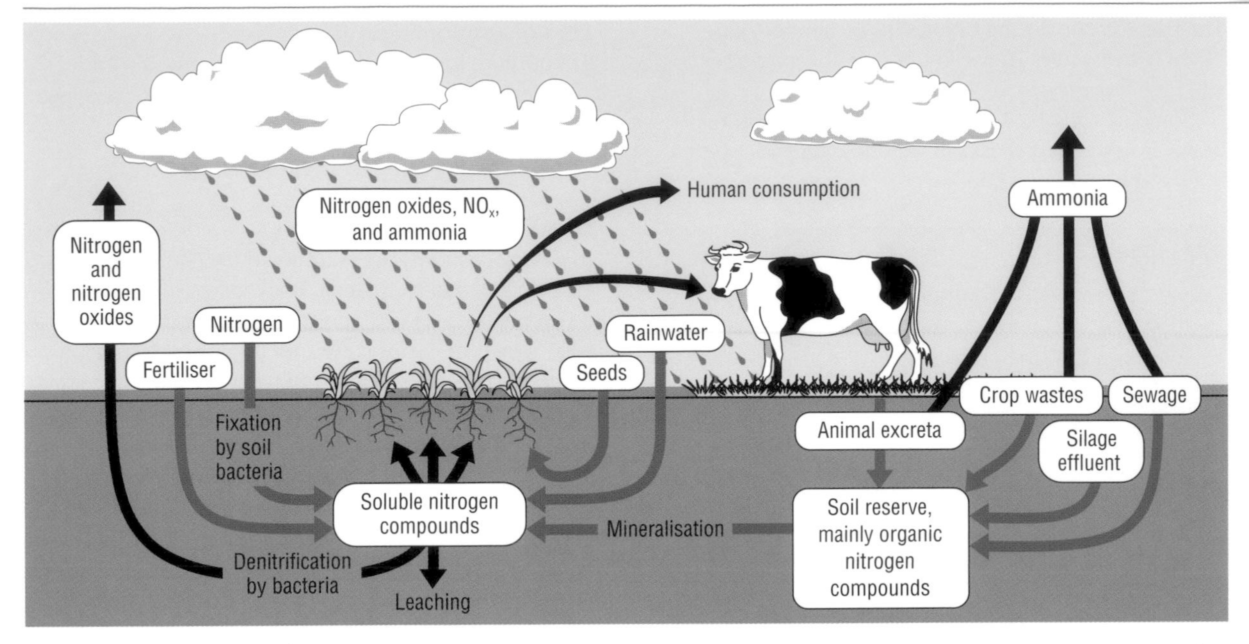

▲ **Figure 14** Nitrogen cycle.

Additions to soil nitrogen

Biological fixation

Some kinds of bacteria in soil, and in root nodules in legumes such as peas and beans, can convert nitrogen gas to ammonium ions. A reducing agent is needed to provide the electrons. The half-equation for the reduction is

$$N_2(g) + 8H^+(aq) + 6e^- \rightarrow 2NH_4^+(aq)$$

Other additions to soil nitrogen

Lightning, burning hydrocarbon fuels and natural fires all produce nitrogen oxides, which are released into the atmosphere and are then deposited on the soil. In Europe about 20–40 kg N ha^{-1} are deposited from the air onto the soil each year. Some of this is in the form of nitrogen oxides and some as ammonium ions, which have come from ammonia emitted by animal excreta.

Transformations in the soil

Mineralisation

Soil bacteria and other microorganisms break down organic nitrogen compounds into simpler molecules and ions. Any nitrogen not needed by the organisms themselves is released into the soil as ammonium ions:

$$\text{organic N} \xrightarrow{\text{several steps}} \rightarrow \rightarrow NH_4^+(aq)$$

The NH_4^+ ions are held by clay minerals as exchangeable cations, but are readily converted to NO_3^- by certain microorganisms.

A well-drained soil is best for mineralisation, and the reaction is much faster at the higher temperatures of tropical soils. Radioactive labelling with ^{15}N shows that only 1–3% of soil nitrogen is mineralised each year.

▲ **Figure 15** Lightning generates enough energy to convert nitrogen and oxygen in the air to nitrogen(II) oxide – $N_2 + O_2 \rightarrow 2NO$.

Nitrification

Ammonium ions can be oxidised by certain aerobic bacteria in the soil – the bacteria carry out the reactions as a means of obtaining respiratory energy. The overall process is called nitrification because the end product is the nitrate(V) ion, formed via the nitrate(III) (nitrite, NO_2^-) ion. Nitrification occurs in several stages. The formation of nitrate(III) can be represented by

$$NH_4^+(aq) + 1\tfrac{1}{2}O_2(g) \rightarrow NO_2^-(aq) + 2H^+(aq) + H_2O(l)$$

The bacteria that do this are called *Nitrosomonas*. The optimum pH is between 7 and 9, and the reaction stops in dry conditions.

The nitrate(III) ion, NO_2^-, produced is rapidly oxidised further. This can be represented by

$$NO_2^-(aq) + \tfrac{1}{2}O_2(g) \rightarrow NO_3^-(aq)$$

The bacteria converting nitrate(III) to nitrate(V) are called *Nitrobacter*. They can tolerate dry conditions and higher acidity than *Nitrosomonas*.

Some nitrification occurs in soils at temperatures down to 0 °C, but it ceases in waterlogged soils where the oxygen content is too low.

Assignment 3

Overall, mineralisation has a first-order rate equation:

$$\text{rate of mineralisation} = k[\text{N}]$$

where k is the rate constant at a particular temperature and [N] is the quantity of organic nitrogen per hectare in the top 20 cm of soil.

The rate constant k varies from 0.01 yr^{-1} to 0.06 yr^{-1}. For a time interval of 1 year, the quantity of organic nitrogen mineralised in the top 20 cm of soil equals k[N] kg ha^{-1}.

a Use the above equation to calculate the quantity of nitrogen mineralised in 1 year (in the top 20 cm of 1 ha) in the three soils A, B and C. The soils have different organic nitrogen contents and different temperatures. (Remember – the value of k depends on the temperature.)

Soil	Soil organic nitrogen/kg ha^{-1}	Rate constant k/yr^{-1}
A	1000	0.01
B	2000	0.03
C	2000	0.06

b Refer to Figure 11 on page 79 for information about the nitrogen content of different types of soil. Think about the factors that affect the rate of mineralisation, and explain why a humid tropical soil has the highest rate of mineralisation.

Losses of nitrogen from the soil

Denitrification

Where oxygen content is low, anaerobic bacteria reduce nitrate(V) ions in the sequence

$$NO_3^-(aq) \rightarrow NO_2^-(aq) \rightarrow NO(g) \rightarrow N_2O(g) \rightarrow N_2(g)$$

In flooded soils, like those used in rice cultivation, losses by denitrification can be high.

Leaching

Nitrogen is lost by leaching, mainly as the nitrate(V) ion, NO_3^-, which is not held by clays or humus in temperate soils. The quantities lost depend on the soil structure and the amount of rainfall, as well as the nitrate(V) concentration in the soil.

Loss of ammonia gas

Ammonium ions are converted into ammonia under alkaline conditions – the ammonia then disperses into the atmosphere.

$$NH_4^+(aq) + OH^-(aq) \rightarrow NH_3(g) + H_2O$$

Uptake of nitrogen by plants

In natural systems, the quantity of nitrogen removed each year is relatively small. For example, a coniferous forest takes up 25–78 kg nitrogen per hectare each year. However, crops cultivated to give high yields need much more nitrogen, 100–500 kg nitrogen per hectare each year. This nitrogen is used to make plant protein necessary for growth. Soil nitrogen cannot be mineralised to ammonium ions and nitrate(V) ions fast enough to meet this demand, even when the reserves of organic nitrogen are high.

In order to maintain or increase crop yields, nitrogen must be added to the soil in a form that plants can readily use – this is where organic and conventional farming differ.

In **Activity AI2.1** you can use the information in this section to investigate the nitrogen cycle in more detail.

In this module you have encountered many types of structure – **Chemical Ideas 5.8** summarises all the ideas on structure and bonding that you have encountered on this course.

In **Activity AI2.2** you can investigate the properties and structures of materials.

In **Activity AI2.3** you can match the properties of materials with their structures.

It may help you to revise earlier work on electronic structure in **Chemical Ideas 2.4**, bonding in **Chemical Ideas 3.1** and the shapes of molecules in **Chemical Ideas 3.2**.

Organic farming

Organic farmers are governed by clear guidelines in the ways that they can add nitrogen to the soil. (Pesticide use is also limited – this is covered in section **AI4**.) The activity of organic farmers in the UK is governed by European laws on production. These laws require foods sold as 'organic' to come from growers who are registered and approved by organic certification bodies – in the UK , the Soil Association is one of these. The certification bodies are in turn registered by the Department for Environment, Food and Rural Affairs (DEFRA). If farmers want to sell their produce as organic under the Soil Association's banner, then they must adhere to the Soil Association's standards during production.

▲ **Figure 16** Organic foods bearing the Soil Association's logo must have been produced in accordance with their strict standards.

The Soil Association does not allow the use of:

- sewage sludge, effluents or sludge-based composts
- residues or manure from livestock systems that do not meet their standards
- chemically synthesised fertilisers.

Instead, organic farmers must maintain fertility and nitrogen content of soil by using some or all of the following methods.

1 Growing crops that add nutrients to the soil as part of a crop rotation.

Different crops have different mineral requirements, and growing one crop continuously in a particular space will lead to a depletion of certain minerals – so it makes sense to change or *rotate* the crop grown each year. Moreover, some crops (those with nitrogen-fixing root nodules such as peas and beans) will actually *increase* the nitrogen content of a soil.

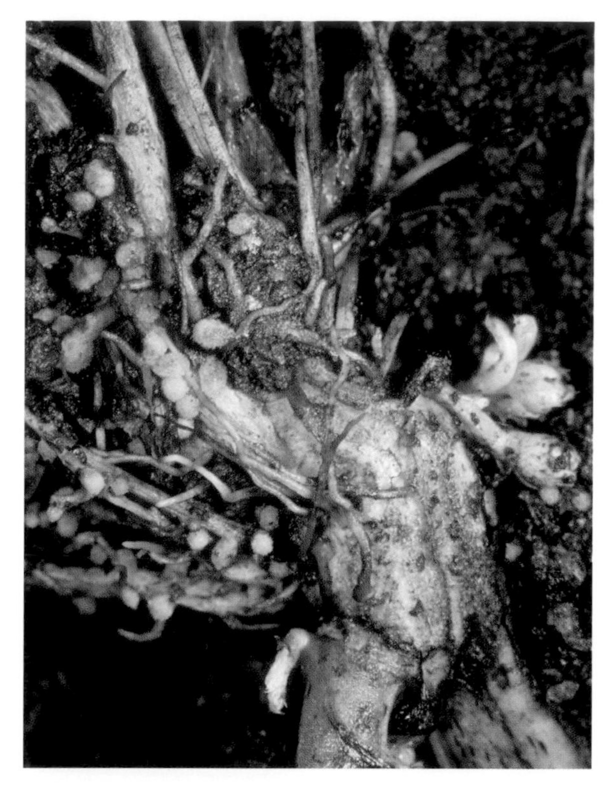

▲ **Figure 17** Nitrogen-fixing root nodules on broad bean roots.

Therefore, crops are grown in sequence so that those with a high nitrogen demand (e.g. brassicas) follow crops that increase soil nitrogen content (e.g. peas and beans). Crops with similar nutrient requirements do not follow one another on the same land, so the land has time to regenerate the depleted nutrients. Green manures (see page 84) are also used to increase soil fertility.

Year 1

Field A	Field B	Field C	Field D
LEGUMES (peas, beans)	BRASSICAS (cabbage, kale, broccoli, cauliflower)	ROOTS (carrots, parsnips etc.)	POTATOES or GREEN MANURE

Year 2

Field A	Field B	Field C	Field D
POTATOES or GREEN MANURE	LEGUMES (peas, beans)	BRASSICAS (cabbage, kale, broccoli, cauliflower)	ROOTS (carrots, parsnips etc.)

▲ **Figure 18** An example of a typical four-year crop rotation – here you can see the first two years of the cycle.

If the same 'family' of crops is grown year after year in one field, there is a likelihood of pests and diseases related to that crop building up. For example, cabbage root fly and club root can be prevalent where brassicas have been planted for many growing seasons. Crop rotation discourages the build-up of such pests and pathogens in the soil.

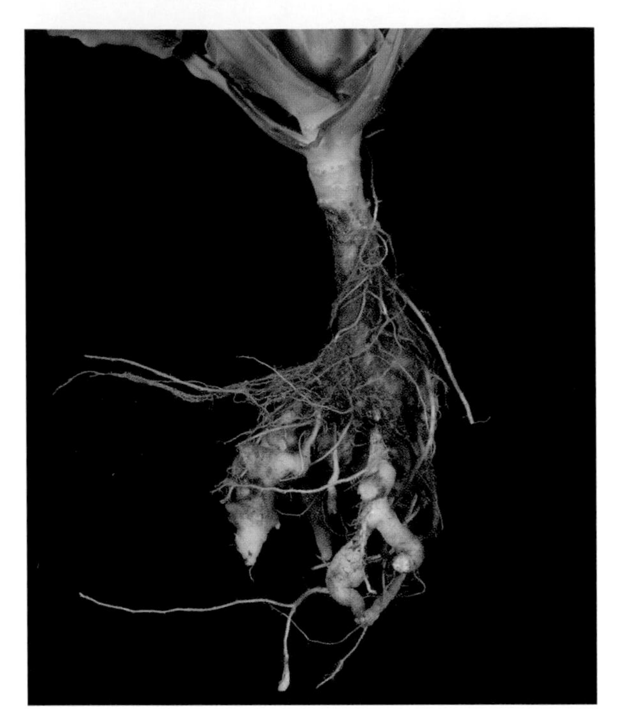

▲ **Figure 19** Cabbage with club root – note the swollen root system.

2 Growing crops that are ploughed back into the soil (green manures)

'Green manures' are crops such as clover or mustard that are grown for their ability to fix nitrogen. These are usually grown through the winter months and their benefits are twofold. Firstly, they increase the nitrogen content of the soil through nitrogen fixation. Secondly, the minerals taken up from the soil by these plants cannot be leached from the soil over the winter, but are returned to the soil when the crop is ploughed into the soil in the spring.

▲ **Figure 20** Green manures are grown to increase the nitrogen content of the soil and prevent the leaching of minerals.

3 Applying composts and farmyard manures

Manure or composted household waste increases the nitrogen content of the soil directly. Long-term experiments have been conducted at the Rothamsted Experimental Station in Hertfordshire since 1843, investigating the effects of adding nutrients to the soil. They have shown that yields can be increased, and high productivity maintained, by adding nutrients to the soil and by controlling weeds, pests and diseases.

After a century and a half, the application of farmyard manure to plots of land has more than doubled the organic nitrogen content of soil, increasing its nitrogen reserve. In addition, farmyard manure has encouraged flourishing populations of soil microorganisms and small animals such as earthworms. These assist in converting organic nitrogen into forms usable by plants. Table 4 gives the nutrient content of some farmyard manures. It must also be remembered that if a farm is to maintain its organic status, guidelines concerning the quantity and source of manure must be adhered to.

Table 4 The nutrient content of some farmyard manures.

	Manure produced/$t\,yr^{-1}$	Nutrient content/$kg\,t^{-1}$		
		N	P	K
1 dairy cow	23	4.7	0.6	4.4
10 pigs	21	6.3	1.5	2.9

4 Applying certain permitted mineral fertilisers and supplementary nutrients

Certain minerals (e.g. sulfur) and trace elements such as boron, copper or sodium can be applied to soil. The source of these is tightly regulated and an organic farmer has to apply for special permission before they can be applied.

▲ **Figure 21** Applying manure to soil will increase its nitrogen content.

Assignment 4

A farmer plans to add 144 kg nitrogen per hectare to a small field of wheat. The area of the field is 7 ha.

a Calculate the total mass of nitrogen needed.

b Calculate the mass of cow manure needed to supply the nitrogen for the field.

AI3 *The fertiliser story*

The agricultural revolution

At one time, all farming was essentially organic. Farming changed very little from medieval times through to the early 1700s. Most of the land around a village was split into open or common fields. 'Farmers' did not think about how to make money from their land – they just grew enough food for their families and the village. A basic crop-rotation system was used, with one field left fallow each year and manure used to maintain soil fertility (see section **AI2**).

▲ **Figure 22** People grazed their animals on common land in medieval times and just grew enough food for their family and the village. These monks are producing food for their monastery.

In the 1700s the population started to increase and this system began to change. Between 1750 and 1825 the population of Britain almost doubled and the number of people living in towns – who could not grow their own food – increased rapidly, so Britain's farmers needed to grow more food.

Several changes in agricultural technology made this possible:

- Turnips and clover were introduced into the crop rotation system. Both crops were ideal as animal feed. Additionally, clover plants have root nodules containing bacteria that are able to fix nitrogen and increase the nitrogen content of the soil. A fallow year was no longer needed and productivity increased.

▲ **Figure 23** Over time Britain became increasingly urbanised. People living in towns and cities could not grow their own food.

- As the iron industry grew, tools and equipment improved and became cheaper – many tasks could be completed more quickly and were less labour-intensive.
- Fields were enclosed following the 1785 Enclosure Act – this led to land being used more efficiently.
- Nitrogenous fertilisers in the form of imported guano (bird droppings) and sodium nitrate imported from South America were used to improve soil fertility – this alone trebled the yield of certain crops where it was used.

▲ **Figure 24** Advances in the iron and steel industry made it possible for metal agricultural tools to be manufactured.

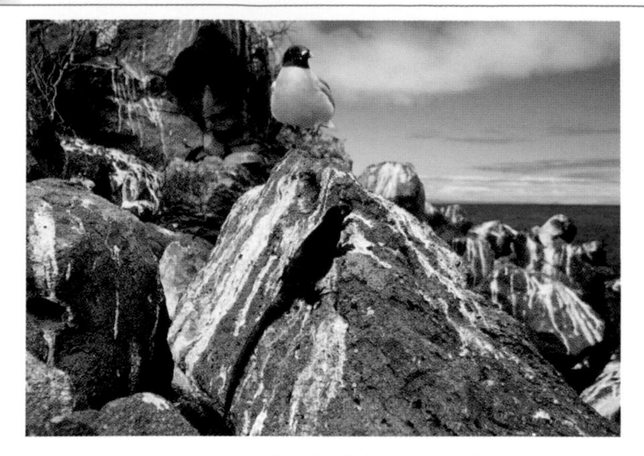

▲ **Figure 25** A guano island off the coast of South America.

The population of Britain increased from 11 million to 40 million between 1750 and 1900. Larger and larger supplies of food were needed for this growing population, not just in Britain but all over the world – particularly in Europe and North America. At the same time, the chemical industry also required increasing quantities of nitrogen compounds to make nitric acid for dyes and explosives such as TNT and dynamite.

Consequently, agriculture and industry were competing for dwindling supplies of nitrogenous raw material. By 1900 Peruvian guano had already been worked out and it was clear that supplies of sodium nitrate from Chile would soon be exhausted.

An alternative supply of nitrogen in the form of ammonium compounds, ammonia or nitrate had to be found, or the chemical industry would stagnate and the world's growing population would be at risk of starvation. Ironically, it was during war preparations in Germany between 1909 and 1914 that a solution to the problem was found. Military leaders in Germany realised that, once war was declared, their country would be subjected to a strict blockade and the importation of goods and materials from the rest of unoccupied Europe and America would cease. German industry, therefore, had to be able to meet the country's requirements for nitrogenous fertilisers, and the tremendous demand for explosives and therefore nitric acid, that a war would create.

Fritz Haber's discovery

In 1909, the leading German chemical company Badische Anilin und Soda Fabrik (BASF) turned its research expertise and financial resources towards the development of ammonia manufacture from atmospheric nitrogen. In the previous year Fritz Haber, a young German research chemist, had discovered that nitrogen and hydrogen would form an equilibrium mixture containing ammonia in the presence of a suitable catalyst. By 1909 he had managed to produce 100 g of ammonia using the apparatus similar to that shown in Figure 26.

The process was scaled up by Carl Bosch, a chemical engineer employed by BASF near Mannheim in

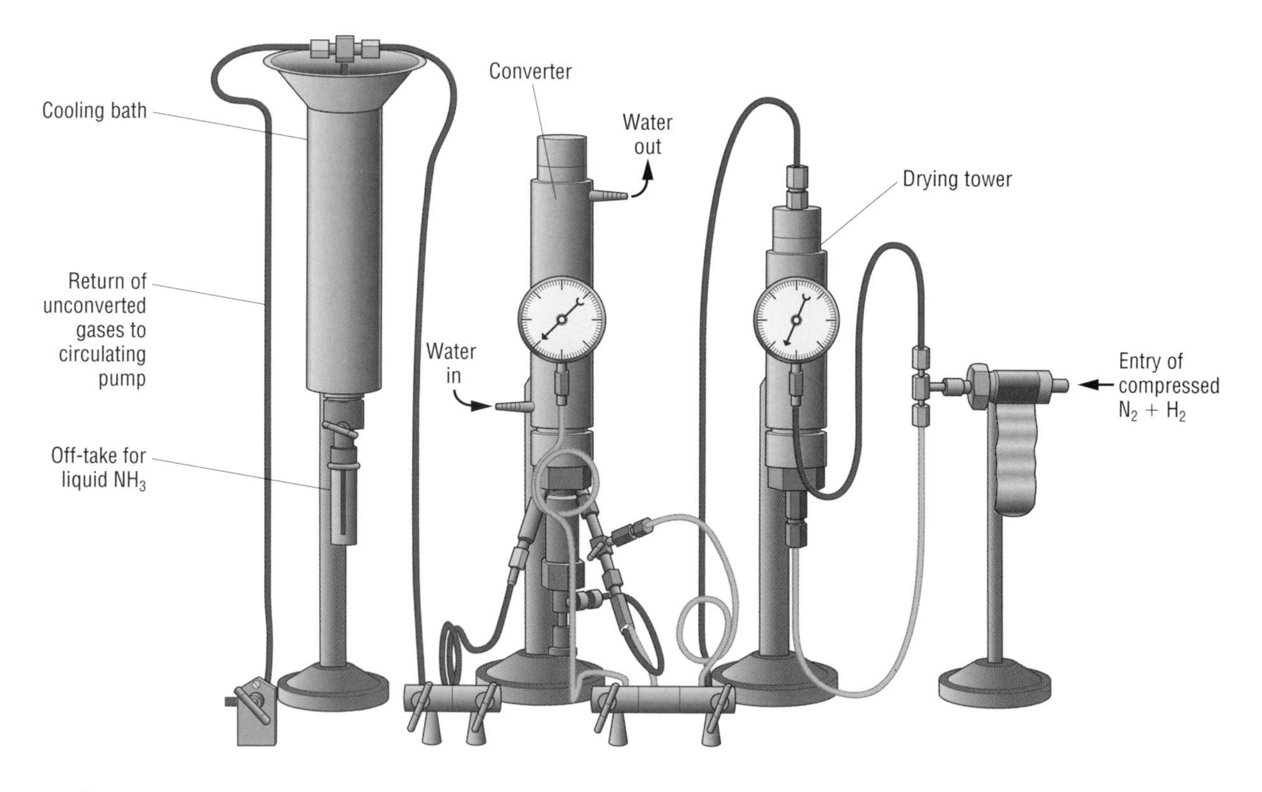

▲ **Figure 26** Diagram of Haber's apparatus for the synthesis of ammonia.

Germany. In 1913 the first industrial plant went into production, with a capacity of 30 tonnes of ammonia per day. Modern plants use the same basic design principles, but with capacities of about 1500 tonnes per day.

About 6500 experiments were carried out between 1910 and 1912 to discover an effective catalyst. Today, finely divided iron is used – small amounts of potassium, aluminium, silicon and magnesium oxides are added to improve its activity.

You can revise the effect of catalysts on the rate of a reaction in **Chemical Ideas 10.5** and **10.6**.

By 1913, German production of nitrogen compounds had reached 120 000 tonnes per year. Without this effort Germany would almost certainly have run out of food and explosives and the war would probably have ended before 1918.

Both Haber and Bosch were awarded Nobel Prizes in Chemistry – Haber in 1918 for his academic work and Bosch in 1931 for inventing and developing the high-pressure technology.

▲ **Figure 27** Fritz Haber was awarded the Nobel Prize in Chemistry in 1918 for his work on ammonia synthesis.

▲ **Figure 28** Carl Bosch was awarded the Nobel Prize in Chemistry in 1931 for developing the high-pressure technology that made large-scale synthesis of ammonia possible.

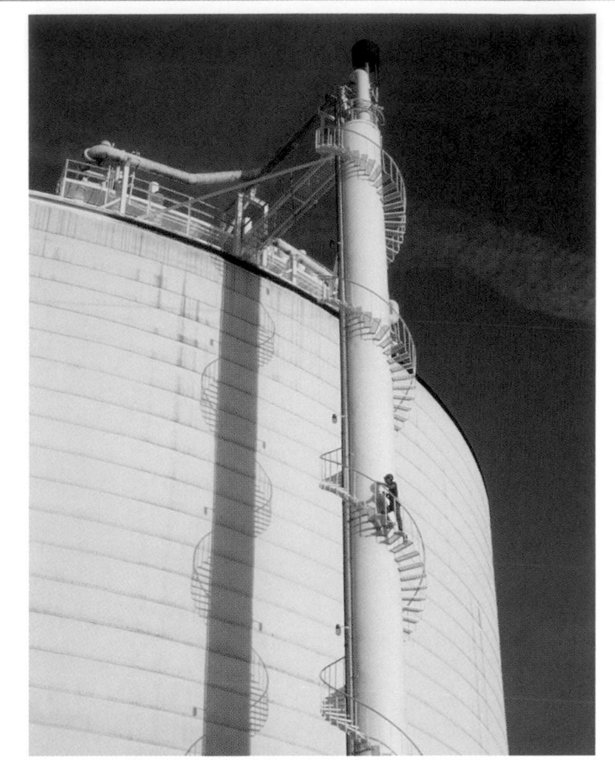

▲ **Figure 29** A man climbing the steps of an ammonia tank at a fertiliser factory.

The modern Haber process plant

In modern factories, the hydrogen needed to make ammonia is usually made by reacting water with natural gas. Nitrogen from the air is purified and mixed with the hydrogen – the heated gases pass over the catalyst where they react to form ammonia:

$$N_2(g) + 3H_2(g) \underset{}{\overset{\text{Fe catalyst}}{\rightleftharpoons}} 2NH_3(g) \quad \Delta H \text{ -ve}$$

The reaction is reversible – ammonia produced in the forward reaction can decompose to nitrogen and hydrogen in the reverse reaction. If a mixture of nitrogen and hydrogen is placed in contact with the catalyst and left for a sufficient time, then an equilibrium mixture containing all three gases is obtained. This is an example of an equilibrium reaction.

You were first introduced to equilibria in **Chemical Ideas 7.1**.

You can find out more about equilibrium reactions in **Chemical Ideas 7.2**.

Activity AI3.1 provides an opportunity for you to check your understanding of chemical equilibria.

Equilibrium constants allow you to calculate product yields in a reaction mixture at equilibrium. In **Activities AI3.2** and **AI3.3** you can determine the equilibrium constant for two different reactions.

To investigate how the highest yields can be obtained, nitrogen and hydrogen in the volume ratio 1 : 3 were mixed and held at different temperatures and pressures. The equilibrium yield of ammonia was recorded. Some results are shown in Table 5.

The equilibrium position depends on the temperature and pressure chosen for the reaction. As you can see from the data in Table 5, a high pressure obviously gives a higher yield of ammonia, but the higher the pressure, the greater the cost and maintenance of equipment. There are also more health and safety issues associated with working at very high pressures. Pressures up to 600 atm have been used, but nowadays a lower pressure is favoured. In contrast to the high pressure required, the temperature must be low to give a high yield of ammonia because the reaction is exothermic. However, at low temperatures the rate of reaction is so slow that it makes the process uneconomical. What this means is that the kinetic and equilibrium considerations are in conflict in the choice of reaction temperature. The greatest yield of ammonia would be obtained at low temperatures, but under these conditions the reaction rate would be uneconomically slow. In practice a compromise temperature of 450 °C is chosen, which is the lowest that can be used without reducing the reaction rate to an unacceptable level.

In **Activity AI3.4** you can investigate the effect of temperature and pressure on an equilibrium reaction.

You can revise the effect of temperature and concentration on the rate of a reaction in **Chemical Ideas 10.2** and **10.3**.

Ammonia is separated from unreacted nitrogen and hydrogen before equilibrium is reached – the unreacted gases are recycled over the catalyst.

Most reactors now operate at pressures between 25 and 150 atm, and at temperatures between 400 and 500 °C. Energy consumption in modern factories is about 35 MJ for every kilogram of nitrogen converted to ammonia.

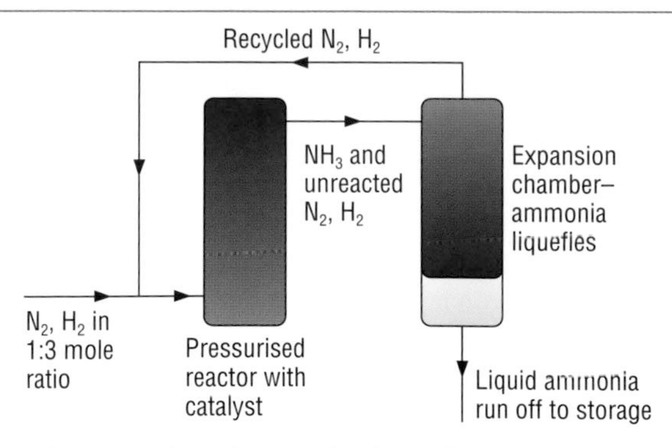

▲ **Figure 30** Flow diagram for the Haber process.

Assignment 5

a The forward reaction in the Haber process is exothermic ($\Delta H^\ominus = -92\ \text{kJ mol}^{-1}$). Explain how the experimental results in Table 8 are in agreement with le Chatelier's principle.

b Pick out the conditions of temperature and pressure from those listed in Table 5 that would give:

i the highest yield of ammonia

ii the fastest rate of conversion to ammonia.

c What practical reasons can you think of for not using very high pressures?

d The boiling points of N_2, H_2 and NH_3 are –196 °C, –253 °C and –33 °C respectively. Explain how the ammonia is separated from unreacted nitrogen and hydrogen.

Fertiliser use today

Today, inorganic fertilisers support about half the world's cereal production. The quantities of inorganic fertilisers being used in the developing world are increasing particularly rapidly – in China, for example, fertiliser applications have risen from 1% of plant nutrients in 1950 to over 60% in 2000.

The range of fertilisers produced includes:

- ammonium nitrate(V)
- ammonium sulfate
- ammonium phosphate

Table 5 Volume percentage of NH_3 in equilibrium mixtures in the reaction $N_2(g) + 3H_2(g) \rightarrow 2NH_3(g)$.

Pressure/atm	NH$_3$ present at equilibrium(%)					
	100 °C	200 °C	300 °C	400 °C	500 °C	700 °C
10	–	50.7	14.7	3.9	1.2	0.2
25	91.7	63.6	27.4	8.7	2.9	–
50	94.5	74.0	39.5	15.3	5.6	1.1
100	96.7	81.7	52.5	25.2	10.6	2.2
200	98.4	89.0	66.7	38.8	18.3	–
400	99.4	94.6	79.7	55.4	31.9	–
1000	–	98.3	92.6	79.8	57.5	12.9

- triple superphosphate – a form of calcium phosphate
- urea – $CO(NH_2)_2$
- potassium chloride.

The compounds are sold individually or mixed to produce a range of products with different N : P : K ratios to meet farmers' needs.

▲ **Figure 31** The Kemira Growhow fertiliser factory at Ince, near Helsby, Cheshire, UK.

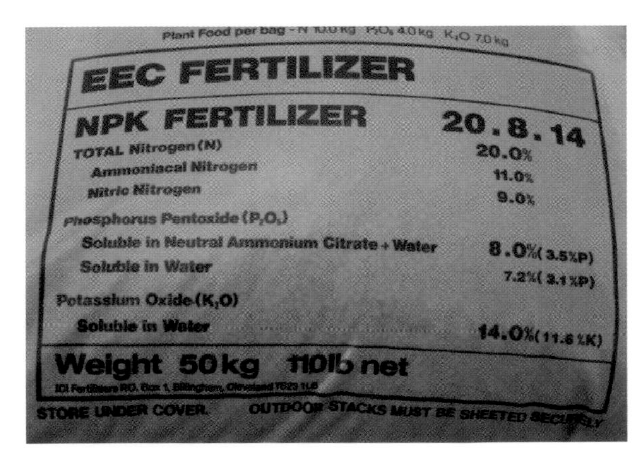

▲ **Figure 32** The percentages of N, P (as P_2O_5) and K (as K_2O) are shown on this bag of fertiliser.

Assignment 6

The major costs involved in distributing and applying nitrogen fertilisers are to do with the mass of material that must be transported for a given mass of nitrogen.

a Calculate the percentages by mass of nitrogen in the fertilisers ammonium nitrate(V), ammonium sulfate and urea. (Make sure that you write the correct formula for each one.)

b List these fertilisers in order of increasing transport costs with respect to nitrogen.

Saving money and protecting the environment

Fertilisers cost money – farmers don't want to waste them. They don't want to apply a nitrogen fertiliser just to have it leached out of the soil as nitrate(V) ions. To avoid wastage, they need to match the addition of fertiliser to the needs of the crop they are growing, and also apply it when the crop is most likely to take up the nitrogen. Any excess leached out of the soil could eventually get into drinking water. Concern over nitrate(V) levels in drinking water has led to an EC limit of 50 mg dm^{-3}, which is considered to be well inside the safety margin.

To reduce the risk of nitrate(V) leaching, winter crops can be grown. Figure 33 shows the soil nitrate(V) levels during a year for a crop sown in the autumn and harvested in August.

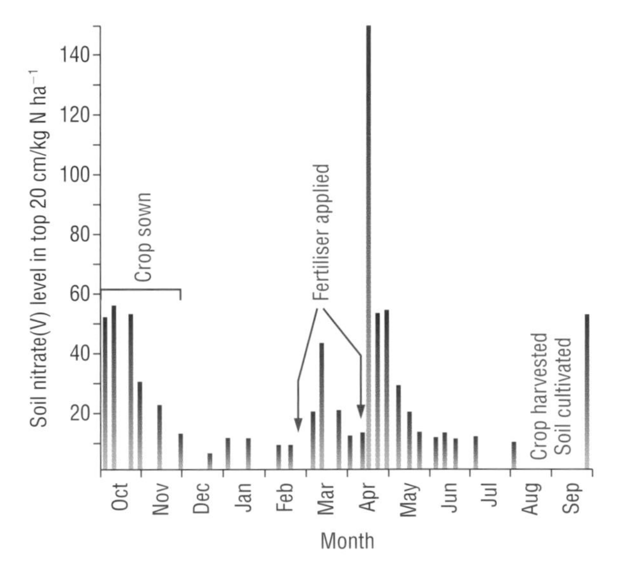

▲ **Figure 33** Soil nitrate(V) levels throughout the year measured at Jealott's Hill Research Station.

Nitrate(V) levels are low in January and February. In spring, the rate of mineralisation increases and fertiliser is added. Higher soil nitrate(V) levels result, but the rapidly growing crop takes up nitrogen and nitrate(V) levels decrease again. After harvesting there is an increase in soil nitrate(V).

Ploughing introduces air into the soil and encourages microbial activity. Warm, moist soil in autumn leads to rapid mineralisation of organic matter – soil nitrate(V) levels rise. As temperatures fall and the soil becomes waterlogged, denitrification causes a decrease in soil nitrate(V) levels. Leaching reduces the levels even further from December onwards.

Figure 34 shows the uptake of nitrogen by winter wheat.

Over 40 kg ha^{-1} of soil nitrogen have been absorbed by late February. Ammonium nitrate(V) fertiliser is then applied in two stages to match the needs of the crop.

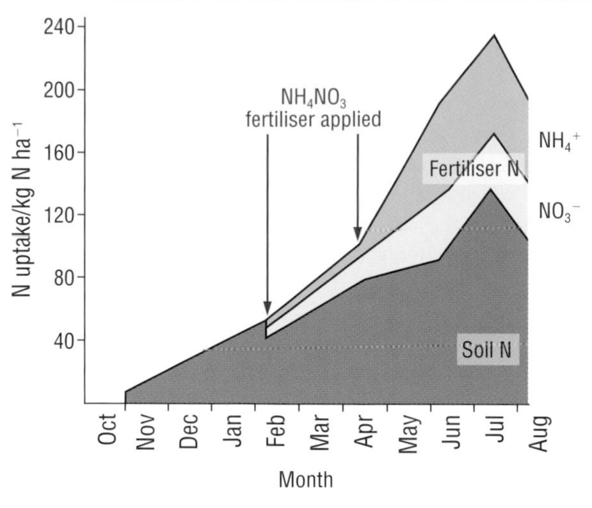

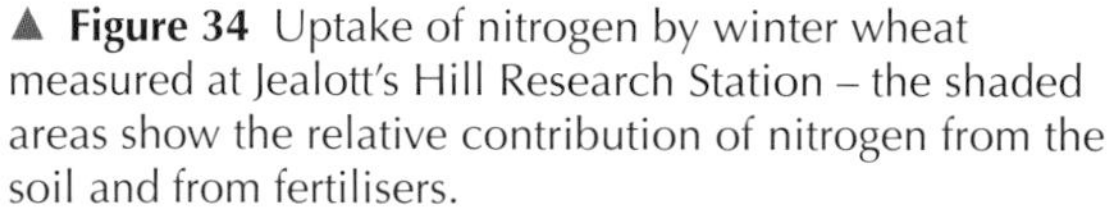

▲ **Figure 34** Uptake of nitrogen by winter wheat measured at Jealott's Hill Research Station – the shaded areas show the relative contribution of nitrogen from the soil and from fertilisers.

The fertiliser used in this study was labelled with the radioactive ^{15}N isotope so that nitrogen entering the crop could be identified as either soil nitrogen or fertiliser nitrogen. Such experiments show that when the correct amount of fertiliser is applied to cereals at the correct time, less than 2% remains in the soil as nitrate(V) by autumn to be at risk of loss by leaching. This means that nearly all the nitrate(V) in the soil at this time of the year comes from the mineralisation of soil organic matter.

Work at Rothamsted has shown that, in plots where no crops have been grown and no fertiliser applied, 45 kg ha^{-1} nitrate(V) are leached from the soil each year. This comes from the large reserve of organic nitrogen in soil, converted to nitrate(V) ions by microbes.

Inorganic fertilisers are designed to contain different proportions of nutrients – this makes it easier for farmers to apply just the quantities needed by the crops. If used correctly, inorganic fertilisers can supplement the nutrient levels in the soil, so that high yields of crops can be maintained without damaging the environment.

Assignment 7

In answering these questions you may need to refer to the data in Figures 33 and 34, as well as the nitrogen cycle in Figure 14.

a What effect does ploughing have on the rate of mineralisation of soil organic matter to inorganic ions? Why does it have this effect?

b In uncropped land, when do you think the greatest loss of nitrate(V) will occur? What two processes account for this loss? What are the environmental consequences?

c Suggest why denitrification tends to be the most important factor in determining soil nitrate(V) levels in early winter.

d Organic farmers grow crops without the use of any fertilisers or pesticides. However, they must be very careful about when they apply farmyard manure. If applied in the autumn, it can lead to far greater nitrate(V) loss by leaching than inorganic nitrogen fertilisers. Explain why farmyard manure increases the rate of nitrate(V) production.

e 'Catch' crops can be grown between main crops. What do you think they are catching? When should catch crops be ploughed into the soil?

Controlling soil acidity

As well as ensuring that the correct nutrients are available for crops, farmers also have to ensure that the soil pH is not too low to inhibit growth. Table 6 shows soil pH values below which plant growth is restricted.

Table 6 Soil pH values below which plant growth is restricted.

Crop	pH below which growth is restricted
Beans	6.0
Oats	5.3
Potatoes	4.9
Wheat	5.5
Lettuces	6.1
Cabbages	5.4
Apples	5.0
Blackcurrants	6.0

Clay soils are made up of sheet structures with negative charges on their surfaces. These use ionic bonding to hold cations on their surfaces. Under natural conditions, H^+ ions from rain, and from plant roots and microbe activity, displace Ca^{2+} and other ions from soil solids. This has two effects – the soil becomes more acidic, and its store of nutrients in the form of exchangeable cations is reduced.

Exchangeable cations are lost for two reasons. Firstly, cations held to the surfaces of the clay layers are replaced by H^+ ions. Secondly, under acidic conditions (at low pH) weathering of clay minerals takes place more quickly (see Figure 9 on page 78).

Aluminium ions are released into the soil and some aluminium oxide is formed. The surface of the aluminium oxide binds H^+ ions, so that it becomes positively charged. It repels cations and holds anions.

The release of aluminium ions from clays at low pH causes another problem – high aluminium concentrations in the soil solution are toxic to crops.

Basic carbonates such as ground limestone or chalk ($CaCO_3$), or bases such as lime $Ca(OH)_2$, can be added to the surface to make a soil less acidic. The amount needed to reach a desired pH depends on the soil's capacity to resist the neutralising action of the base. This is called its **buffering capacity.** A clay soil at pH 5 will need more lime to bring its pH to 7 than the same mass of sandy soil – the clay soil has a higher buffering capacity.

As lime is added to a clay soil, the pH of the soil changes very little at first, and then slowly rises. The soil acts as a **buffer** and is able to resist changes in pH to some extent. It can do this because H^+ ions bound onto the soil solids replace some of the H^+ ions in the soil solution as soon as they are removed.

Figure 35 shows the effect of adding alkali to some clay minerals. Plots like Figure 35 for different soils allow the **lime requirement** of each soil to be calculated.

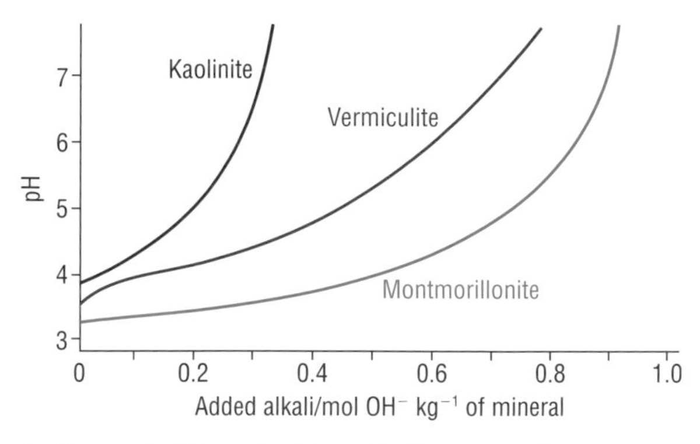

▲ **Figure 35** Effect of adding alkali to some clay minerals.

Assignment 8

Look at Figure 35.

a Which of the three minerals has the lowest buffering capacity?

b Suggest a reason why the mineral you chose has a lower buffering capacity than the other two.

▲ **Figure 36** Farmers add lime to reduce soil acidity.

AI4 *Competition for food*

Increasing the crop yield by improving the concentration of nutrients in the soil is only one of the concerns of an agricultural chemist. Another is to protect the crop – both before harvest and in storage.

Other organisms compete with us for the food we grow. In 2006 the Institute for Plant Diseases in Bonn, Germany, reported that the total global potential loss of crops due to pests varied from about 50% in wheat to more than 80% in cotton production. Overall, weeds were reported to produce the highest potential loss, with animal pests and pathogens being less important.

Control of diseases and pests is now easier because of a variety of scientific advances, such as selective breeding of plant species more resistant to attack, and chemical control using pesticides. Nevertheless, the problem is compounded by the variety of species that can attack crops. For example, there are at least two types of bacteria, five types of fungi and four types of viruses that are responsible for the diseases of rice. By using new plant species and chemical control, a developed country such as Australia can limit losses to these to about 20%, but a developing country not using these techniques, such as Bangladesh, may still suffer losses of up to 70%.

Pesticides (insecticides, herbicides and fungicides) kill insects that eat our crops, weeds that compete with the crops for soil nutrients, and moulds that rot plants and seeds. Many disease-carrying organisms, such as mosquitoes, are also controlled by pesticides.

While there are undoubted benefits from using pesticides, there can also be problems. Many pesticides can be damaging to human health and to the environment if used incorrectly – pesticides may leach into our water supplies. Organisms other than the targets can be killed – if these are predators, which would eat the pests we wish to destroy, then the pests might actually benefit.

Some pesticides that are now banned, such as DDT, can remain in the soil and then build up through food chains, affecting predators such as birds and contaminating human food supplies. Fat-soluble molecules such as DDT can accumulate in the fatty tissues of animals and become more and more concentrated higher up the food chains (Table 7).

Table 7 DDT concentrations up a food chain (data from Long Island, US).

	Concentration of DDT/mg kg^{-1}
sea water	3×10^{-6}
fat of plankton	4×10^{-2}
fat of minnows	0.5
fat of needlefish	2
fat of cormorants	25

▲ **Figure 37** Some common pests: **a** 'rust', a fungal disease; **b** lupin aphids; **c** a cotton bollworm caterpillar; **d** flowering weeds (poppies) in a crop of wheat.

Therefore the challenge to the modern agricultural chemist is an enormous one – to find substances:

- that are *specific* to the target organism,
- that kill at low dosages so that only small amounts need be applied, and
- that do not persist in the environment or travel into the water supply.

Great advances have been made in the last 30 years. Unfortunately, pests can also build up resistance to chemicals, so chemists need to keep finding new products to overcome this.

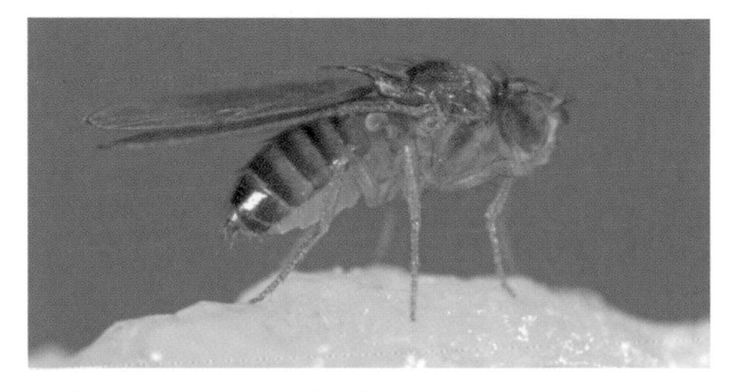

▲ **Figure 38** Pests can develop resistance to chemicals – the fruit fly Drosophila has developed resistance to DDT.

The search for a new pesticide

The research and development involved in producing a new product is lengthy and expensive, requiring the collaboration of a great many scientists – chemists, biologists, toxicologists, chemical engineers and process engineers are all involved.

Large companies may invest sums well in excess of £100 million each year in research and development, from which only one or two new products may result.

When an interesting compound is discovered that is active against pests, chemists will usually try to improve its activity by making systematic changes to the structure, continually testing and working out the 'best' substitutions in various parts of the molecule.

The compounds are tested on target pests and compared with existing products for potency and for the range of pests affected. The compounds that come out best in laboratory tests may be tried out to see if they will work in real field situations. Then hundreds of field trials are conducted on substances chosen for development.

A successful compound is judged on a range of factors. These will include:

- ease of manufacture
- specificity
- persistence in soil
- cost of the final product
- marketability
- leaching losses into drainage water
- toxicity to humans
- comparison with known compounds
- ownership of the patents surrounding the invention.

Patents are important because the company needs to be able to make a profit in return for its investment in research and development.

The pyrethroid story

For many centuries, the dried flower heads of a chrysanthemum, *Chrysanthemum cinerariaefolium*, have been used to ward off insects, particularly mosquitoes. The structures and stereochemistry of the natural insecticides present in the flower heads were worked out between 1920 and 1955. One of them is pyrethrin 1:

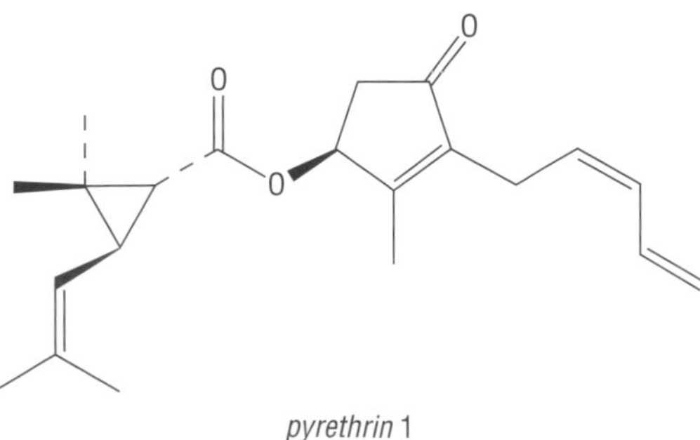

pyrethrin 1

Pyrethrins have some of the qualities of the ideal insecticide – they are powerful against insects, but are harmless to mammals under all normal circumstances. However, natural pyrethrins are unstable in light – a photochemical oxidation reaction occurs. This limits their use in agriculture because they break down so quickly.

▲ **Figure 39** Pyrethrum flower.

Assignment 9

a A carbon atom with four different groups attached to it is described as a chiral centre. (If the carbon atom is part of a ring, and the structure of the ring is different on each side of the carbon atom, the ring counts as two different groups.) Copy the structure of pyrethrin 1 and identify the chiral centre(s) by marking them with an asterisk (*).

b List the functional groups present in pyrethrin 1.

Michael Elliott and his co-workers spent many years working at Rothamsted on the synthesis of pyrethroids – compounds related to natural pyrethrins. They were looking for substances like the natural compounds that would be active against insects, but more stable in light and air.

They made an important breakthrough in 1977 with the synthesis of Permethrin, the first pyrethroid sufficiently stable to be used widely in agriculture. Permethrin is a mixture of stereoisomers – one isomer, biopermethrin, is shown below:

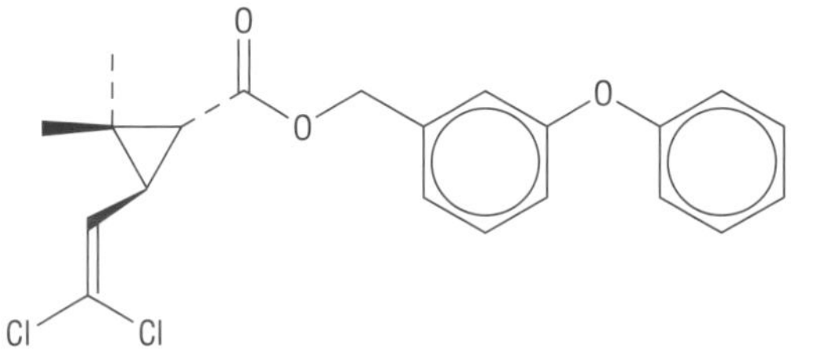

biopermethrin

A later discovery, biocypermethrin, is a more active insecticide so smaller quantities can be applied to achieve the same effect. One isomer is shown below:

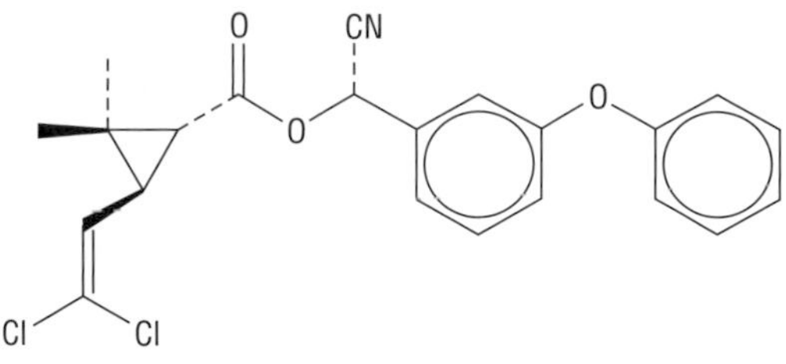

biocypermethrin

This is a classic example of the way in which both new agrochemicals and new pharmaceuticals can be found. Something in nature is known to have special properties – chemists then determine which substance it is that confers these properties. The next step is to find a means of synthesising it to test the effectiveness of the pure form. Chemists may then develop compounds that have an even greater effect, using variations on the structure of the natural product. At this stage, extensive trials are made and methods of transferring the synthesis from laboratory to manufacturing scale are explored. No wonder it is so expensive to develop an agrochemical or a medicine.

You can consider some of the issues involved in the production of a chemical on an industrial scale in **Activity AI4**.

You can revise some of the important stages in the development and production of a new agrochemical or pharmaceutical in **Chemical Ideas Chapter 15**. This will help with **Activity AI4**.

What happens to pyrethroids in the environment?

In mammals, pyrethroids are rapidly broken down into polar products, either by oxidation or by hydrolysis of the ester group. These polar products are not attracted to the fatty membranes, but remain in aqueous solution and are excreted before they can reach sensitive sites in the body.

Synthetic pyrethroids persist on crops for 7–30 days (Figure 40). Any pyrethroid residues reaching the soil are attracted into the soil organic matter. Once there they are rapidly hydrolysed and oxidised by routes like those in mammals. The products are inactive polar compounds, so residues of active, non-polar compounds do not build up in the environment.

Assignment 10

This assignment requires you to make use of chemical ideas met in earlier parts of the course. You may wish to refer to **Chemical Ideas 13.5** for the reactions of esters, **Chemical Ideas 7.3** for thin-layer chromatography and **Chemical Ideas 10.3** for information about using half-lives to determine order of reaction.

Biopermethrin is safe to mammals because enzymes, called esterases, catalyse the hydrolysis of the ester linkage.

a Look at the structure of biopermethrin on page 93, and draw the structural formulae of the hydrolysis products. Explain why these are more soluble than biopermethrin in water.

b The course of the hydrolysis can be followed by thin-layer chromatography. The R_f values in the eluting solvent used in one experiment are shown below:

	Biopermethrin	Alcohol derivative
R_f value	0.6	0.15

The acid derivative did not move much above the base line.
Describe in outline the procedure you would use to follow the progress of the reaction.
Sketch how the chromatograms would look
i when the hydrolysis reaction was incomplete and
ii when it had reached completion.

c The hydrolysis of biopermethrin in the soil is a first-order reaction. Calculate the half-life of biopermethrin if there is 2% of the insecticide left in the soil 2 months after application.

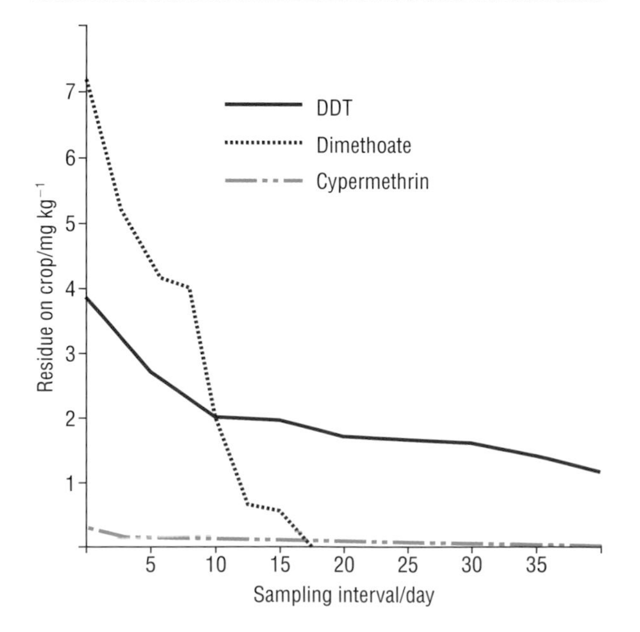

▶ **Figure 40** Decay of pesticides – the graph shows the large reduction in pesticide residues on crops when newer pesticides such as cypermethrin replace older ones such as DDT and dimethoate; less pyrethroid is needed, and it decays faster to inactive products.

▲ **Figure 41** Measuring the effects of an agrochemical.

Herbicides

Herbicides can increase crop yields by destroying weeds. There are two main types of herbicide – **total herbicides** and **selective herbicides**. Total herbicides destroy all green plant material and are used in fields before a crop is planted.

One example of a total herbicide is *paraquat* – pure paraquat is highly toxic, but in the concentrations applied to kill weeds it is relatively harmless to humans and is rapidly inactivated on contact with soil. The structure of the paraquat ion is

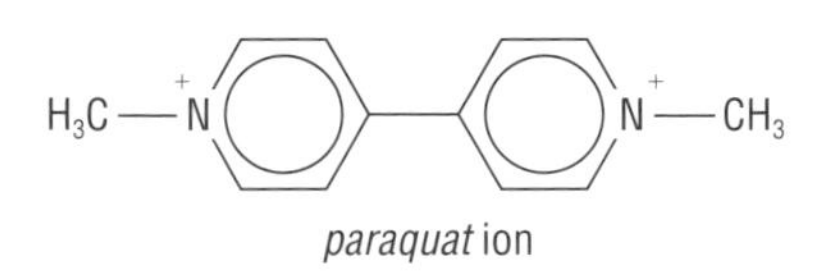

paraquat ion

As soon as the positive paraquat ions make contact with the soil they are removed by adsorption onto the soil solids. This is a particularly useful property because it means that paraquat is inactivated as soon as it reaches the soil – it kills only those plants whose *leaves* it touches.

Each type of soil is capable of holding a particular amount of paraquat irreversibly – this amount is called the **strong adsorption capacity** of the soil. Some values are given in Table 8. If the paraquat concentration rises above the strong adsorption capacity of the soil, paraquat can be displaced into soil water and can damage growing plants.

ADSORPTION AND ABSORPTION

Paraquat is deactivated when it comes into contact with soil because it is adsorbed onto the soil solids. A substance is **adsorbed** when it is bound to the *surface* of another substance.

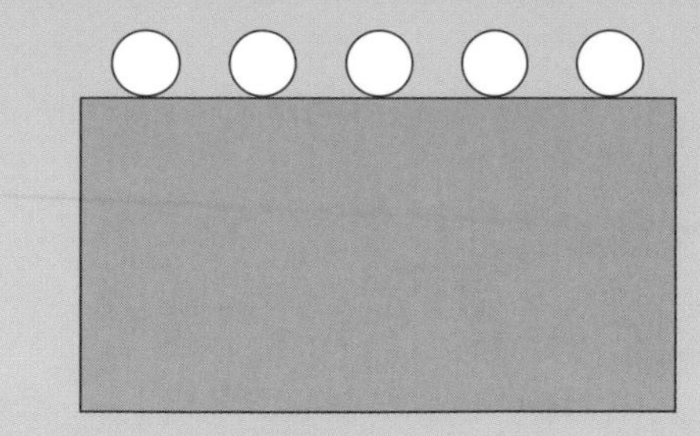

▲ **Figure 42** Adsorption.

Be careful not to confuse this with **absorption** – in this the absorbed substance diffuses *into the bulk* of another substance. In coloured plastics, for example, dye molecules are absorbed into the bulk of the plastic.

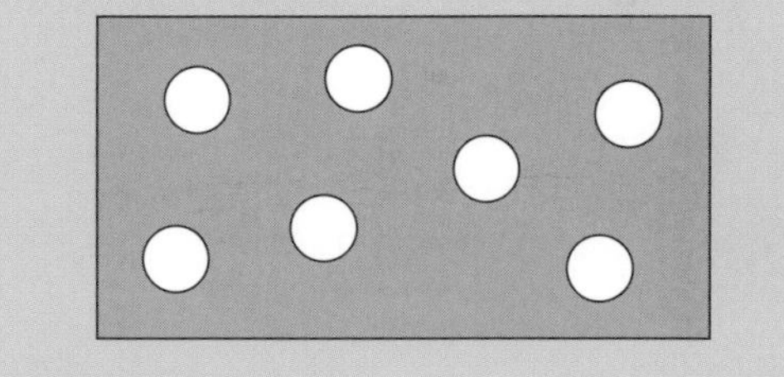

▲ **Figure 43** Absorption.

Paraquat can be used as an alternative to ploughing as a means of destroying weeds before a crop is sown, but farmers also need to destroy weeds in growing crops without harming the crops themselves. Therefore, they want a range of selective herbicides. The selectivity of some herbicides relies on naturally occurring differences in the plants they are applied to. For example, because of their narrow leaf shape, grasses are generally more resistant to herbicides than broadleaved plants – so herbicides have been developed that can kill broadleaved plants without damaging grasses or grass-type crops such as rice. Other selective herbicides have been developed that work by exploiting differences between the metabolisms of different plants.

Table 8 Strong adsorption capacities of some soils.

Source	Soil type	Composition			Strong adsorption capacity/mol kg^{-1}
		% sand	% silt	% clay	
Newark	clay	34	24	42	0.137
Oakham	clay loam	33	37	29	0.047
Jealott's Hill	sandy loam	49	33	18	0.041
Sutton Coldfield	sandy loam	78	15	8	0.0078
Bagshot	loamy sand	85	11	4	0.0019

ATOM ECONOMY AND AGROCHEMICALS

Insecticides with brand names such as Furadan and Temik belong to a group of organic chemicals that contain the carbamate functional group:

Carbamates have been produced using a chemical called phosgene. However, pollution control legislation has limited the use of phosgene. This has necessitated finding an alternative synthetic route.

Industrially one of the most important carbamates is phenyl N–phenyl carbamate:

Initially, chemists looked at producing this in one of two ways:

- Substituting an O for an NH group:
- Substituting an NH for an O group:

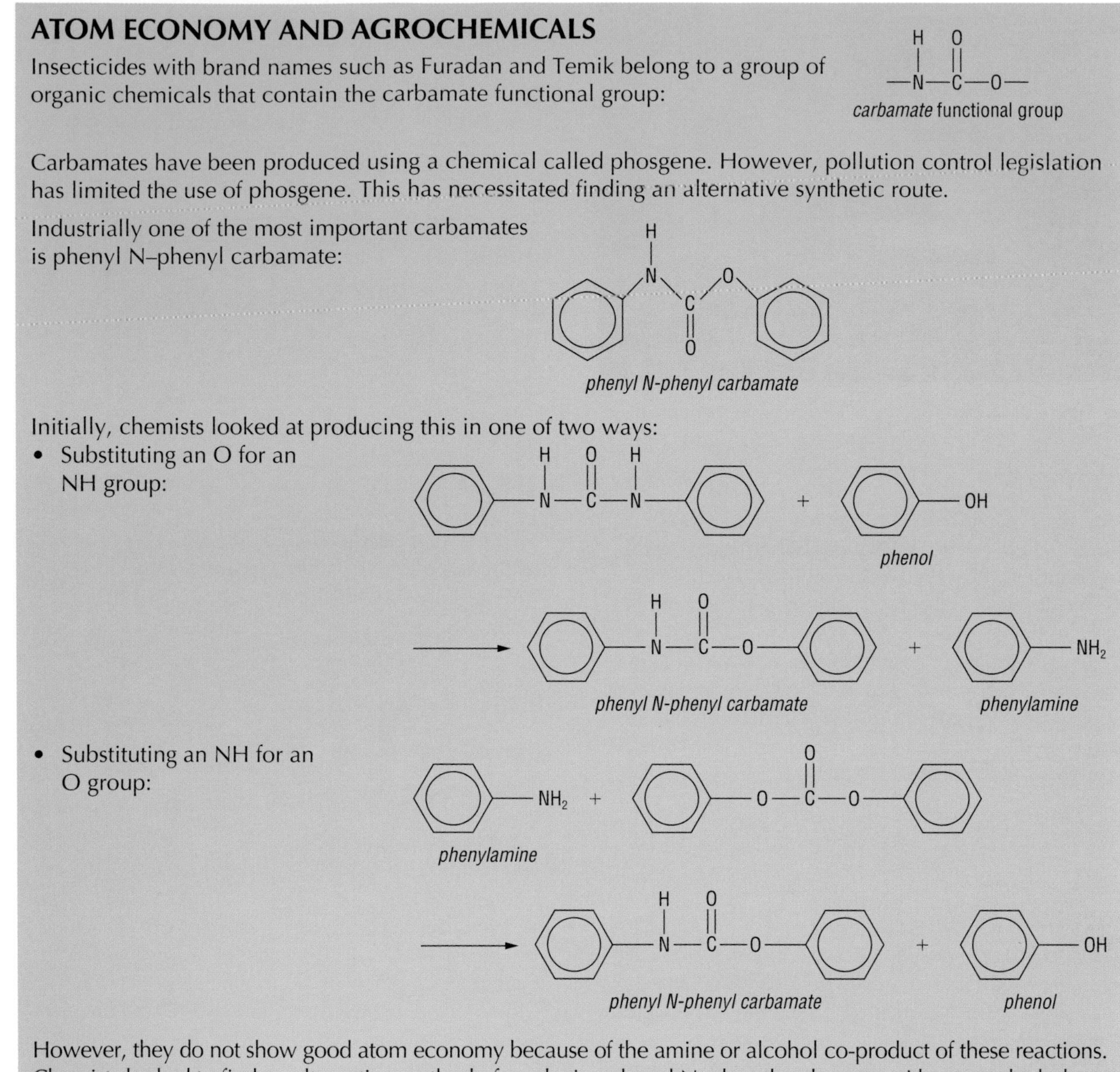

However, they do not show good atom economy because of the amine or alcohol co-product of these reactions. Chemists looked to find an alternative method of producing phenyl N–phenyl carbamate without an alcohol- or amine-leaving group. In the course of their research they became interested in the following reaction:

Carbamate synthesis via an addition reaction:

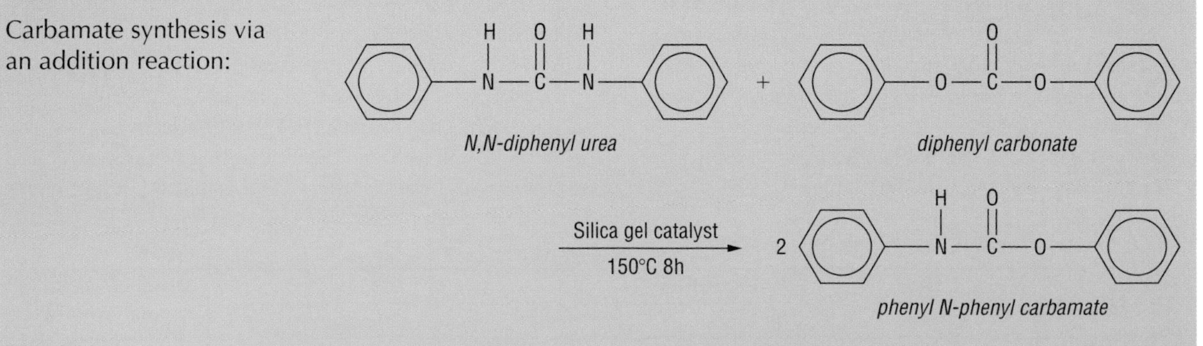

The initial experiment without a catalyst yielded only trace amounts of the target molecule (phenyl N–phenyl carbamate), but several catalysts were screened and silica gel was found to be very suitable. Not only did it produce a yield of 81%, it is also low cost and easily recycled. This reaction can also be easily scaled up for industrial production.

Genetic engineering can also be used to produce plants that are resistant to certain herbicides – then a crop can be sprayed that kills weeds without damaging the crop itself. However, with genetically engineered plants there is a danger that the resistant gene could be transferred to other plants through cross-breeding and this could lead to the production of 'superweeds' – weeds that cannot be destroyed with herbicide.

Assignment 11

a Calculate the atom economy of the three reaction methods for producing phenyl N–phenyl carbamate given in the 'Atom economy and agrochemicals' purple box on page 96.

b Why does silica gel make such a good catalyst for this reaction?

Controlling pests organically

Unfortunately, organically farmed crops are not immune to attack from pests, but the ways in which they can be controlled are very different from conventional farming. Organic farmers claim that their crops have greater natural resistance to pests and disease, because they grow more slowly and have thicker cell walls that provide a stronger barrier to pests. For example, aphids are pests that damage crops by feeding on plant sap – however, a plant with thick cell walls is harder for aphids to feed on. Organic farmers also endeavour to prevent pests from becoming a problem by choosing plant varieties that show resistance to disease. In organic farming, 'prevention rather than cure' is the key to success in controlling pests. The following are some ways in which organic farmers prevent pests from becoming a problem on their land.

Crop rotation

The golden rule in organic farming is to not grow the same crop in the same soil in consecutive years. This is not just to prevent nutrient loss from the soil, but also to prevent pests from building up in the soil. Removing the crop removes the pest organism's habitat, which keeps its numbers to a minimum.

Physical barriers

Covering crops with netting prevents predators from eating or laying their eggs on a crop. For example, the cabbage white butterfly lays its eggs on vegetables such as purple sprouting broccoli – for the very reason that these plants will provide food for the young caterpillars. Covering the crop with fine mesh netting prevents this from happening.

Encouraging predators

Some pests, such as greenfly, are a food source for another garden insect – the ladybird. If farmers take measures to increase the numbers of these predators, then they will see a decrease in the numbers of pests.

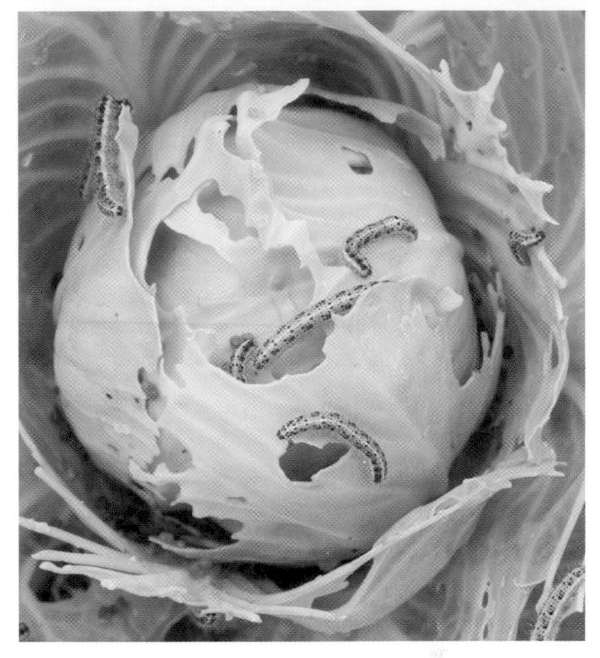

▲ **Figure 44** Young caterpillars feed on brassicas such as cabbage and can destroy the crop.

▲ **Figure 45** Some crop pests have predators – if the numbers of these predators can be increased then pest numbers will reduce. Ladybirds (shown here with ants) are natural predators of aphids (greenfly).

Weeding

Instead of using herbicides, organic farmers need to remove weeds (plant pests) mechanically. This can be done before a crop is planted by ploughing and some weeding can be done by machine, but often there is no alternative to labour-intensive hand weeding. Growers of organic crops claim that this is one of the major contributors to the higher cost of organic vegetables.

Limited use of certain pesticides

Soil Association growers in the UK can use small quantities of four pesticides – copper, Rotenone (sometimes referred to as Derris), soft soap and sulfur. These are used to control insect pests and fungi – no herbicides are permitted. Of these the use of copper and Rotenone are restricted and can only be used with the specific consent of the Soil Association – if the farmer provides evidence of a threat to the crop and where no alterative is available. In general, organic pesticides are used mainly to control the fungal disease potato blight, which can have devastating effects.

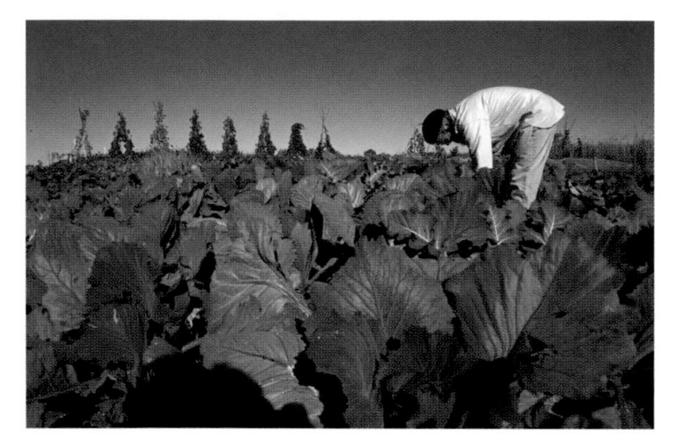

▲ **Figure 46** Organic farmers' pesticide and herbicide use is limited, so often there is no alternative to labour-intensive hand weeding.

Companion planting

This is a system where two or more plants are grown in close proximity so that one or all of them may gain some form of benefit. For example, African Marigolds produce a chemical called thiopene that repels nematodes (a type of worm), making them a good companion plant for root crops prone to attack by nematodes. Marigolds also reduce attacks on tomatoes by whitefly – it is thought that this is because the smell of the marigolds masks the attractive (at least to whitefly) smell of the tomatoes. Borage deters tomato worms and garlic repels aphids – there are many more examples.

◀ **Figure 47** Potato blight is widespread in the UK and can have devastating effects.

Organic and conventional farming exist side by side in our world today. They employ different techniques to maximise crop yields, but ultimately the aim of both is the same – to provide enough food to meet the demands of a growing world population.

AI5 *Summary*

This module has introduced you to some of the issues involved in developing and maintaining a system of agriculture that is able to feed the world's rapidly increasing population. There is a constant dilemma concerning how to do this without damaging the environment.

Between 1960 and 2000, the world's food supply increased threefold, without increasing the area used for farming. However, between 2000 and 2050 the population is likely to increase from 6 billion to over 9 billion. Furthermore, it is hoped and expected that the quality of life for the poorest people will increase significantly during this period. This means that agricultural output has to be increased even more, raising an inevitable question about the future of agriculture.

To understand these issues, you first needed to learn about the composition and structure of soil, and to understand the relationship between the properties of a substance and its structure and bonding.

Understanding the processes that occur as plants grow and decay is vital to understanding how nutrients are cycled. Central to this are the nitrogen cycle and the redox chemistry of nitrogen.

You looked at how crop production is managed in organic farming and conventional farming. The use of inorganic fertiliser is one way of maintaining soil fertility and high crop yields. The Haber process for the manufacture of ammonia from nitrogen and hydrogen is the key step in the production of nitrogen fertilisers. An understanding of equilibrium reactions and the effects of temperature and pressure, both on the position of the chemical equilibrium and on the rate of the reaction, is necessary to select the optimum conditions for this process. The use of a catalyst is also important.

In the last section of the module, you studied some methods of pest control. A study of the action of pyrethroids, modern insecticides that do not persist in the environment, provided a setting in which to revise some of the issues involved in the development and production of an agrochemical. You also looked at the pest control methods available to organic farmers. Organic and conventional farming exist side by side in our current world – who knows where the balance of these methods will take us in the future?

Activity AI5 will help you to check your knowledge and understanding of this topic.

COLOUR BY DESIGN

Why a module on 'Colour by Design'?

Coloured compounds are everywhere. The ability to make synthetic dyes and pigments to colour an enormous variety of things is one of the great achievements of modern chemistry.

The first part of the module describes how some coloured pigments were discovered by accident. These pigments are d-block compounds and you will develop ideas to explain why such compounds are coloured.

The second part of the module describes work carried out in the Scientific Department of the National Gallery, London, on the conservation of two paintings – *A Wheatfield, with Cypresses*, painted by Vincent van Gogh in 1888, and *The Incredulity of Saint Thomas*, painted by Cima da Conegliano in 1504.

The chemists at the National Gallery use a variety of analytical techniques to investigate pigments and paint media. This provides an ideal setting in which to learn about ultraviolet and visible spectroscopy, atomic emission spectroscopy and gas–liquid chromatography. Analysis of the drying oils used by Cima da Conegliano in his sixteenth-century Italian altarpiece requires an understanding of the structure of oils and fats. This links in well with your previous work on acids, alcohols and esters.

The third part of the module traces the development of synthetic dyes for fabrics. In this section you will extend your knowledge of organic chemistry further and learn about the special structure of the benzene ring and the types of reaction that arenes undergo. You will also see why structure is important in determining the colour of an organic compound.

Finally, the module provides an opportunity to draw together some ideas you have met earlier to understand how dyes can be made to dissolve and to adhere strongly to fabrics.

Overview of chemical principles

In this module you will learn more about ideas introduced in earlier modules in this course:

- the interaction of radiation with matter (**Elements of Life, The Atmosphere**, **Polymer Revolution** and **The Steel Story**)
- spectroscopy (**Polymer Revolution, What's in a Medicine?**)
- atomic emission spectra (**Elements of Life** and **The Steel Story**)
- aromatic compounds (**Developing Fuels** and several other storylines)
- esters (**What's in a Medicine?** and **Materials Revolution**)
- reaction mechanisms (**Elements from the Sea, The Atmosphere** and **Polymer Revolution**)
- intermolecular bonds (**Polymer Revolution**)
- the relationship between structure and bonding, and properties (**Agriculture and Industry** and several other storylines).

You will also learn new ideas about:

- why compounds are coloured
- ultraviolet and visible spectroscopy
- gas–liquid chromatography
- oils and fats
- the structure of benzene and the reactions of arenes
- the chemistry of dyes.

CD

COLOUR BY DESIGN

 *Ways of making colour*

The natural world is full of colour. Some colours, like the blue of the sky or the colours in a rainbow, are produced by the scattering or refraction of light. But in most cases colour is due to the presence of *coloured compounds* and arises from the way these compounds interact with light.

▲ **Figure 1** Cave painting in Lascaux in France thought to date back 17 000 years.

From the earliest times people have used the natural substances around them to colour themselves and their possessions. We know that some Neanderthal tribes roaming Europe 180 000 years ago prepared their dead for burial by coating them with *Red Ochre* (iron(III) oxide). For tens of thousands of years, humans made colouring agents from minerals they found in rocks, so the colours produced were mostly dull and earthy. These mineral **pigments** were mixed with oil or mud to form a paste that would stick to surfaces.

You can revise why some compounds are coloured by reading **Chemical Ideas 6.7**.

Many of the early dyes came from crushed berries or plant juices. The early Britons used a blue dye that they extracted from the woad plant (*Isatis tinctoria*). The main coloured component of woad is *indigo* – the same dye is used today to colour blue denim jeans.

PIGMENTS AND DYES

The way a coloured substance is used determines whether it is called a *dye* or a *pigment*. You can dye your hair or your clothes, but when you paint a picture you are using pigments. The main thing to remember is that *dyes are soluble substances*, whereas *pigments are insoluble*.

Pigments can be spread in a surface layer (as in a paint or a printing ink) or mixed into the bulk of a material (as when making a coloured plastic bowl).

Dyes are always incorporated into the bulk of a material and they attach themselves to the molecules of the substance they colour. This attachment can be the result of **hydrogen bonding** or of weaker intermolecular bonds, such as **instantaneous dipole–induced dipole bonds** or other **dipole–dipole bonds**. Sometimes stronger ionic or covalent chemical bonds are involved.

You will find that there is another general distinction – most *dyestuffs are organic compounds*, whereas *pigments can be organic or inorganic*.

This was fine until people learned to weave and make fabrics. When the paste-like pigments were applied to fabrics, the cloth became stiff and the colouring material soon fell out – pigments were no good for colouring cloth. Cloth could only be coloured by soaking it in a solution of a *dye*.

Dyes can be made into pigments by coating particles of an inert solid material with the dye – these pigments were called **lakes**, e.g. *Crimson Lake*.

▲ **Figure 2** Modern dyes and pigments make life very colourful.

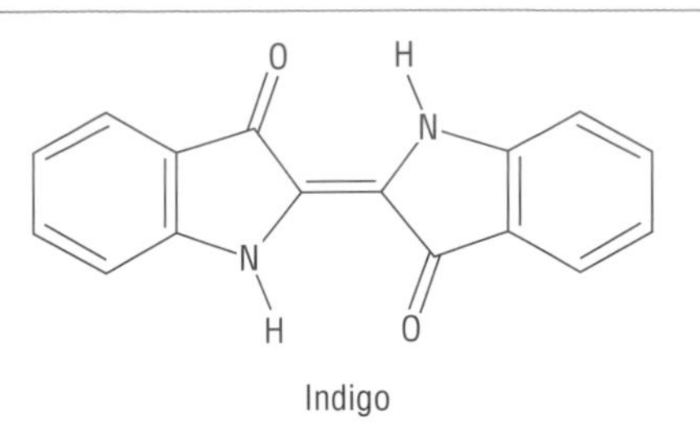

Indigo

Some dyes came from animals. Mexican dyers around 1000 BC discovered the red dye *cochineal*. They extracted it from small insects that live on the *Opuntia* cactus. Only female insects produce the dye, and they had to be collected by hand – about 150 000 insects were needed to make 1 kg of dye! It was the Spaniards who brought cochineal to Europe in 1518 AD. It was used until 1954 to produce the bright red jackets of the Brigade of Guards. You may have eaten it as a food colouring.

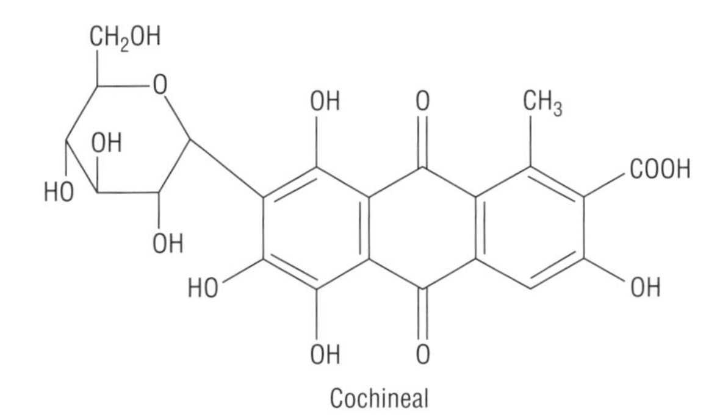

Cochineal

Until about 150 years ago, dyes and pigments were expensive and colours were mainly for the wealthy. Ordinary people wore clothes dyed with cheap vegetable dyes – the colours were often drab and faded quickly.

The great breakthrough came when chemists learned how to make coloured substances in the laboratory. The starting materials for the new *synthetic* dyes were cheap. They came from coal tar, an unwanted by-product of the new coal-gas industry. In the second half of the nineteenth century, as towns and cities were lit up by gas lamps, chemical companies producing dyes from coal tar flourished – and Europe exploded in a riot of colour!

Not only did chemists learn to copy natural colours, they also used the compounds they obtained from coal tar to make a whole range of entirely new coloured compounds.

Modern colour chemists now have a vast range of coloured substances available to them. They need to understand not only why compounds are coloured, and which structures lead to particular colours, but also how to bind coloured substances to different types of fibres and surfaces. Once the chemistry is understood, it becomes possible to *design* a coloured molecule for a particular purpose – colour by design.

▲ **Figure 3** Late nineteenth-century dyehouse.

▲ **Figure 4** Gas holder built by the Imperial Gas Company at Bethnal Green in London.

COAL TAR

'Coal gas' was made by heating coal in the absence of air. Coke, coal tar and a liquid rich in ammonia were also produced.

The coal-gas industry provided the raw materials for the production of synthetic dyes – and later for the production of pharmaceuticals, plastics, perfumes and explosives. Coal tar was then replaced by oil as the source of organic chemicals.

In **Activity CD1** you can look at ways of changing colours chemically.

Assignment 1

a Look carefully at the structure of cochineal on this page. Make a table showing the different functional groups in the molecule – give the name and full structural formula of each.

b What structural feature present in both indigo and cochineal do you think could be responsible for their colours?

CD2 Colour by accident

Once in a blue moon!

Many of the advances in colour chemistry have been made as a result of intensive research and painstaking 'trial and error'. Every so often, though, an important step forward seems to happen by chance. The development of blue pigments illustrates the role of 'serendipity' in scientific progress.

From ancient times blue has been considered a 'noble' colour, synonymous with royalty. It is cool and calming, reminding us of matters spiritual and infinite, indicating stability. Blue is a popular colour for corporate logos.

Despite the abundance of the colour blue in nature, the sky and the sea providing a rich palette of various deep-blue shades, the colour is rare in minerals. After grinding, minerals were used as the source of colour in pigments, inks and glass. Only the semi-precious stone *Lapis lazuli*, beloved by the ancient Egyptians and Persians, was of a strong enough colour. Unfortunately, its rarity and difficult extraction made it far too expensive to be used widely as an artist's material.

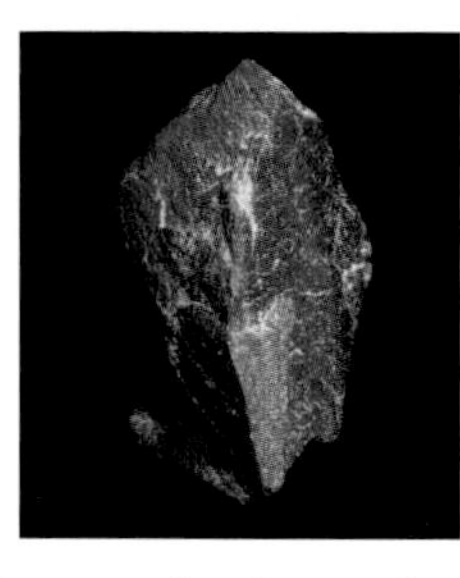

◄ **Figure 5** Lapis lazuli is a rare, deep-blue gemstone formerly used as a source of blue pigment.

However, by chance the ancient Egyptians discovered how to make the first synthetic pigment by heating together *natron* (a mineral containing mainly sodium carbonate), copper oxide, lime and sand. The pigment was known as *Egyptian Blue Frit*. For some 4000 years, until the nineteenth century, this was the darkest and most stable blue pigment available at a reasonable cost, but it had a rather coarse texture.

In the Middle Ages, a few artists used the blue pigment *Ultramarine*, derived from Lapis lazuli, but the cost was generally prohibitive. A cheaper, less stable and less intense material based on the copper mineral *azurite* was used by most painters. In the early nineteenth century the French Government offered a large prize to anyone who could synthesise Ultramarine. All attempts failed until a French factory accidentally produced a deep-blue colour as a by-product of heating lime in a kiln. John Freidrich Gmelin, professor of chemistry at the University of Tübingen in Germany, carefully analysed the process in detail and came up with a workable process for synthesising Ultramarine, thus winning the prize. Artificial Ultramarine is now known as *French Ultramarine* and is one of the cheapest artist's pigments available.

▲ **Figure 6** King Ramses III (1170 BC) – the blue helmet with a golden serpent was the symbol of royalty.

Another important blue pigment, often used in printing inks, is *Prussian Blue*. This was discovered in 1705 by a German dyer, Heinrich Diesbac, who was attempting to make a red lake from cochineal, iron(II) sulfate and potash (potassium hydroxide). However, he tried to save money by using potash supplied by an alchemist, Johan Dippel, in whose laboratory he was working. The potash was contaminated with cattle blood and was supposed to be thrown out! Instead of producing the expected deep-red colour, the lake was very pale. On heating it to make it more concentrated, it turned purple and then a deep blue – he had accidently synthesised the first blue pigment since the ancient Egyptians.

Prussian Blue is a complex containing iron (a transition metal). In the second half of **Chemical Ideas 6.9** you can find out why pigments containing transition metal ions are coloured.

Assignment 2

Prussian Blue was used for many years before the exact structure in the solid was determined. The formula of the complex is often given as $Fe_4(Fe[CN]_6)_3$.

a Give the oxidation state of each type of iron atom in the complex.

b Draw the structure of the complex iron present in Prussian Blue, showing its shape.

c The absorption spectrum of Prussian Blue has a strong peak centred at 680 nm. Explain why the pigment is blue.

d Using ideas of energy levels and ligands, explain why transition metal complexes in solution are coloured.

Accidents will happen …

Monastral Blue is one of the best blue pigments ever made. It is widely used to colour plastics, printing inks, paints and enamels because it gives a very pure blue. The paint on most blue cars, for example, contains Monastral Blue. Its discovery arose from a chance observation at a Scottish dyeworks. Fortunately, the people who made the observation took the trouble to investigate further.

▲ **Figure 7** Monastral Blue is the pigment in the blue paint on this classic steam locomotive, the 'Mallard'.

In 1928 in Grangemouth, Scotland, the firm Scottish Dyes (which later became part of ICI) was making phthalimide, a compound needed for dye manufacture. In this process, a white substance, phthalic anhydride, is melted in a glass-lined iron vessel, and ammonia is passed into it:

ammonia(g) + phthalic anhydride(l) → phthalimide(l)

The product is white too, but on this occasion they found that some batches contained traces of a blue substance. The blue compound seemed to have formed where the reaction mixture came into contact with a part of the vessel where the lining was damaged.

Later, when the blue substance was analysed, all samples were found to contain 12.6% iron. The structure of the blue substance, although complex, was interesting for a number of reasons – not least because it was closely related to that of some naturally occurring substances called *porphyrins*.

At first it was thought that the new blue compound, named *iron phthalocyanine*, would be of academic interest only, but further investigation suggested that it might be a useful blue pigment.

Assignment 3

Read the account of the discovery of iron phthalocyanine.

a Why did the blue pigment form?

b What is the significance of the fact that *all* samples of the blue pigment contained 12.6% iron?

PORPHYRINS EVERYWHERE

You already know of an important example of a coloured substance that contains a porphyrin ring system. *Haemoglobin* is responsible for the red colour of blood and is involved in transporting oxygen around the body. *Chlorophyll* a gives leaves their green colour and is responsible for harvesting light and initiating photosynthesis. It too is a porphyrin pigment, but with magnesium as the central metal atom.

The structure of the blue compound

The blue compound is made up of large, flat molecules. Each molecule contains a 16-membered ring of alternating carbon and nitrogen atoms, with an iron atom at the centre. Look carefully at Figure 8, showing the structure of iron phthalocyanine, and pick out the 16-membered ring surrounding the iron atom.

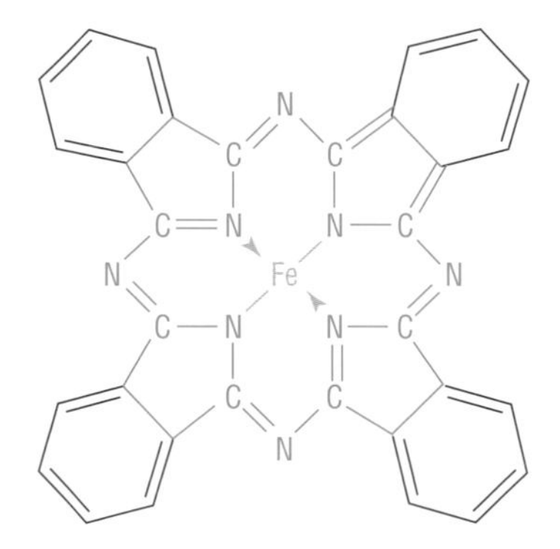

▲ **Figure 8** Structure of iron phthalocyanine – the porphyrin ring system is shown in blue.

Don't be put off by the size of the molecule – you won't be expected to remember its structure, but you should be able to recognise some of its important features.

Four of the nitrogen atoms act as *ligands* to form a planar complex with the iron atom. You will remember from **The Steel Story** that complexes of d-block metals are often highly coloured – you can revise this in **Chemical Ideas 11.6.**

The 16-membered ring in iron phthalocyanine contains alternate single and double bonds. This type of arrangement is called a *conjugated system* of bonds. For each double bond, one pair of electrons is not confined to linking two particular atoms, as in a single bond, but is spread out or **delocalised** over the whole conjugated system. The lone pairs of electrons on two of the nitrogen atoms are also involved, so the delocalisation extends over the whole 16-membered ring – and over the four benzene rings too. The result is a huge delocalised structure.

The more that electrons are delocalised like this, the lower the energy of the molecule. So, iron phthalocyanine is a very stable pigment.

Molecules with extended conjugated systems also tend to be coloured – you will find out more about this in section **CD5**.

It's even better with copper

Molecules with similar structures have similar properties, so it is always worth making a series of related compounds in the hope that you come across one or two with even better properties than the ones you already have.

Once the structure of the blue iron compound was known, a team of chemists led by R.P. Linstead at Imperial College, London, set to work investigating other phthalocyanines. They repeated and modified the original process using other d-block metals instead of iron as the central atom for the molecule.

Within a few years of the original discovery they had shown that a whole range of similar pigments could be made. Copper phthalocyanine was particularly promising. Writing about this in 1934, Linstead said:

> '... it is even more stable than the other compounds of this series, and in this respect must be classed among the most remarkable of organic compounds. It resists the action of molten potash (potassium hydroxide) and of boiling hydrochloric acid ... It is exceptionally resistant to heat, and at about 580 °C it may be sublimed ... in an atmosphere of nitrogen or carbon dioxide ... The vapour is deep pure blue and the crystals, which have the usual purple lustre, may be obtained up to 1 cm in length.'

By 1939, copper phthalocyanine was on the market as a pigment with the trade name Monastral Blue (Colour Index Pigment Blue 15: CI 74160).

The commercial importance of Monastral Blue and other phthalocyanine pigments is due to a combination of factors:

- they have very beautiful bright blue to green shades and high colour strengths – which means that only a relatively small amount of pigment is needed to give a good colour
- they are very stable pigments and have excellent fastness to light – they don't fade.

THE COLOUR INDEX

Most pigments and dyes are known by the name originally given to them by the manufacturer, rather than by a systematic name. So it's quite common for these compounds to have several names.

To avoid confusion, the Society of Dyers and Colourists in Bradford have drawn up a Colour Index in which coloured compounds are classified according to their application and colour, and given a generic name, e.g. CI Pigment Blue 15. All commercial products having the same structure are classified under a generic name. Where the structure is known, the compound is also given a Constitution Number (e.g. CI 74160).

Yellow can also be improved

Artists are always on the lookout for new and better pigments from which to make their paints – by the middle of the eighteenth century a variety of yellow pigments were being used. *Orpiment* (arsenic(III) sulfide) and *Yellow Ochre* (hydrated iron(III) oxide) had been known since earliest times. *Massicot* (lead(II) oxide) was introduced in the sixteenth century. But there was still a need for a pigment capable of giving a bright, lemon-yellow colour.

The French chemist Louis Vauquelin discovered chromium in 1797 while investigating a mineral called Siberian red lead spar from the Beresof gold mine in Siberia – another accident! The colours of samples of the mineral varied from orange–yellow to orange–red. We now know that the mineral was a form of lead chromate(VI) called crocoite, but at the time its composition was unknown.

▲ **Figure 9** Louis Vauquelin (1763–1829) discovered metallic chromium, Chrome Yellow and other artists' pigments.

COLOURFUL COMPOUNDS

A contemporary of Vauquelin described how the new metal, chromium, got its name (*chroma* is the Greek word for colour):

> '[The new element] ... had the property of changing all its saline or earthy combinations to a red or orange colour. This property, and that of producing variegated and beautiful colours when combined with metals, induced him to give it the name "chrome".'

By 1809, a new source of crocoite had been found in the Var region of France and the mineral became readily available in Europe. Within a few years, the enormous potential value of lead chromate(VI) as a yellow or orange pigment became clear. Painters particularly valued the dark lemon-yellow colour that lasted better than the pigments previously available.

In 1809, Vauquelin carried out a series of investigations to determine the best conditions for making the lead chromate(VI) pigment in the laboratory. He obtained it by **ionic precipitation**, using solutions of a lead salt and a chromate(VI), for example:

lead nitrate(V)(aq) + sodium chromate(VI)(aq) →
sodium nitrate(V)(aq) + lead chromate(VI)(s)

Chemical Ideas 5.1 will remind you about ionic precipitation.

Assignment 4

a For the reaction between lead nitrate(V) and sodium chromate(VI), write:
  i a full balanced equation
  ii an ionic equation that shows only the ions involved in the precipitation reaction.
  Include state symbols in all your equations.

b Another soluble lead salt is lead(II) ethanoate. Write equations, as in part **a**, for the precipitation reaction between potassium chromate(VI) and lead(II) ethanoate.

WHAT IS A PAINT?

A paint is usually made up of two main parts:

- the colouring matter – the pigment
- a liquid that carries the pigment, allows it to spread and helps to bind it to the surface.

In modern paints, the liquid is made up of a polymer or resin (known as the *binder*) and a solvent that can evaporate. In spray paints, the volatile organic solvent evaporates quickly once the paint has been applied.

In watercolours, the finely ground pigment and a binder are suspended in water. The paint dries as the water evaporates and the pigment and binder are absorbed into the paper.

In oil paints, the drying process is slow. Oil paints usually contain some solvent, such as white spirit, to give the paint a workable consistency. The oil is the binder (sometimes called the *binding medium*) and does not evaporate. Instead, it slowly hardens as it reacts with air to produce a flexible film. The oil protects the pigment and helps to bind it to the surface being covered.

Chrome Yellow was the first of a range of metal chromates containing lead, barium, strontium or zinc that could be used to produce pigments with different shades of yellow. Chrome Yellow is still widely used today in paints and printing inks.

▲ **Figure 10** Yellow road markings contain Chrome Yellow pigment.

Pigments can be poisonous

Many of the artists' pigments used in the past were hazardous in one way or another, but the artists were almost certainly unaware of the risks they were taking.

Some pigments, such as *Red Ochre* and *Yellow Ochre* (both forms of Fe_2O_3) and many organic pigments, are harmless – others, like the green and yellow arsenic pigments that were once used, can be deadly. We now know that lead, cadmium and mercury compounds are highly toxic, and certain soluble chromates have been shown to be *carcinogenic* (can induce cancers) and are now rarely used in paints. So, like any chemist, handle pigments with care and follow the safety instructions.

Modern paints tend to be much less toxic. Lead compounds, for example, are no longer included in household paints. In many cases inorganic pigments are being replaced by less toxic organic ones.

A famous wheatfield

During 1888, the Dutch artist van Gogh painted three pictures of *A Wheatfield, with Cypresses*. One of these (Figure 11) is in the National Gallery in London – if you are in Trafalgar Square you can go and see it.

▲ **Figure 11** *A Wheatfield, with Cypresses* by Vincent van Gogh.

During the 1980s this picture was cleaned and remounted. This gave scientists at the National Gallery a welcome opportunity to examine the materials and techniques that van Gogh used.

The picture was photographed in daylight, ultraviolet and infrared light. These photographs gave information about areas in the painting where particular pigments have been used. For example, *Zinc White* (zinc oxide, ZnO) fluoresces under ultraviolet light. *Emerald Green* (a pigment containing copper and arsenic) absorbs in the infrared region, so areas painted with this pigment appear dark in infrared light.

The pigments on the surface of paintings are coloured because they absorb some of the white light falling on them. **Activity CD2** will help you to understand this.

Scientists took very small samples of different colours of paint from the edges of the painting and analysed them to find out which elements were present. One way they did this was to look at the lines in the **atomic emission spectrum** from each sample of paint. You will learn more about this technique later in the 'Laser microspectral analysis' green box on page 112. They also used a scanning electron microscope to look at the tiny individual crystals in the paint (see Figure 12).

These investigations, combined with historical records, showed that van Gogh used Chrome Yellow, mixed with Zinc White and other pigments, to create the different shades of yellow in the wheatfield. Table 1 shows an extract from the data obtained by the scientists at the National Gallery.

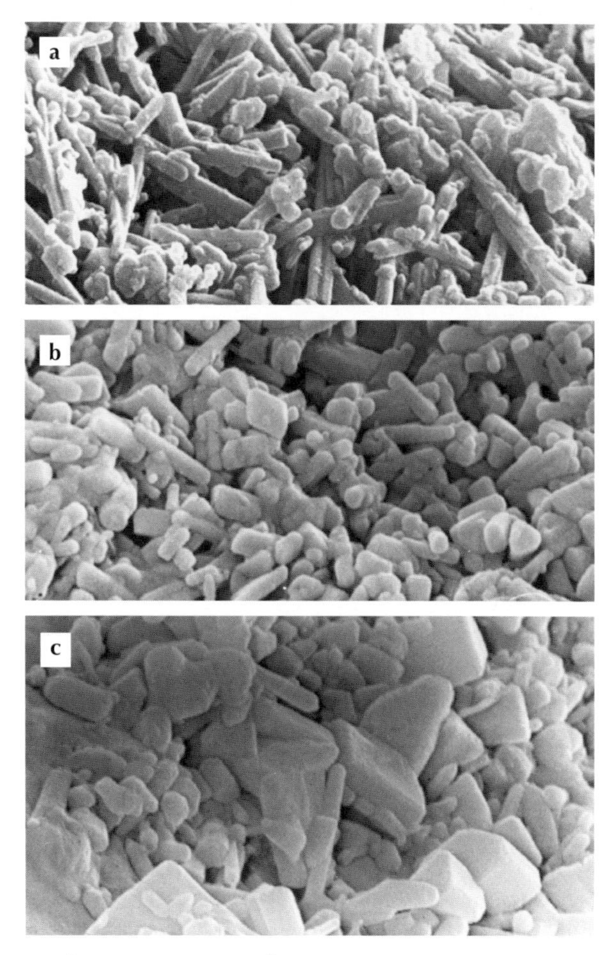

▲ **Figure 12** Scanning electron microscope pictures of pure lead chromate(VI) and pigment samples from the wheatfield (the scale at the top of each picture shows 2 μm, i.e. 2×10^{-6} m):
a pure lead chromate(VI);
b from the dark yellow in the wheatfield;
c from the light yellow in the wheatfield.

Table 1 Pigment mixtures used in *A Wheatfield, with Cypresses* by Vincent van Gogh.

Paint sample	Pigment
darkest yellow of wheatfield	Chrome Yellow
mid-yellow of wheatfield	Chrome Yellow + Zinc White
lightest yellow of wheatfield	Zinc White + Chrome Yellow
dull yellow of wheatfield	Chrome Yellow + Zinc White + small amounts of Emerald Green
pale green bushes	Zinc White + Chrome Yellow + Viridian (a green pigment)

Assignment 5

a How do the electron microscope pictures in Figure 12 support the conclusions of the scientists shown in Table 1 concerning the pigments used in the dark yellow and light yellow areas of the wheatfield?

b Van Gogh used Cobalt Blue and Zinc White for the distant mountains. Both these pigments reflect infrared light strongly. He used Emerald Green to produce the dark green of the cypresses. In infrared light, the trees appear dark and the mountains cannot be distinguished from the sky. Explain why this happens.

CD3 *Chemistry in the art gallery*

The chemists involved in identifying the pigments used by van Gogh are called *analytical chemists* – their job is to find out which compounds are present in different substances. Some of the problems involved in restoring a painting can be very challenging. Solving them can involve a great deal of patience and ingenuity, a knowledge of chemistry and the application of some sophisticated techniques.

The Incredulity of Saint Thomas

In 1504 an Italian artist, Cima da Conegliano, put the finishing touches to his altarpiece *The Incredulity of Saint Thomas* – the painting now hangs in the National Gallery in London. It is considered to be a masterpiece and is highly prized.

By the standards of the time, Cima used an exceptionally wide range of pigments in creating the altarpiece. Some of the substances he used will be familiar to you. For example, to produce certain shades of green, he used *malachite* ($CuCO_3 \cdot Cu(OH)_2$, basic copper carbonate). For reddish brown, he used *haematite* (an ore containing Fe_2O_3, iron(III) oxide).

Many of the pigments, such as the blue Ultramarine, would have been quite expensive. In fact, there are several areas of blue paint on the altarpiece. These vary widely in colour from the rich turquoise of the ceiling to the pale-blue drapery worn by the apostle to the left of Jesus Christ.

How do we know which blue pigments Cima used to achieve these colours nearly 500 years ago? One way to find out is to shine light on different areas of the painting and examine the wavelengths present in the reflected light. Different pigments absorb different wavelengths from the incident light, so the resulting **reflectance spectrum** is characteristic of a particular pigment.

If you are ever fortunate enough to stand in front of Cima's painting in the National Gallery and look at its vast array of bright colours, you will find it difficult to believe how badly disfigured and fragile the painting was when restoration was started in 1969 (Figure 13). Layers of dirt and old varnish masked or dulled the colours. The paint was badly blistered and was flaking off in many places.

▲ **Figure 13** *The Incredulity of Saint Thomas* by Cima da Conegliano in 1969, before restoration.

Over the next 15 years a great deal of laborious restoration took place. Much of this was made possible by scientific analysis – not only of the pigments involved but also of the binding medium used to make the paint and of the foundation coating beneath the paint layer.

But first we will look at the story of how the painting got into such an awful state.

You can read about ultraviolet and **visible spectroscopy** in **Chemical Ideas 6.8**.

In **Activity CD3.1** you can find out how chemists can help in deciding which pigments Cima used to produce the different areas of blue in the altarpiece.

The history of the painting

The start of the trouble (1504–1870)

Cima's painting *The Incredulity of Saint Thomas* was originally hung above an altar in a church in the Italian town of Portogruaro, about 50 miles north-west of Venice. For some reason the paint began to flake and blister.

What may have been the first in a series of restorations was attempted in 1745. Unfortunately, the condition of the painting continued to deteriorate. Eventually it was sent to the Academy of Fine Arts in Venice where one of the foremost experts of the time, Professor Giuseppe Baldassini, restored it as best he could.

Between 1822 and 1830 the painting was stored in a ground-floor room of the academy while arguments were taking place over Professor Baldassini's bill. This led to the most disastrous episode in the troubled history of the painting – a sudden and unusually high tide flooded the room and the picture was submerged in salt water for several hours.

More flaking and blistering occurred and there were several further attempts at restoration. Despite these problems, the painting caught the eye of Sir Charles Eastlake in 1861 while he was on a picture-buying expedition for the National Gallery. After some haggling over the price and a lengthy legal dispute about the ownership of the painting, it was bought for the National Gallery in April 1870 for £1800.

1870 – the National Gallery, London

At this point the gallery authorities made the emphatic recommendation that 'no restoration work should again be attempted'. But paint continued to flake off and there were repeated treatments over the next 80 years to reattach this flaking.

During the Second World War, the whole of the National Gallery collection of art was housed in large artificial caves in a quarry near Bangor in North Wales. The conditions of stable humidity and temperature in the caves were far superior to those in the gallery, and the problems of warping wooden panels and blistering paint were much reduced.

It was when the collection was returned to London in 1945 that serious problems arose. The effect was particularly bad in the very cold winter of 1947, when the heat had to be turned up high to maintain a tolerable temperature for visitors to the gallery. The low humidity caused many paintings to dry out, resulting in cracking and flaking. Cima's altarpiece flaked more than almost any other painting.

When an oil painting is produced on a wooden panel, the wood is first coated with a stable inorganic substance bound together with an animal glue. In Cima's painting, gypsum (hydrated calcium sulfate, $CaSO_4{\cdot}2H_2O$) was used for this. This coating layer is called the *gesso* (the Italian word for gypsum).

In 1969 investigators found that the repeated flaking was due to the gesso layer becoming detached from the wooden support, and not to the paint coming away from the gesso. Worse still, it was found that the wooden panel had dry rot and woodworm!

The challenge

In the light of all this, experts at the National Gallery decided to give the picture the most drastic form of treatment used in restoration. The delicate and priceless layers of paint and calcium sulfate were to be transferred from the wooden panel to a new support.

You might get some idea of the scale of the problem if you consider that the picture measures approximately 3 m × 2 m and that the thickness of the brittle paint and gesso together is only a fraction of a millimetre.

The first step was to cover the painting with layers of special tissue using an organic resin as an adhesive. It was then placed face-down on a temporary support and the original 5 cm-thick wooden panel was removed from the back. No high-tech approach here – it was a slow, laborious process achieved by hand using small gouges, chisels and finally scalpels! What remained was a priceless film of paint and calcium sulfate.

▲ **Figure 14** The painting in 1986, after cleaning and restoration.

Next, a heat-activated adhesive was used to secure the back of the painting to a new support. This consisted of a fibreglass plate secured to an aluminium honeycomb. Now the layers of tissue on the front of the painting could be carefully peeled away. After 500 years the painting was finally attached to something from which it was unlikely to come unstuck – at least not for a few more centuries.

However, after all the years of hard work the painting was still unsuitable for exhibition (see Figure 14). The last stage was to repair the damaged paint layer.

So far, the scientific department at the National Gallery had played a relatively small role. In the final restoration, the *retouching* of the paint layer, the work of analytical chemists was crucial. When retouching the painting, the restorers used modern pigments and modern *binding media* to make up the paints they used. They had to choose these carefully to make sure of two things:

- that any restorations could easily be removed without damaging the original work
- that the colours used in retouching were a good match for the original colours.

The first of these principles applies to the restoration of any work of art in any country. It ensures that a future generation can remove the work of the restorer if they decide that it is inappropriate or if scientific advances make better restorations possible.

Restorations can only be removed with ease if the modern binding medium dissolves in organic solvents more readily than the one used by Cima 500 years ago – so it is important to know exactly which binding medium Cima used.

THE BINDING MEDIUM

The medium used to bind the particles of pigment together to form a paint must be viscous enough to prevent the paint from running as it is applied. But it must not be *too* sticky, otherwise the artist's freedom would be restricted during painting.

Once the paint has been applied, the medium must then dry and become hard in order to produce a durable painted surface.

What medium did Cima use?

Throughout the history of painting, artists have experimented with different types of medium. Egg yolk (known as *egg tempera* when used as a binding medium) was widely used in European paintings in the Middle Ages. Egg yolk is an emulsion of globules of fat and protein in water. It dries and hardens quickly as the water evaporates. You will be able to confirm this if you have ever had to wash up plates coated in dried egg.

▲ **Figure 15** The painting after cleaning and transfer to a new support – but before repairing the damaged paint layer.

EMULSIONS

An emulsion is a mixture of two liquids that do not dissolve in one another, such as oil and water.

For example, egg yolk is an oil-in-water emulsion. The oil droplets are so small that they do not settle out, and the mixture appears uniform and opaque.

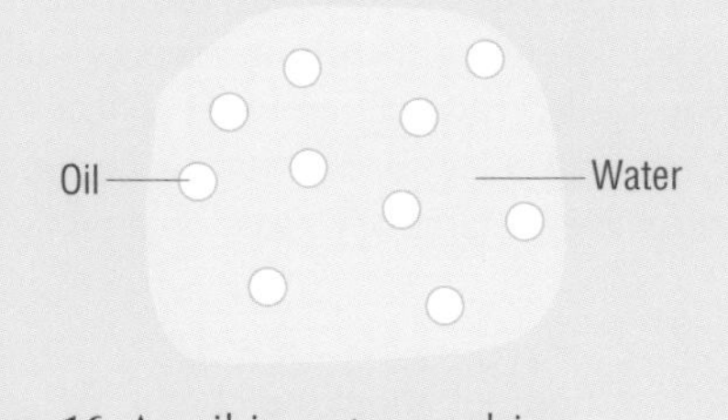

▲ **Figure 16** An oil-in-water emulsion.

At the time Cima was working on *The Incredulity of Saint Thomas*, paints were commonly prepared using oils as the medium. Oils dry more slowly than egg yolk and give the artist more freedom. Natural oils, such as linseed or walnut oil, that dry and harden to develop a protective coating are suitable – they are called *drying oils*.

The drying oils that Cima might have used all contain significant amounts of *triesters* based on two carboxylic acids, *palmitic acid* and *stearic acid*. Different oils contain the palmitate and stearate esters in different ratios. So, if the palmitate and stearate ratio can be measured, we can tell with some confidence which oil Cima used to bind his pigments.

This ratio can be found using an analytical technique called **gas–liquid chromatography (g.l.c.)**. Figure 17 shows the gas–liquid chromatogram obtained from a small sample of paint from Cima's painting.

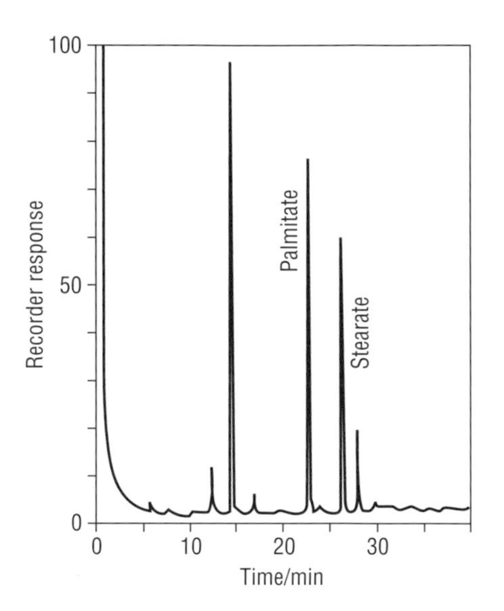

▲ **Figure 17** Gas–liquid chromatogram from a sample of paint from Cima's painting.

The area under each peak in the chromatogram is proportional to the amount of each substance present. So you can work out the ratio of palmitate to stearate esters in the paint, and then compare this ratio with those obtained from known oils.

The paints used in the restoration were made using a synthetic polymer as the medium, rather than oil. The polymer is soluble in hydrocarbon solvents – all the recent retouchings could be removed, if necessary, with a solvent-soaked swab in just a few hours, leaving behind only what remains of Cima's original work.

It will help to find out about the chemistry of oils and fats – you can do this by reading **Chemical Ideas 13.6**.

You can read about gas–liquid chromatography in **Chemical Ideas 7.3**.

In **Activity CD3.2** you can use a computer drawing package to construct structures for oils and fats.

You can work out which oil Cima used by doing **Activity CD3.3**.

What were the pigments?

You have already met some of the methods that chemists can use to identify a pigment. One of the methods mentioned earlier is a form of **atomic emission spectroscopy** (using laser light as a source) called laser microspectral analysis (LMA). It proved particularly useful when analysing paint from Cima's altarpiece. It allows chemists to identify the elements present in a sample of paint. From this they can usually say which pigments are present.

You can remind yourself about atomic emission spectroscopy by reading **Chemical Ideas 6.1**.

LMA is just one of the analytical techniques used to identify the elements present in a pigment. A newer and more versatile technique called *energy dispersive X-ray fluorescence* (EDX) is now commonly used within the scanning electron microscope.

Assignment 6

The LMA emission spectrum obtained from a paint sample consists of a complex arrangement of lines corresponding to radiation of different frequencies.

a Explain why excited atoms emit only certain frequencies of radiation.

b Why are these frequencies different for atoms of different elements?

c How do you think chemists work out which elements are present in the paint sample from the lines in the LMA spectrum?

Scientific evidence is always used along with information from other sources. Art historians, for example, can often provide additional information by studying contemporary manuscripts and by using what is known about other paintings from the same period.

Once the composition of the original pigment is known, a modern substitute can be chosen to give a good match for use in restoration.

Now you can try your hand at finding matches for two of the blue pigments in Cima's painting, in **Activity CD3.4**.

▲ **Figure 18** At home again! Cima's great painting hanging in the National Gallery, London.

LASER MICROSPECTRAL ANALYSIS

In this technique, a pulse of laser light is focused onto a small sample of paint. Laser light is a high-energy beam of light of a single wavelength. The energy of the pulse is high enough to vaporise the high-boiling-point metal compounds present in the pigments. A small plume of vapour rises from the paint sample into a region between two electrodes. The atoms and ions in the vapour are then excited to higher electronic energy levels by an electrical discharge between the two electrodes. Each chemical element present gives rise to a characteristic **emission spectrum**.

The method is very sensitive and can be carried out on very small quantities of material – the sample of paint taken from a valuable picture is usually of the order of 1×10^{-5} g. Of this, a minor pigment component might comprise only 10^{-7}–10^{-6} g.

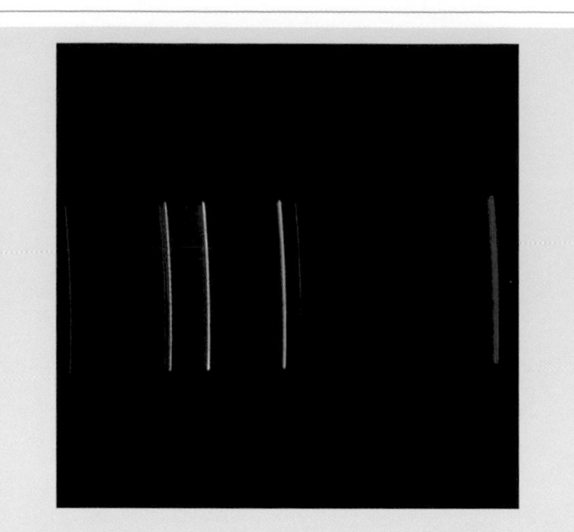

▲ **Figure 19** Emission spectrum of cadmium.

CD4 *At the start of the rainbow*

During the nineteenth century, many new pigments became available as chemists investigated more and more compounds and learned how to imitate the properties of natural substances.

The nineteenth century was also a time of great innovation in the production of dyes for cloth. Today, many of our more obviously high-tech industries are based on electronics – then, it was the colourist who enjoyed the regard we now have for software engineers and system analysts.

The story of the development of dyes is closely linked to the development of organic chemistry. Colourists learned about both the methods of dyeing and the principles of chemistry as they moved about Europe selling their skills and knowledge.

▲ **Figure 20** William Perkin (1838–1907).

To understand this section you will need to know about the special nature of benzene and the type of reactions it undergoes – you can find out about this by reading **Chemical Ideas 12.3** and **12.4**.

Three young entrepreneurs

Perkin

William Perkin was only 18 years old when he invented a method of synthesising a dyestuff, later known as *Mauve*. He was a student at the Royal College of Chemistry in London and in those days the study of plant extracts dominated organic chemistry.

August Hofmann, director of the college, encouraged Perkin to try to make quinine, a natural product used to treat malaria. Perkin tried to do this during the Easter holiday of 1856 in his home laboratory.

His starting material was a complex organic amino compound obtained from coal tar. Little was known of the structure of quinine, so Perkin's chance of success was minimal – and sure enough the experiment failed.

But Perkin was not deterred. He repeated the process with a simpler starting material called aniline (phenylamine), also obtained from coal tar. The reaction produced a purple solution – but again nothing related to quinine.

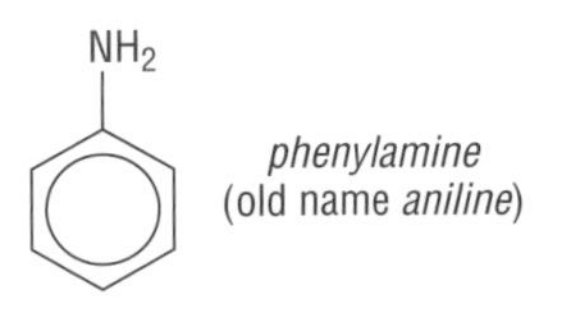

This type of reaction was already known to analytical chemists, who often used reactions involving colour changes to test for compounds. Perkin realised that the purple substance might have other uses. Purple was a popular colour in fashion – the new colour was brilliant and held fast to a piece of cloth.

He decided to test it as a dyestuff – the results were promising and Perkin filed a patent for his discovery. It was the first *synthetic* dye. With the help of his father and brother, he built a factory to manufacture the dye – the difficulties were immense. For one thing, he had to devise a way of producing aniline on a large scale, but very little was known about the behaviour of reactions when scaled up. Figure 21 shows the type of equipment Perkin used.

The venture was an overwhelming success. The colour was in great demand from the world of fashion and both Queen Victoria and the Empress Eugénie in France wore dresses made from material treated with the new dye. The dye was originally called *Aniline Purple*, but it was later named *Mauve* – after the French for the mallow flower.

Perkin's discovery led to increased interest across Europe in aniline as a starting material. Other synthetic aniline dyes in a variety of colours soon followed.

You can investigate some typical reactions of arenes in **Activity CD4.1**.

Activity CD4.2 will help you to understand how **substitution** reactions of arenes take place.

Levinstein

Perkin was not the only teenager to realise the potential of his college experiments. A German student, Ivan Levinstein, worked on the new aniline dyes at the Gewerbe Akademie in Berlin.

In 1864, with the support of his family, he opened a factory there to make *Aniline Green*. The following year, aged 19, he moved to England and settled near Manchester. Here he eventually started manufacturing *Aniline Red* in a row of cottages at Blackley near Manchester (and thus at the heart of the textile industry so important in the great Industrial Revolution) and gradually built up the biggest British company in the field. Since 1926 it has been incorporated, first into ICI, and is now known as Avecia.

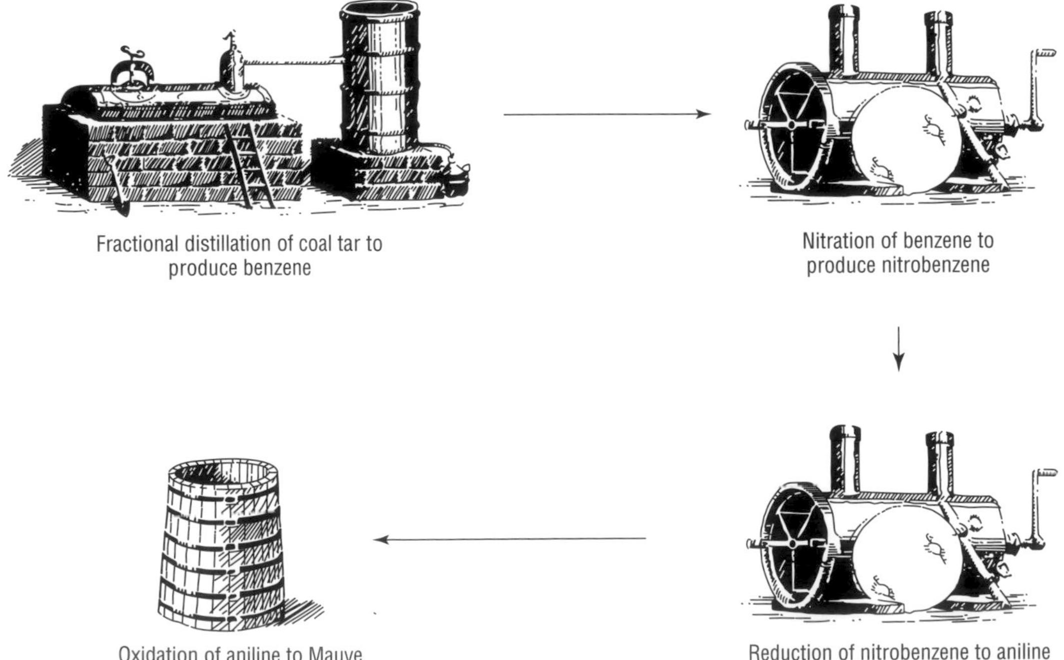

▲ **Figure 21** Manufacture of Mauve showing the type of equipment Perkin used.

▲ **Figure 22** Advert for dyes similar to those made by Ivan Levinstein's company. Manufacturers like Levinstein often started in their own homes using household appliances – to this day, the main building in any modern dyeworks is called the dyehouse.

Caro

Another young German entrepreneur, Heinrich Caro, saw the economic potential of dye manufacture and made his way without family support.

Caro was also a student at the Gewerbe Akademie before spending seven years in Manchester as a colourist and plant manager in a dyeworks. He returned to Germany in 1866 and two years later joined the newly formed Badische Anilin und Soda Fabrik (BASF).

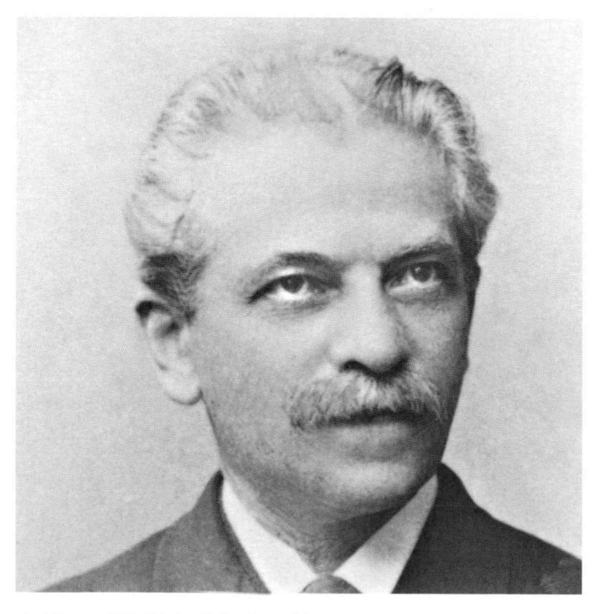

▲ **Figure 23** Heinrich Caro (1834–1910).

ALDEHYDE GREEN

This was the first pure green dye. A mixture of yellow and blue was used previously to get a good green colour, but this mixture looked dull in artificial light. Empress Eugénie dazzled everyone when she went to the opera wearing the dress.

▲ **Figure 24** Empress Eugénie, wife of Napoleon III, in the 1860s wearing a dress dyed with Aldehyde Green – a synthetic 'coal tar' dye.

The dye provided the German company Hoechst with a product that enabled it to become one of the largest chemical companies in the world.

During the late 1860s, work in Germany centred on a systematic study of organic chemistry, as well as on possible commercial developments. Caro's skills in chemistry and chemical engineering enabled him to work out a key step in the synthesis of the red colour of the natural madder dye obtained from the madder root (Figure 25). Like Perkin and Levinstein, he was able to transform laboratory experiments into successful commercial processes.

But was it chemistry?

It is important to bear in mind the state of chemical theory at the time these young entrepreneurs were making their discoveries. Ideas that we now take for granted about atoms, atomic masses and periodicity were just beginning to emerge.

For example, the idea that carbon can form four bonds and is capable of forming long chains of carbon atoms emerged in 1858. The German chemist August Kekulé first reported his ideas about a ring structure for benzene in 1865. It was not until much later (the 1880s) that chemists began to write the structural formulae of dyes in their scientific papers as a matter of routine.

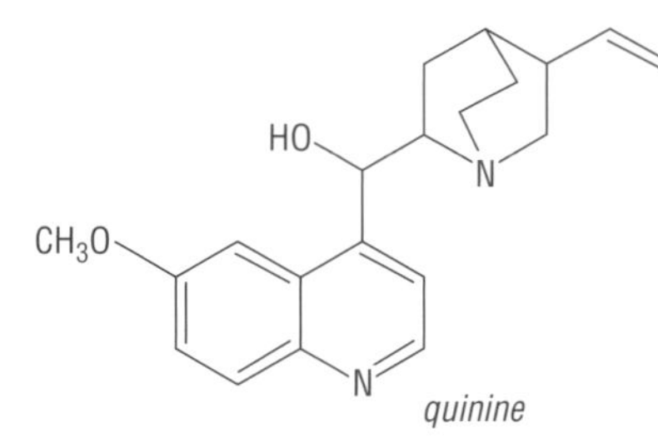

When Hofmann challenged Perkin to make quinine, it was thought that the formula of quinine was $C_{20}H_{22}O_2N_2$ – this was actually two hydrogens short. He therefore proposed the following synthesis, using sodium dichromate(VI) as the oxidising agent:

$$2C_{10}H_{13}N + 4O \rightarrow C_{20}H_{22}O_2N_2 + 2H_2O$$

This approach to tackling a synthetic problem now seems ridiculous, but at the time it seemed sound chemical theory! When the structure of Mauve was unravelled in the late 1880s, it turned out to be a mixture of compounds. The most important one, mauveine, is a complex organic salt with the structure

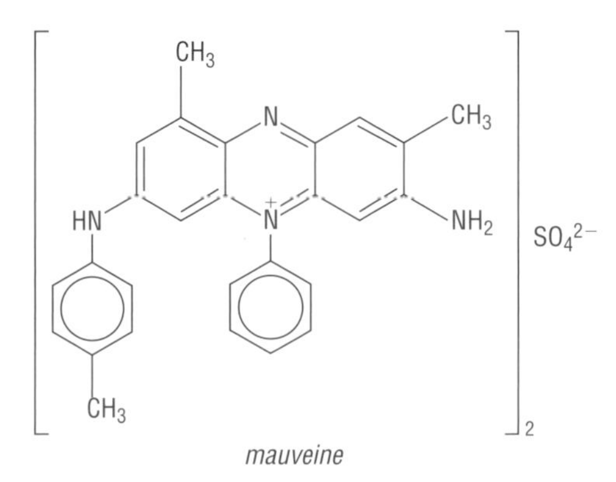

So it was quite amazing that Perkin should stumble across this molecule by chance – particularly as it was shown later that the formation of Mauve depended on the presence of impurities in the starting aniline. This also explained why Mauve was made up of more than one purple compound.

Alizarin

Until Perkin made his discovery, dyes had come mainly from plants and vast areas around the world were given over to cultivation for the dyestuffs industry.

The indigo plant was used to make blue dyes – red dyes came from the madder root.

▲ **Figure 25** The root of the madder plant.

The colouring matter from the root of the madder plant is called *Alizarin*. We now know that it has the structure

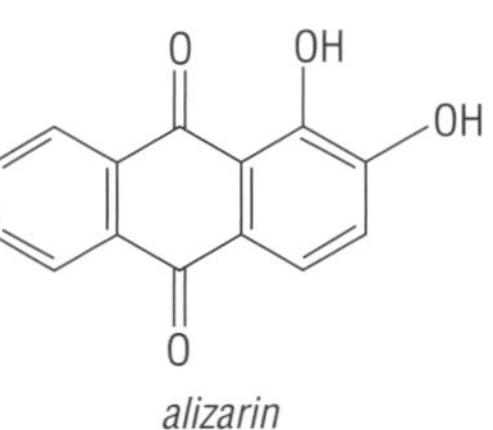

Alizarin sticks fast only to cloth that has been impregnated with a metal compound such as aluminium sulfate. The process is called *mordanting*. The colour of the dyed cloth depends on the metal chosen – when Alizarin is used with an aluminium mordant, the cloth is dyed red; when tin(II) ions are used the cloth is dyed pink; and iron(II) ions give a brown colour.

▲ **Figure 26** The nineteenth-century Turkey Red dyeing and printing process used Alizarin as the red dye – Turkey Red fabric (above) was famous for its bright colour and distinctive patterns.

The mordanting takes place under alkaline conditions, so that the metal hydroxide is precipitated in the fibres. The metal ions attach themselves to the cloth firmly and then bind the dye molecules by forming chelate rings, as shown in Figure 27.

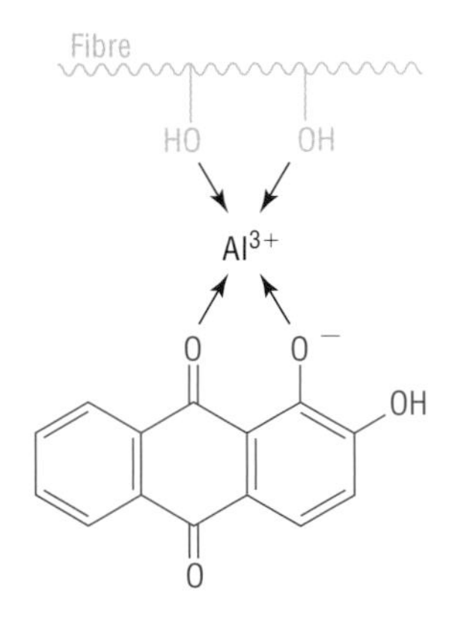

▲ **Figure 27** Chelate of Alizarin with the metal ion Al^{3+} – the two remaining ligand sites above and below the Al^{3+} ion could be taken up by OH^- ions.

Assignment 7

a Write an ionic equation, with state symbols, for the precipitation of aluminium hydroxide by the reaction of an aluminium compound with alkali.

b What type of reaction takes place when mordanted cloth is dyed with Alizarin? What is the role of the Alizarin in this reaction?

c In terms of intermolecular bonds, explain how the use of a mordant holds the dye in the fibres.

Alizarin had been used as a dye for thousands of years, but no one had any idea of its chemical nature until Carl Graebe and Carl Liebermann, two talented students of Adolf Bayer at the Gewerbe Akademie in Berlin, decided to find out. Bayer had recently developed a method for converting aryl compounds into their parent hydrocarbons by heating them with zinc dust. For example:

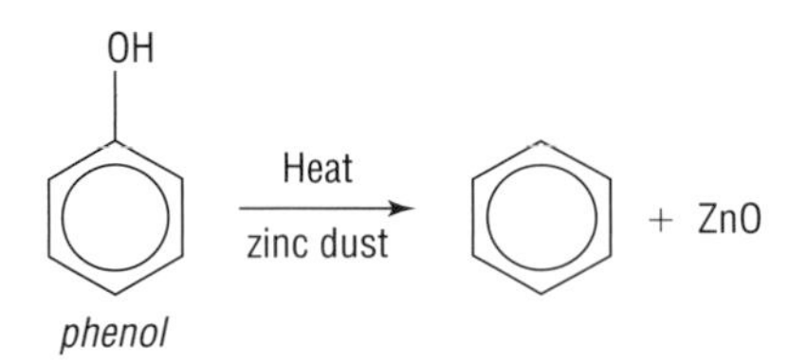

In January 1868, Graebe and Liebermann used this technique to show that Alizarin is derived from anthracene, a minor component of coal tar.

Graebe and Liebermann then set out on a quest to find a method for synthesising Alizarin. They eventually devised the route shown in Figure 28 (route 1). Their method worked in the laboratory and they patented it in 1868, but the yield was very poor – in addition, a route based on bromine was too expensive at that time for a larger-scale process.

The last step of their route, in which the dibromo compound is heated with solid KOH, only added to the confusion of those later trying to work out the structure of Alizarin – this is because an isomerisation reaction takes place at the same time!

Once Graebe and Liebermann had announced their success, the race was on to find a cheaper commercial route to synthetic Alizarin – a route that was expected to bring rich returns to the first people to patent the process. Both Perkin and Caro joined in the competition.

The race turned out to be a tie – the competing chemists independently hit on the same solution in May 1869! It's the reaction scheme shown in route II in Figure 28 – but remember that the structures were not known at the time.

Perkin filed his patent in London on 26th June 1869. But an almost identical patent had been deposited at the Patent Office on the previous day on behalf of Caro, Graebe and Liebermann working for BASF. Chemists at the Hoechst company in Germany also solved the problem at the same time, but they didn't patent their process in London.

Assignment 8

Graebe and Liebermann knew the molecular formula of anthracene. Its properties suggested that it was some kind of condensed aromatic system. Below are the two structures that they suggested for anthracene in 1869:

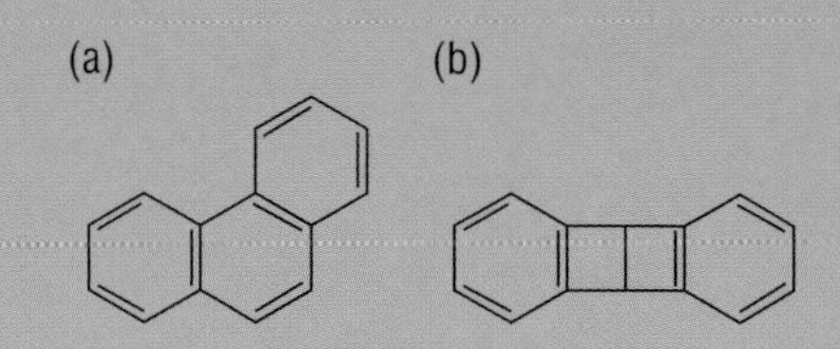

Compare structures A and B with C, the modern structural formula of anthracene:

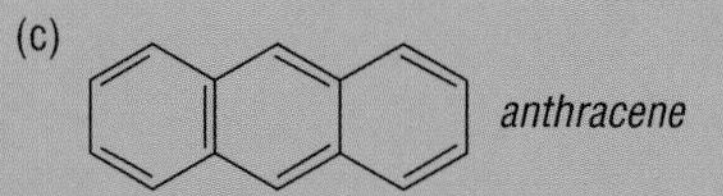

a What is the molecular formula of anthracene?

b What is the relationship between the three structures A, B and C?

c Use your knowledge of bonding in organic compounds to explain why structure B is the least likely possibility.

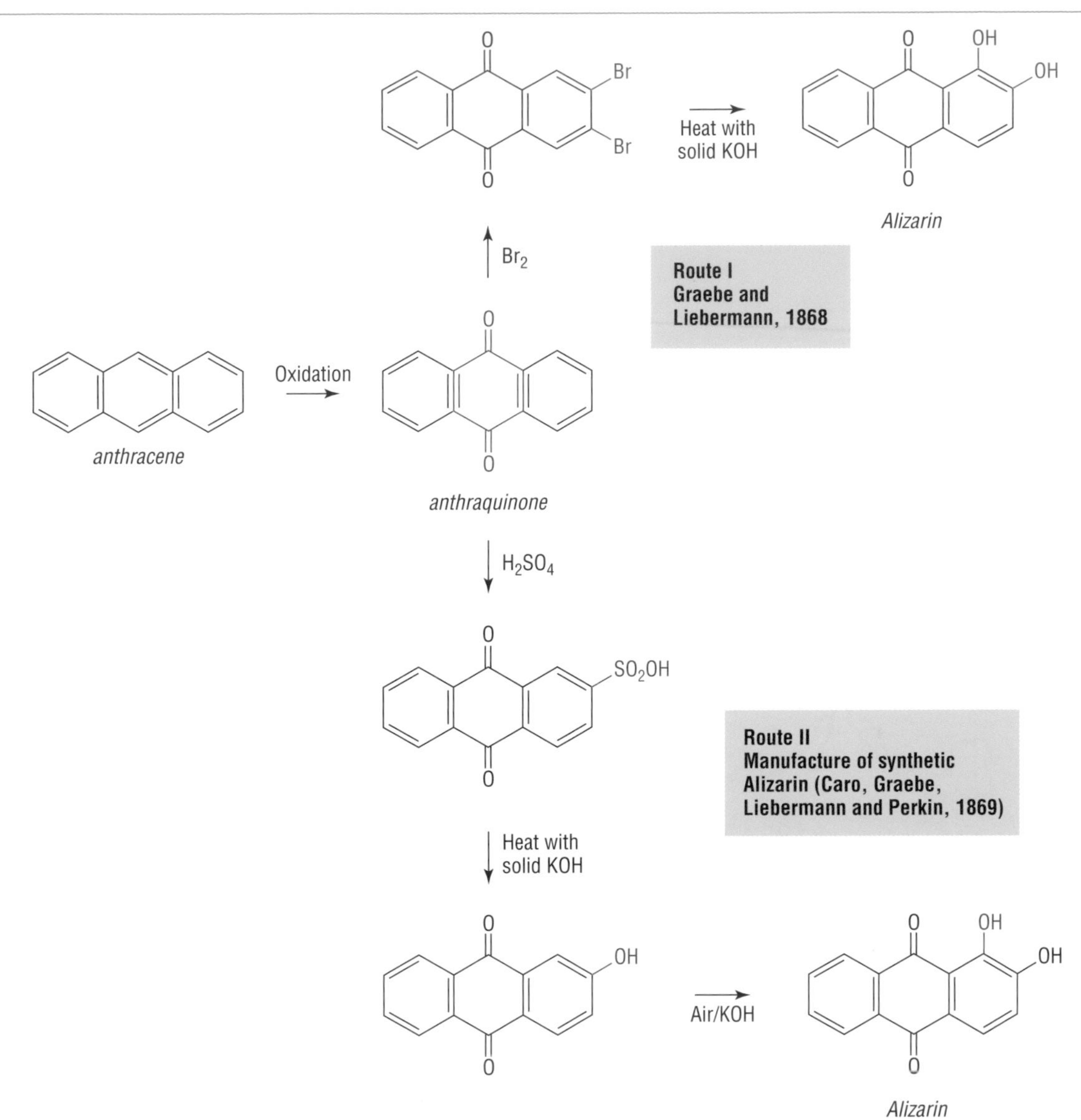

▲ **Figure 28** Conversion of anthracene to Alizarin – the red colour shows the changes for each of the steps.

The situation was resolved amicably later in the year when Perkin and BASF agreed to share the market in Alizarin – Perkin retained the UK trade while BASF was to dominate Europe and the USA.

The discovery of a route to synthetic Alizarin had a devastating effect on the madder industry. Hundreds of thousands of acres in southern Europe and eastwards towards Asia had been devoted to growing madder. In 1868 a crop of 70 000 tonnes of madder root was processed to produce around 750 tonnes of Alizarin. By 1873, just five years later, the madder fields had disappeared – Perkin's company alone produced 430 tonnes of Alizarin that year.

Assignment 9

Compare the two reaction schemes (routes I and II) for the synthesis of Alizarin in Figure 28.

a What is the essential difference between the two routes?

b Why was the second route a commercial success when the first was not? Would the same apply today?

c What type of reaction is involved in the reaction of anthraquinone in each route?

d Which of the following would be better to show that Alizarin contains a phenol group – addition of neutral iron(III) chloride or infrared spectroscopy? Explain your answer.

CD5 Chemists design colours

Figure 29 shows a cartoon of Kekulé's ideas about the structure of the benzene ring, in which monkeys appear instead of carbon atoms. This appeared in a hoax pamphlet put out by two of Kekulé's young disciples. One of them was Otto Witt, a Swiss-trained chemist, who was working for a dyemaking company in Brentford near London in 1875.

Witt was working on a theory that related colour to structure. He wanted to know *why* certain structures led to coloured substances, and how small changes to the structure led to changes in colour.

To do this he was investigating the **diazo reaction**, known since 1858, which led to strongly coloured products that were insoluble in water.

To understand this section you will need to know about the structure and preparation of azo compounds – you can find out about these in **Chemical Ideas 13.10**.

▲ **Figure 29** Witt's cartoon lampooning Kekulé's ideas about the structure of the benzene ring – monkeys form single bonds by holding hands and double bonds by linking tails as well.

The first azo dyes

The first azo dyes were made by coupling a **diazonium salt** obtained from phenylamine with one of a variety of **coupling agents**. Figure 30 shows one example of the reaction involved, in which the coupling agent is another molecule of phenylamine.

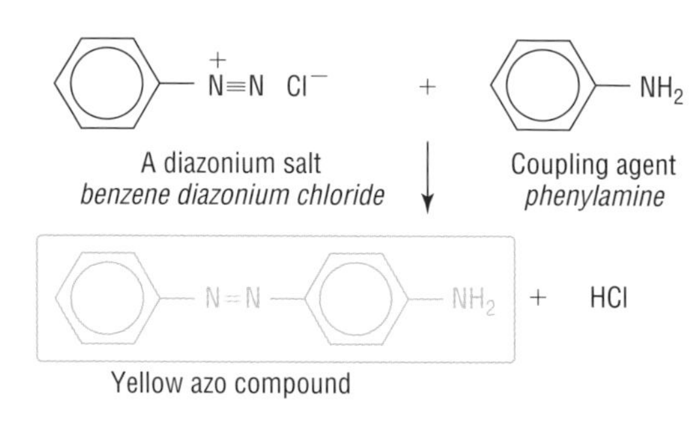

▲ **Figure 30** Azo coupling reaction.

Witt knew about this reaction. He also knew that the corresponding reaction using triaminobenzene as the coupling agent in place of phenylamine gave a brown azo compound:

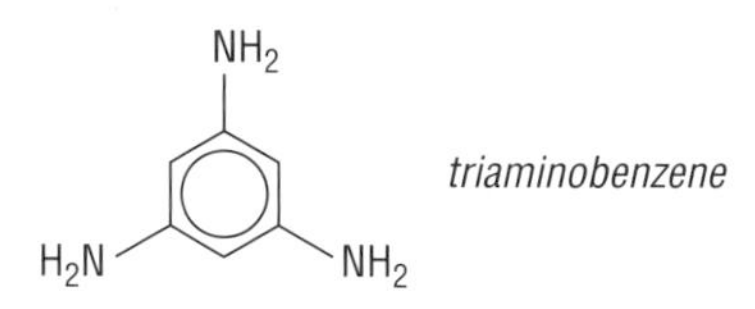

Witt thought that the colour of an azo compound was related to its structure. He predicted that the missing azo compound in the series, the one with *two* amine groups on the benzene ring, should be intermediate in colour. Sure enough, when he made this compound he found it was coloured orange, midway between yellow and brown. It proved to be a successful dye for cotton and was marketed as *Chrysoidine* – it was the first commercially useful azo dye.

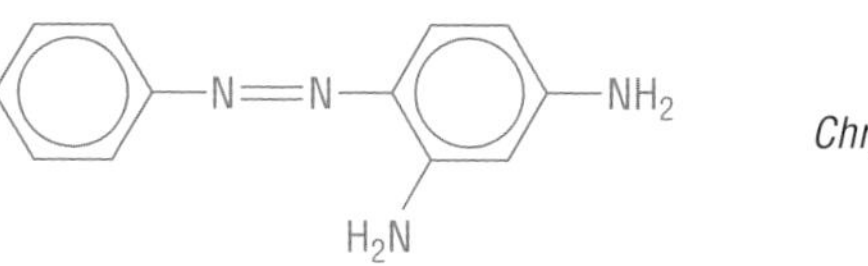

Ivan Levinstein manufactured a range of azo dyes at Blackley, but many other British manufacturers retired around this time. William Perkin sold his factory in 1874 having made his fortune. Meanwhile, the German companies Agfa, BASF, Bayer and Hoechst flourished. They set up research teams with links to the universities to develop new colours. By 1913, Germany was exporting around 135 000 tonnes of dyestuffs per year, while Britain was exporting only 5000 tonnes.

The German companies were also using azo chemistry to produce pigments. Even today, azo pigments dominate the yellow, orange and red parts of the spectrum. Azo pigments are used to colour paints, plastics and printing inks. They provide the yellow and magenta standards used in the three- and four-colour printing processes.

▲ **Figure 31** Many azo dyes are still widely used today.

Assignment 10

a Write an equation for the azo coupling reaction that Witt used to make Chrysoidine.

b In the early years, azo dyes could only be manufactured satisfactorily in winter. How do you account for this?

c Chrysoidine is a basic dye. Which group in the molecule is responsible for its basic properties?

How does structure affect colour?

Witt's work on azo dyes helped him to put forward a theory of colour in dye molecules that still influences our thinking today. A dye molecule is built up from a group of atoms called a **chromophore** and it is this that is largely responsible for its colour.

We now know that chromophores contain unsaturated groups such as C=O and –N=N–, which are often part of an extended delocalised electron system involving arene ring structures.

For Chrysoidine, the chromophore is the delocalised system shown in the box in this diagram:

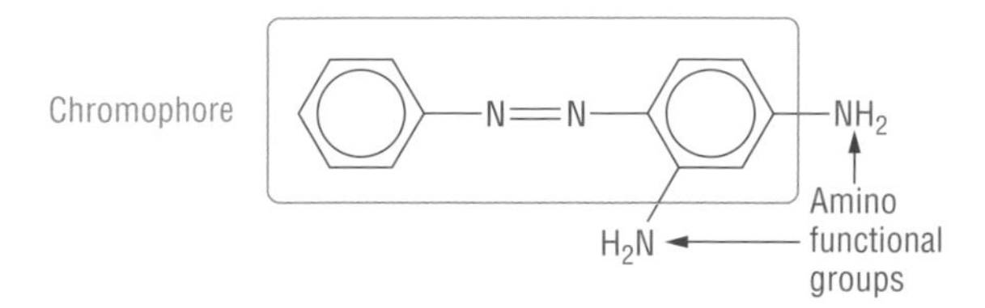

Attached to the chromophore in Chrysoidine are two $-NH_2$ **functional groups** with lone pairs of electrons – these interact with the chromophore to produce the orange colour.

Other functional groups may be added that can:

- modify or enhance the colour of the dye
- make the dye more soluble in water
- attach the dye molecule to the fibres of the cloth.

All azo dyes contain the X–N=N–Y arrangement. Chemists worked to make as many different XY combinations as possible to find new dyes with good colours, which were fast to fabrics and commercially viable. As they did this they began to understand the effect of chromophores and functional groups on the colour and properties of the dyes they produced.

A vast range of azo dyes is now available, produced by coupling one of 50 diazonium salts with one of 52 coupling agents to give a whole rainbow of colours – although these are mostly yellow, orange or red with relatively few blues and greens.

In the first half of **Chemical Ideas 6.9** you can read more about chromophores and why some organic compounds are coloured.

In **Activity CD5.1** you can make a range of azo dyes using different diazonium salts and coupling agents.

Activity CD5.2 will help you learn the reagents and conditions used in some reactions of aromatic compounds.

CD6 *Colour for fabrics*

The search for fast dyes

At the turn of the century a Carlisle weaver called James Morton made a trip to London. Standing outside Liberty's window in Regent Street, he examined a display of cotton tapestries he had designed. He was horrified by what he saw – the colours had faded after only one week in the shop window.

Immediately, Morton set out to find dyes that were fast to light and began manufacturing them – you can see an advert for some of his dyes, called 'Sundour' dyes, in Figure 32.

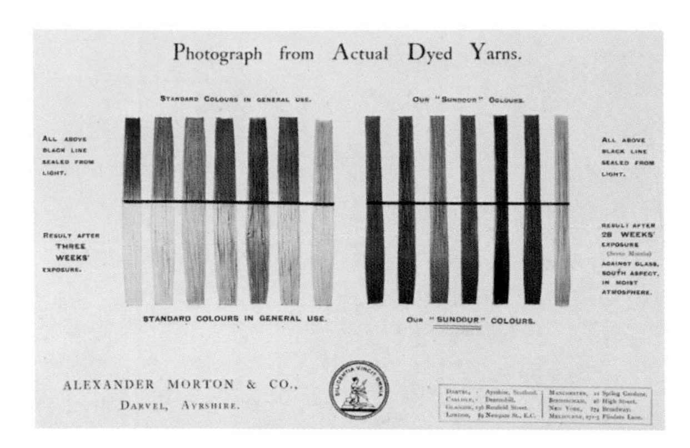

▲ **Figure 32** Page from a leaflet advertising Sundour unfadable colours.

Morton's success helped to re-establish the reputation of the dye industry in Britain; but cotton presented a problem in other ways. Dyes must be fast to washing and rubbing as well as to light – many of the dyes that were fast to wool and silk did not bind at all well to cotton.

How do dyes stick to fibres?

Protein-based fibres such as wool and silk have free ionisable –COOH and $-NH_2$ groups on the protein chains that can form electrostatic bonds with parts of the dye molecule. For example, a sulfonate ($-SO_3^-$)

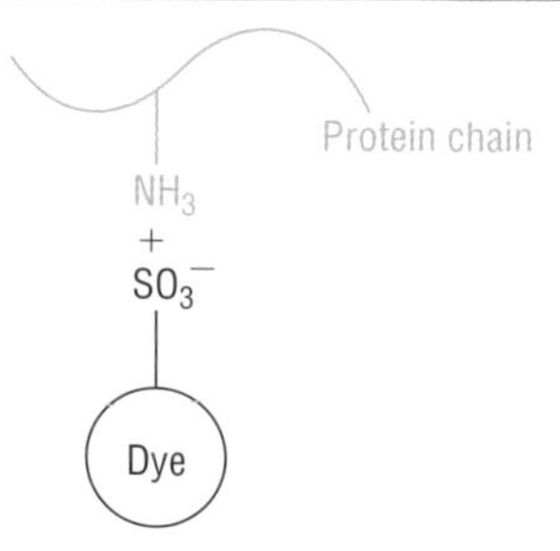

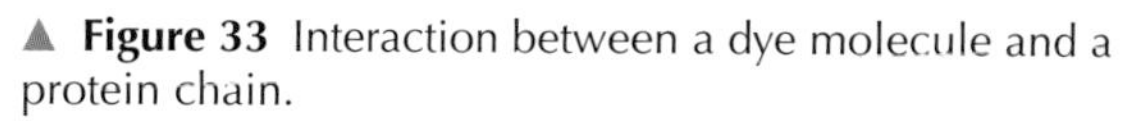

▲ **Figure 33** Interaction between a dye molecule and a protein chain.

group on a dye molecule can interact with an $-NH_3^+$ group on a protein chain, as shown in Figure 33.

Cotton, on the other hand, is a cellulose fibre consisting of bundles of polymer chains with no readily active parts. The polymer is a string of glucose units joined as shown in Figure 34.

Dyes such as Alizarin are bound to the fibres using a mordant (look at Figure 27 on page 116). Indigo, which is used to dye denim jeans, is a *vat dye* – the cotton is soaked in a colourless solution of a reduced form of the dye. This is then oxidised to the blue form of indigo that precipitates in the fibres.

Many azo dyes for cotton are trapped in the fibres because they are insoluble. In contrast, **direct dyes**, such as *Direct Blue 1* (see Figure 35), are applied to the cotton in solution and are held to the fibres by hydrogen bonding and instantaneous dipole–induced dipole bonds. Hydrogen bonds are weak compared with covalent bonds and so these dyes are fast only if

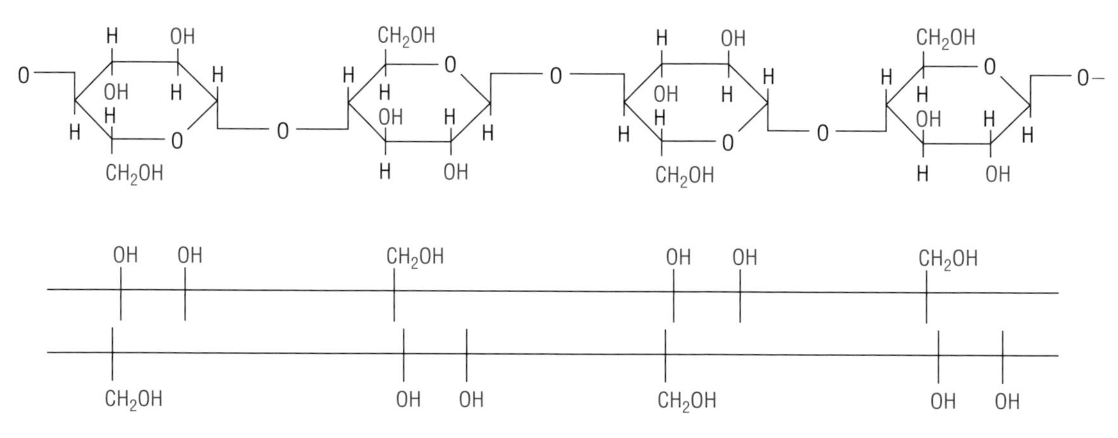

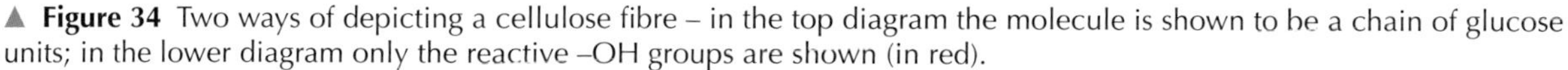

▲ **Figure 34** Two ways of depicting a cellulose fibre – in the top diagram the molecule is shown to be a chain of glucose units; in the lower diagram only the reactive –OH groups are shown (in red).

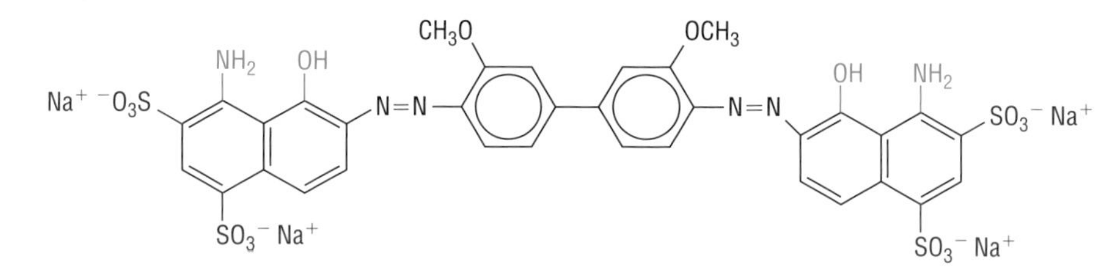

▲ **Figure 35** Structure of Direct Blue 1 (CI 24410).

the molecules are long and straight. They must be able to line up with the cellulose fibres to form multiple hydrogen bonds. Ionic groups on the dye also help to improve solubility in water.

In order to understand why different dyes attach to different fibres, you might like to revise bond polarity using **Chemical Ideas 5.3** and **3.1**.

It would also be useful to revise intermolecular bonding using **Chemical Ideas 5.3** and **5.4**.

Assignment 11

Look at the structure of Direct Blue 1 in Figure 35.

a Which groups of atoms do you think are responsible for the fact that a compound with such large molecules is soluble in water?

b Which groups of atoms would you expect to form hydrogen bonds with cellulose fibres?

c Draw part of the dye molecule and part of a cellulose fibre (in the style of the lower diagram in Figure 34) to show how these hydrogen bonds form. Show appropriate polarities.

A dyemaker's dream

For many years chemists dreamed of developing dyes that would be held to fibres by strong covalent bonds instead of by weak intermolecular bonds. They knew that such dyes would be very fast to washing because they would react with the textile materials and become chemically part of the polymer molecules.

The story starts in the early 1950s with a group of chemists working at ICI's research laboratories in Blackley trying to find better dyes for wool. William Stephen was part of the group. He started with azo dyes and modified the molecules by adding reactive groups, which he hoped might combine with the amino groups of proteins in wool. One idea was to modify an azo dye containing an amino group by reacting it with *trichlorotriazine*, as shown in Figure 36.

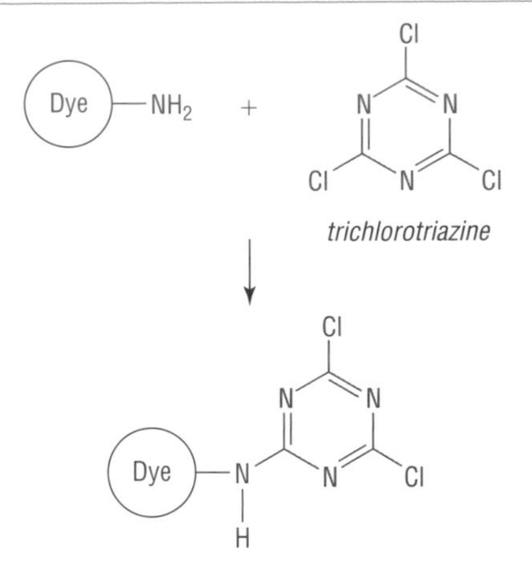

▲ **Figure 36** Building a reactive dye to react with wool.

Stephen took some samples of his modified dyes to Ian Rattee, the senior technician in the wool-dyeing section of the dyehouse. He hoped that the new dyes would react with wool as shown in Figure 37 – however, the results were poor and Rattee was not impressed.

The dream come true

One morning in October 1953, Stephen and Rattee were discussing their experiments with reactive dyes. Stephen pointed out that the reaction of the new dye with the wool would be much more likely to happen under alkaline conditions. Unfortunately, the alkali would damage the wool at the same time.

Assignment 12

Look at Figure 37, which shows the reaction Stephen was hoping to bring about between the modified dyes and wool.

a Why are there free $-NH_2$ groups in the protein molecules in wool?

b What is the second product of the proposed reaction?

c Why might this reaction be expected to go better in alkaline solution?

d Suggest an explanation for the fact that wool is damaged by alkali while cotton is not.

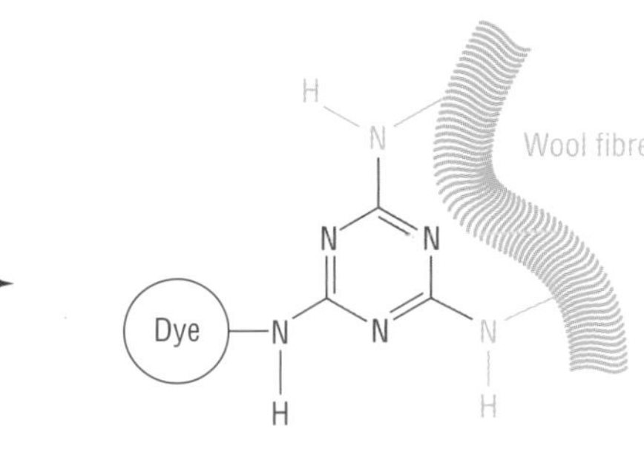

▲ **Figure 37** The planned reaction of the new dye with amino groups in wool fibres.

Stephen suggested that it might be worth trying the dyes with cotton, because cotton is not damaged by alkali in the way that wool is. The reaction that they were hoping to use to bind the dye to the fibres can take place with hydroxyl groups on cotton as well as with amino groups.

Rattee saw the importance of this idea. Dyeing cotton was not part of his responsibility, but that afternoon he took samples of Stephen's dyes and showed that, as predicted, they would react with the hydroxyl groups on cotton when alkali was added.

▲ **Figure 38** Research into dye manufacture is ongoing even today – this chemist is developing dyes used in the manufacture of Lacoste clothing.

The discovery had been made! The idea had at last become a reality – but it was far from certain that it could be made to work in textile mills. A great deal of work lay ahead to develop the innovation, demonstrate its potential and prove that it could be exploited successfully.

Assignment 13

Draw a diagram in the style of Figure 39 to show what happens when a fibre reactive dye reacts with cotton in alkaline solution.

The first two *fibre reactive dyes* are shown in Figure 39. Fibre reactive dyes are now so common that it is hard to recall the feverish excitement and energy released when it was realised that a search that had been going on for around 60 years might be coming to a climax.

You can investigate how different dyes are used to dye different fabrics in **Activity CD6**.

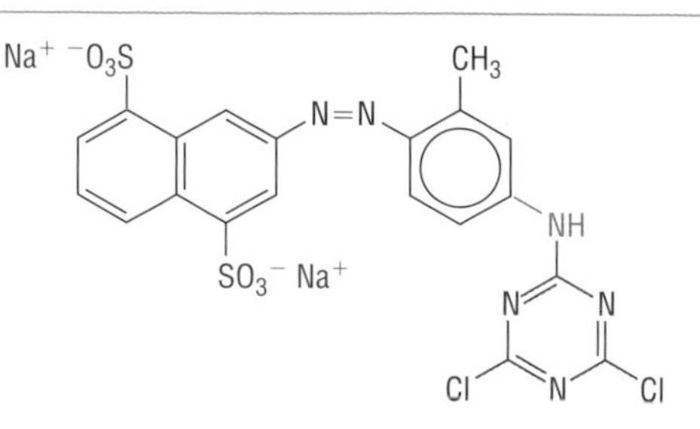

Procion Yellow RS

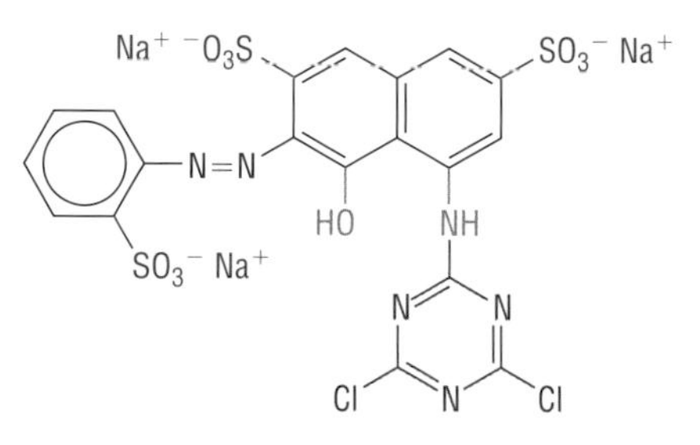

Procion Brilliant Red 2BS

▲ **Figure 39** The first fibre-reactive dyes.

Assignment 14

The two dyes in Figure 39 have the same chromophore.

a Copy the structure of Procion Yellow RS and draw a ring round the chromophore.

b Which functional groups in the structure of Procion Yellow RS help to make the dye soluble in water?

c Which part of the molecule of Procion Yellow RS helps to attach the dye to the cloth? Explain how you decide on your answer.

d What are the differences in the structures of the two dyes which might account for the difference in colour?

Paradox and problem

The new dyes were colour fast because of the reaction with the hydroxyl groups in cotton, but this reactivity was itself a problem because there are hydroxyl groups in water too. The reactivity of the dyes was destroyed by hydrolysis.

This problem had to be overcome, because commercial success with the textile industry depended on the possibility of using the dyes when dissolved in water. Stephen developed systems of *buffers* that kept the pH of the solution within strict limits and kept the hydrolysis reaction under control.

The action of buffers is explained in **The Oceans**.

The decision was finally taken to launch the new dyes in March 1956 – exactly 100 years after Perkin discovered his Mauve dye. There was an explosion of activity in all departments – research, patents, production planning, engineering, the dyehouse, costing, sales and publicity. The dream had finally become a reality.

The work of Stephen and Rattee heralded a new era of bright modern dyes that are fast to light and washing. Other types of reactive dye soon followed, using different groups to bond the dye to the fibres. The 1980s saw the development of a range of extremely fast, brightly coloured dyes for polyester.

Colour chemists can now produce colours to order for a particular application. It is possible to have the exact shade of the season's fashion colour reproduced perfectly in a range of different fabrics – real 'colour by design' (Figure 40).

One of the most exciting recent developments has been in the modification of dyes for high-tech uses, for example dyes used in high-speed inkjet printers and for electronic photography.

▲ **Figure 40** It is even possible to colour leather so that it coordinates with colours of other materials.

CD7 *Summary*

This module is really a series of connected stories involving pigments, paints and the development of synthetic dyes. They illustrate well the very diverse roles chemists play in society.

Many of our pigments made from minerals were discovered by accident – in experiments that 'went wrong'. However, as one of the finest of all scientists, Louis Pasteur, pointed out: 'In the field of observations, chance favours only the prepared mind.' It was the 'prepared' minds of the investigating chemists that enabled the 'accidents' to be carefully analysed and useful discoveries to be made.

The chemists at the National Gallery use a variety of analytical techniques to investigate pigments and paint media. For example, ultraviolet and visible spectroscopy and atomic emission spectroscopy provide information about the chemical composition of pigments, while gas–liquid chromatography is used to identify the medium used to make the paint. Drying oils are unsaturated triesters of propane-1,2,3-triol (glycerol).

The development of synthetic dyes over the last 150 years has been closely linked to the development of organic chemistry – each contributed to the growth of the other. In this part of the module you found out about the special structure of benzene and the types of reaction that arenes undergo. An understanding of the relationship between the structure of a dye molecule and its function, and of the bonds that bind dyes to fibres, is essential when designing new dyes.

Substances appear coloured because they absorb radiation in the visible region of the spectrum. The energy absorbed causes changes in electronic energy and electrons are promoted from the ground state to a higher energy level. Coloured organic compounds often contain unsaturated groups such as C=O or –N=N–. These groups are usually part of an extended delocalised electron system called a chromophore. Groups such as –OH and $–NH_2$ are often attached to chromophores to enhance or modify the colour of the molecules, while $–SO_3H$/ $–SO_3^-$ groups are used to increase the solubility of a dye. Many coloured inorganic compounds contain transition metals. Here, the radiation absorbed excites a d-electron to a higher energy level.

Activity CD7 will help you to review the new chemical ideas covered in this module.

O

THE OCEANS

Why a module on 'The Oceans'?

To many people, the term 'ocean' probably conjures up an image of a seemingly endless expanse of water, of some biological interest but chemically inert – yet this is far from true. Oceans play an essential part in the cycling of many chemicals (sulfur and nitrogen compounds, for example) throughout the Earth. The oceans absorb and store carbon dioxide and must be considered together with the atmosphere in any study of the greenhouse effect. The oceans play a further role in controlling our climate through their absorption of solar energy and the consequent production of water vapour and flows of warm water that help to drive currents in the air and the seas.

The oceans help to make the Earth hospitable to life and have kept our planet that way for over 3.5 billion years. Despite their importance, our understanding of ocean processes is far from complete, and they are one of the major sources of uncertainty in scientists' attempts to model future global conditions.

This module attempts to raise awareness of the importance of the oceans to life on Earth, and to bring out some of the fundamental chemistry which lies behind some ocean processes. Major chemical ideas are developed, and linked (perhaps unexpectedly) to familiar objects such as seashells and to the behaviour of water itself.

Overview of chemical principles

In this module you will learn more about ideas introduced in earlier modules in this course:

- dissolving and solubility (**Elements from the Sea**)
- ions in solution (**Elements from the Sea**)
- enthalpy changes (**Developing Fuels** and **The Atmosphere**)
- entropy (**Developing Fuels**)
- intermolecular bonds (**Elements from the Sea** and **Polymer Revolution**)
- chemical equilibrium (**The Atmosphere** and **Agriculture and Industry**)
- interaction of carbon dioxide with water (**The Atmosphere**)
- acids and bases (**What's in a Medicine?**).

You will also learn new ideas about:

- lattice enthalpies and energy changes in solutions
- entropy and the distribution of energy quanta
- entropy changes in a system and its surroundings
- total entropy changes, spontaneity and equilibrium
- the pH scale
- weak acids and buffer solutions.

THE OCEANS

O1 *Third rock from the Sun?*

At a simple level, planet Earth could be considered as being just an almost spherical lump of rock approximately 12 750 km in diameter travelling through space in an orbit some 150 million kilometres from the Sun.

▲ **Figure 1** The planets in the Solar System.

We know, of course, that the Earth is more complex than this. It is the largest of the four rocky inner planets in our Solar System. It has a dense inner core of iron and nickel, the outer part of which is liquid (see the **Elements of Life** storyline). Motion within this liquid core produces the Earth's magnetic field, which extends into space creating a protective envelope that we call the 'magnetosphere'.

The outer surface of the Earth, the crust, is made up of large solid sections called tectonic plates that float on the partially molten mantle beneath. The outermost part of the Earth is surrounded by layers of:

- air – the 'atmosphere'
- living things – the 'biosphere'
- water – the 'hydrosphere'.

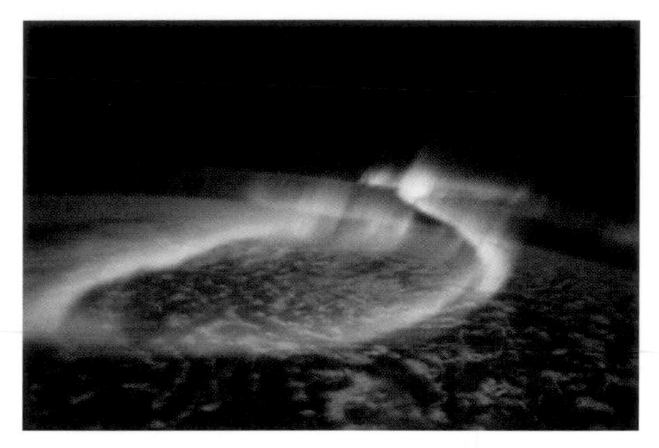

▲ **Figure 2** Earth's magnetosphere is responsible for the spectacular aurorae that occur mainly in the polar regions when charged particles from the Sun become trapped in the Earth's magnetic field.

The contents of these layers are in continuous motion and interact with each other in many ways. You learned in **The Atmosphere** module how the atmosphere protects the surface of the Earth from excessive solar radiation. The density of Earth's atmosphere, together with the favourable distance of Earth from the Sun, is also responsible for the temperature on Earth being maintained at just the right level for water to exist as a liquid! We know now that the atmosphere and hydrosphere *together* also provide the vital mechanisms for distributing energy around our planet.

The blue planet

Planet Earth certainly lives up to this name when viewed from space (Figures 3a and 3b). The Pacific Ocean is the largest of all the oceans and covers one-third of the surface of the Earth. The Atlantic Ocean is the second largest of the oceans and also the youngest, having been formed around 200 million years ago when the continents drifted apart. The Arctic Ocean is the smallest ocean.

Despite the vast areas covered by the oceans, it is surprising to note that the volume of all the water in them (1.37×10^9 km^3) is still only around $^1/_{1000}$ of the total volume of the Earth (1.08×10^{12} km^3).

▲ **Figure 3a** The blue planet – a view of the Earth from space.

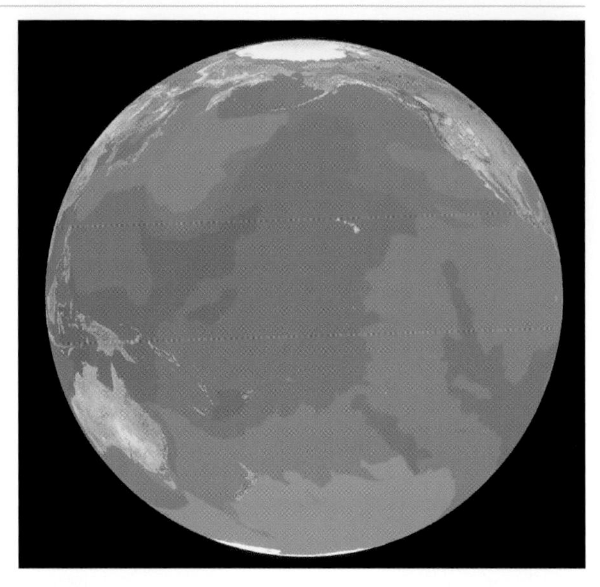

▲ **Figure 3b** The watery planet – a view of the Earth showing the vast expanse of the Pacific Ocean.

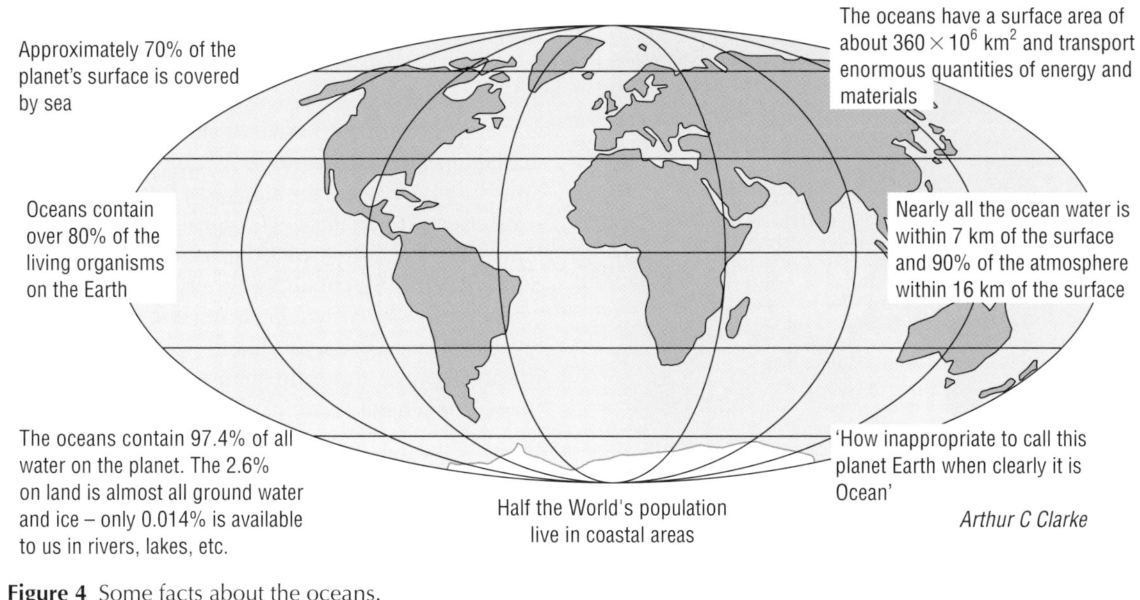

▲ **Figure 4** Some facts about the oceans.

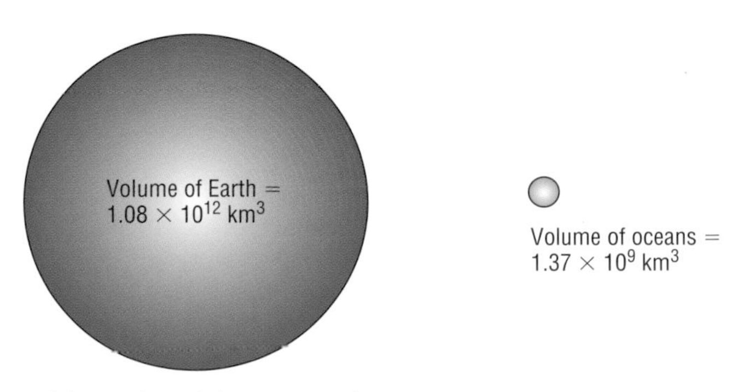

▲ **Figure 5** Relative volumes of the Earth and the water in the oceans.

Surveying the seas

Although humans have been exploring the globe throughout the ages, it is only relatively recently that detailed investigations of the deep oceans have taken place. The historic HMS *Challenger* voyage (1872–1876) was the first-ever dedicated marine science expedition. The small wooden warship sailed out of Portsmouth in December 1872. She had undergone extensive alterations, including the construction of two laboratories on board – one for chemistry and one for biology. Her crew of around 240 men included a team of six scientists and was assigned to investigate 'everything about the sea'.

▲ **Figure 6** HMS *Challenger* – her voyage from 1872 to 1876 was the first systematic study of the oceans.

Three and a half years later, on 24th May 1876, HMS *Challenger* returned home. During a journey of almost 130 000 km, the expedition had collected information from most of the oceans, as well as samples of water, sediments and marine life. More than 4000 new species of marine animals had been discovered, the extent of the Mid-Atlantic Ridge had been measured (although its significance was not recognised until the 1960s), valuable nodules rich in manganese, copper, nickel and cobalt had been found on the sea floor, and the scientists had even managed to measure the depth of the Marianas Trench using only a heavy weight tied to hemp rope! The HMS *Challenger* scientists measured the depth of the trench at 8200 m, which is reasonably close to modern measurements that put it at about 11 000 m. However, some of the most important discoveries have arisen from the measurements the *Challenger* scientists made of the temperature and salinity of deep-ocean water. Their data provided vital evidence for working out how water circulates around the oceans and how the oceans control the global climate (see section **O5**).

The voyage of HMS *Challenger* (see Figure 7) was the first systematic study of the oceans and provided the scientific basis of modern oceanography. The team on HMS *Challenger* took samples and a range of measurements (such as water depth, temperature at different depths, and direction and speed of currents) at 360 sites during the voyage. Even today an ocean survey ship, equipped with much more sensitive equipment, usually makes measurements at only one site per day, travelling 400 km between sites. But accurate information may be of little use if the system you are studying has changed significantly by the time you have finished collecting your data. Some of the

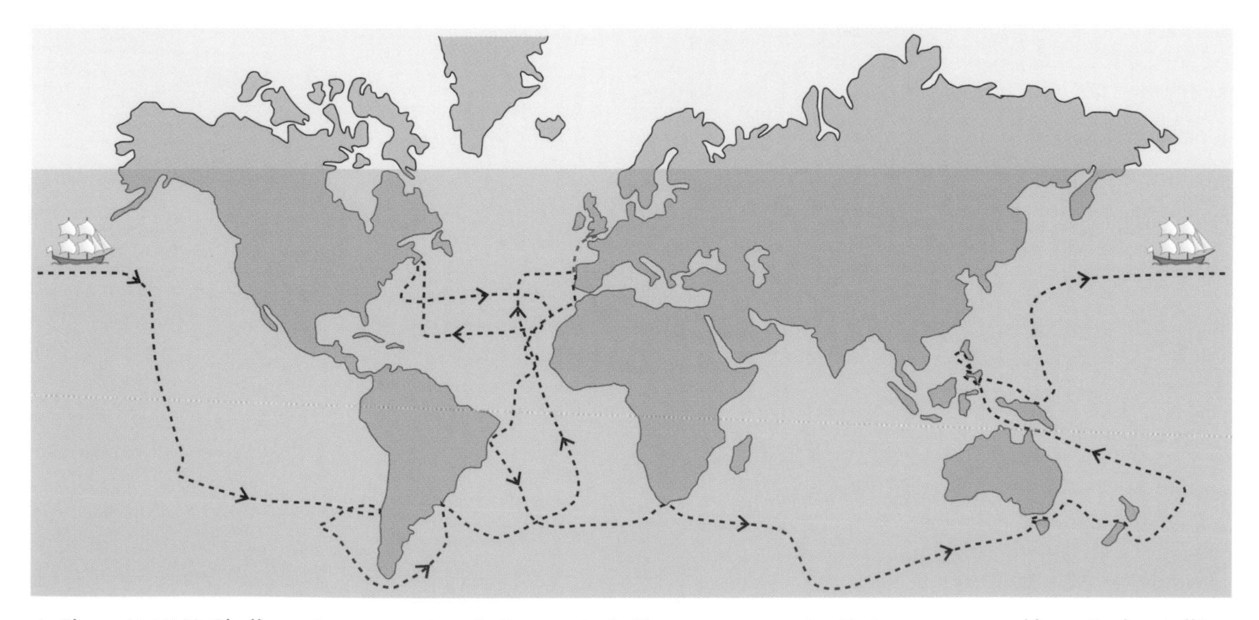

▲ **Figure 7** HMS *Challenger*'s voyage – it took three and a half years, compared with the area covered by a single satellite over a period of 10 days (shaded area).

events being studied are over in a matter of days or a few weeks. A satellite's greater speed is often a considerable advantage (see Figure 7), even though the measurements it makes may sometimes be less accurate.

Surveys have shown that the depth of the oceans is far from constant. In fact, underwater landscapes are more extreme than those we see above the water (see Figure 8). Mount Everest rises 8.85 km above sea level but there are several parts of the ocean that are deeper than 10 km. The average height of the land is only around 840 m above sea level, whereas the average depth of the oceans is 3700 m.

As we have learned more about the seas, we have realised that their role as a storehouse of food and chemicals is almost a sideline compared with the part they play in controlling our climate. Together with the atmosphere, the oceans are at the centre of the system that controls global conditions – the conditions in which we live and under which life has evolved for billions of years.

Perhaps we should have realised much sooner that the oceans have a regulatory role – they are so huge that it would be naive to think they did *not* influence what happens on Earth. But it is much harder to understand *how* the oceans work. Their vastness makes them difficult to study and it is only recently that our knowledge of them has really begun to grow.

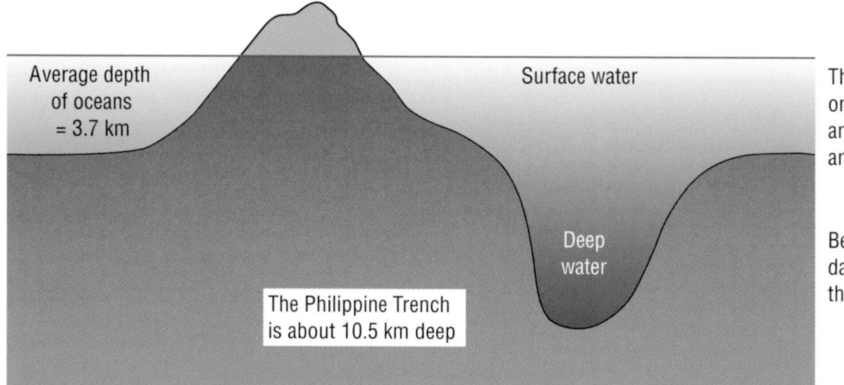

▲ **Figure 8** Underwater landscapes are more extreme than those we see above the water.

Assignment 1

Making accurate measurements on the oceans can be difficult, and errors can become significant when values are multiplied millions of times to scale them up to ocean-sized quantities. This assignment is about one example of such an error.

Fritz Haber was a brilliant German chemist, who developed a process for making ammonia (see section **AI3**). Ammonia can be turned into explosives, and Haber's process was used to supply Germany's munitions factories during the First World War. When the war was over, Haber decided that his country's war debts could be paid off by another of his ideas – extracting gold from the sea.

Gold compounds are present in solution in sea water. Their concentration is very low, but Haber calculated that vast quantities of the precious metal could be extracted by special devices fitted to ships. During the 1920s, German ships sailed around the world hoping to return laden with gold.

They didn't – Haber's figure for the gold concentration was hopelessly high, and our estimate has been falling ever since. For example, a survey carried out between 1988 and 1990 set a new maximum level at $1 \times 10^{-11}\,g\,dm^{-3}$; before 1988 it was thought to be $4 \times 10^{-9}\,g\,dm^{-3}$.

a Convert the volume of the oceans (Figure 5) into dm^3 units. You should now be able to see the 'big number aspect' of calculations on the oceans.

b Estimate the maximum value for the total mass of gold thought to be in the oceans today.

c If all the gold could be extracted from the sea and shared equally among the Earth's population of 6.65×10^9 people, approximately what would be your share today?

d The price of gold varies from day to day, but in 2008 it was worth about £20.0 g^{-1}.

- **i** How much would your share of the gold be worth?
- **ii** Explain why you would be unlikely to receive as much as this, and could even make a loss.

O2 *Salt of the Earth*

Everyone knows that the sea is salty. Salts are ionic compounds and over 99% of all the dissolved substances in sea water are ionic. While they remain dissolved, of course, the ions are free in solution so – strictly speaking – the sea contains not salts, but a mixture of ions.

Figure 9 shows the abundance of these ions. The proportion of one to another is remarkably constant whichever part of whichever sea you choose to sample.

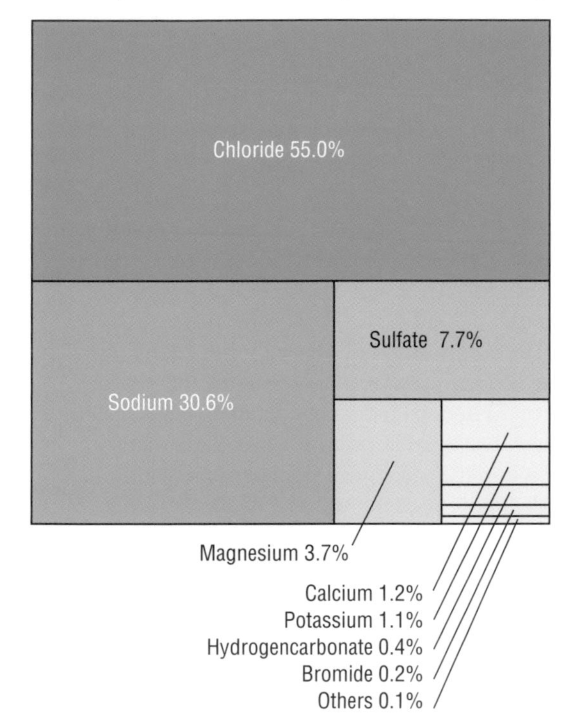

▲ **Figure 9** Percentages by mass of different ions in sea water.

Assignment 2

Use the information in Figure 9 to work out:

a the amount in moles of positive charge present in 100 g of sea water

b the amount in moles of negative charge present in 100 g of sea water. (Remember that 1 mole of doubly charged ions carries 2 moles of charge.)

Comment on the relative magnitudes of the answers you obtain in parts **a** and **b**.

You may find it useful at this point to refer back to the section on writing the formulae of ionic compounds in **Chemical Ideas 3.1**, and the section on the chemistry of ions in solids and solution in **Chemical Ideas 5.1**.

You can revise ionisation by reading **Chemical Ideas 2.5**.

Salt sellers

Sea salt has almost the same composition wherever you get it from. However, the concentration of the salt in the water does vary from place to place. Where large volumes of fresh water enter the sea, the salinity (total salt content) is low. So those parts of the sea that are close to estuaries, abundant in icebergs or in areas of high rainfall are not very salty. On the other hand, in places where the rate of evaporation is high, the sea is saltier than the average. Water evaporates quickly in a hot, dry climate or where it is windy.

The two most abundant ions in sea water are sodium (Na^+) and chloride (Cl^-). These ions are also present in the body fluids of all land creatures – an adult human contains about 300 g of dissolved sodium chloride (common salt). Some of this is excreted each day in urine and sweat, and must be replaced if the body is to function normally.

Our need for salt has been recognised for thousands of years and the substance has entered the language, superstition and history of many societies. The word *salary* means, literally, a payment in salt – as was the custom in the Roman legions. In medieval times, important guests sat 'above the salt' at a banquet – and the saying lives on. Spilling salt brings bad luck – a measure of its value hundreds of years ago.

Not only is it important to eat enough salt, salt was also used to preserve food in the days before refrigerators and freezers had been invented.

The obvious place to get salt from was the sea. Along the east coast of Britain around 600 BC was a string of small saltworks. From the remains of their equipment it would seem that these early salters evaporated sea water in shallow pottery dishes set on brick stands over fires.

When the Romans occupied much of Britain, they imported salt from France (Gaul) where it was possible to evaporate the water by the heat of the Sun. After the Romans left, the British salt industry thrived again.

By medieval times, salt was big business. The Guild of Salters controlled the importation of foreign salt into Britain – mainly from the Bay of Biscay – and there were strict rules governing its handling. Salt could only be measured out by officials called 'salt meters' and carried by 'salt porters'. Large profits were made and the investments continue to produce income even today – the modern Salters' Company uses much of this income to support science teaching in schools.

Although most of our salt now comes from underground deposits, one firm in Essex continues to extract salt from sea water.

Essex has a long coastline with numerous shallow inlets from which water evaporates, leaving sea water with a high salinity. First the water is filtered, and then it is heated by natural gas in large stainless steel pans. As the sea water comes to the boil, some of the

▲ **Figure 10** Salters' Company coat of arms – the motto means 'Salt savours all'.

▲ **Figure 11** Underground salt mine.

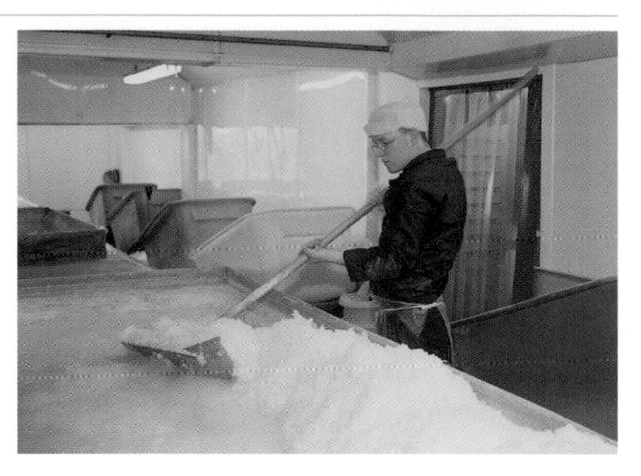

▲ **Figure 12** Raking salt crystals to the side of a pan at the Maldon Crystal Salt Company Ltd.

impurities rise to the surface as a froth and are skimmed off. Then the solution is allowed to simmer. Calcium sulfate is one of the first salts to crystallise out and forms a hard scale deposit on the sides of the pans. As the solution becomes more concentrated, crystals of sodium chloride begin to form. This is partly because sodium and chloride ions are most abundant, and partly because the solubility of sodium chloride is relatively low.

After about 15 hours, the pile of accumulated crystals reaches the surface of the liquid and heating is stopped. The crystals are raked to the sides of the pan and removed with shovels. It is important to do this before the bitter-tasting magnesium salts crystallise out and spoil the flavour. Luckily, they are present in smaller quantities and are more soluble than sodium chloride, so they remain in solution for longer. The early salt makers must have learned by 'bitter experience' not to evaporate sea water to dryness.

The composition of sea water has been known for over 100 years, but it is only recently that we have begun to understand in any detail how the sea became salty. It has been known for a long time that some of the salt in the sea comes from the land. Rainwater leaches salts from the soil and rivers wash them into the sea. But the seas contain, in abundance, some elements that are not found to any great extent in river water.

The source of elements such as chlorine, bromine and sulfur in sea water remained a mystery until we learned more about the structure of the ocean floors.

There is an old Norse myth that tells of a magic salt mill grinding away at the bottom of the ocean, making the sea salt. Strangely enough, as you have already read in the **Elements from the Sea** module, this notion is not so far from the truth. Underneath the sediments on the ocean floor are lavas, generated by long, thin, underwater volcanoes called *mid-ocean ridges*. The gases given off from these volcanoes are rich in compounds containing chlorine, bromine and sulfur. Also, as molten lavas meet cold sea water they solidify and shatter. Water streams down through cracks, scouring out soluble minerals. The superheated solutions that re-emerge through *hydrothermal vents* are much richer sources of elements like chlorine, bromine and sulfur than crustal rock. Figure 13 shows the sources of the dissolved ions in sea water.

Scientists think that the water itself escaped from these deep-seated rocks and that sea water was salty right from the beginning. The balance of ions is kept constant by a complicated geochemical cycle that scientists are only just beginning to understand properly.

Assignment 3

Bottled mineral water contains dissolved ions from the soil and rock through which the water has percolated. Most brands of bottled water give an analysis of the contents on their labels. The contents of typical bottled water (still rather than fizzy (carbonated)) are shown in the table below. Compare the contents with the analysis of sea water.

a What is the total mass of dissolved solids in 1 dm^3 (1 litre) of:
  i sea water
  ii bottled mineral water?

b Compare the analyses for the individual ions in the mineral water with the figures given for sea water. Comment on any differences. Which of the ions in sea water cannot be accounted for by run-off of water from the land?

Contents/mg dm^{-3}		
Ion	**Sea water**	**Bottled water**
Cl^-	19 000	75
Na^+	11 000	62
SO_4^{2-}	2 500	25
Mg^{2+}	1 300	28
Ca^{2+}	400	60
K^+	400	5
HCO_3^-	100	350
Br^-	70	–
Others	40	2

Assignment 4

Human activities remove only a tiny percentage of salt from the sea. Suggest reasons why, when rivers, volcanic gases and hydrothermal vents continually supply the sea with dissolved ions, the composition of sea water has remained constant for at least the last 200 million years.

Chemical Ideas 4.5 tells you more about dissolving and solutions and the use of enthalpy cycles.

Why do many ionic compounds dissolve so readily in water? **Activity O2.1** allows you to investigate the relationship between a solvent and the substances it dissolves.

In **Activity O2.2** you can find the enthalpy changes of solution for some ionic compounds.

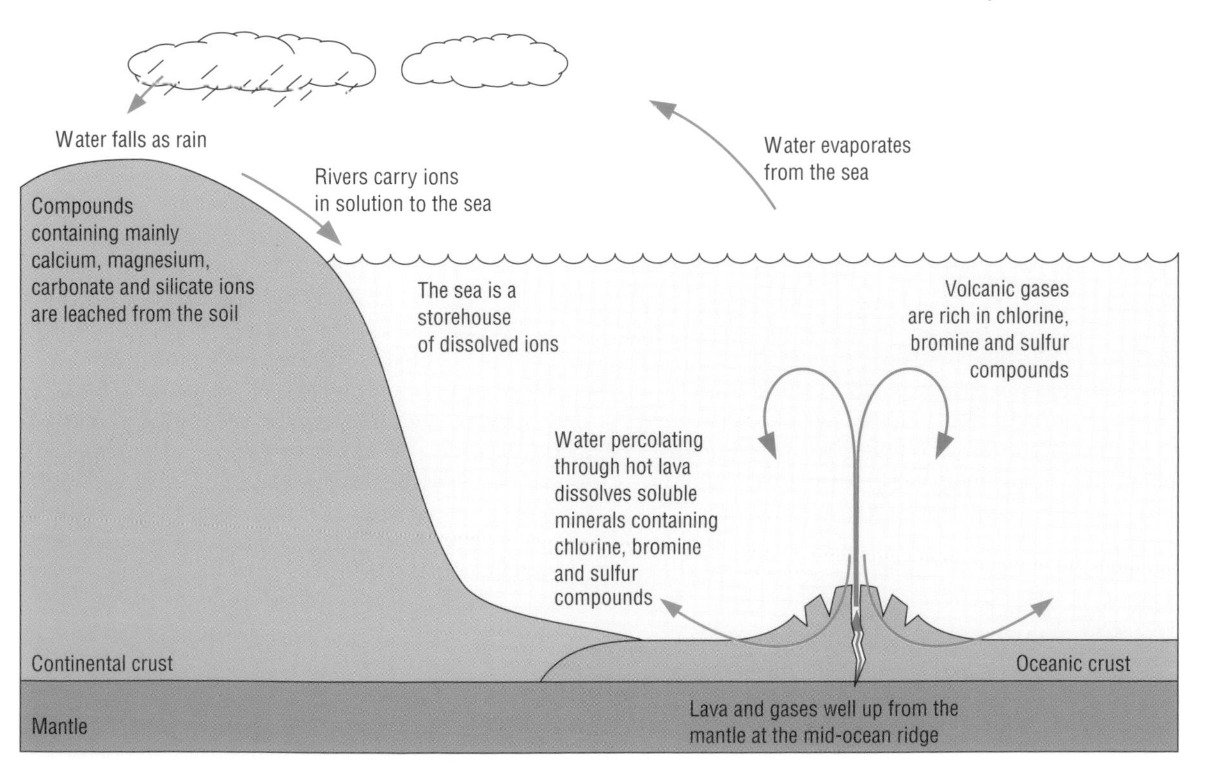

▲ **Figure 13** Sources of dissolved ions in sea water.

O3 *The smell of the sea!*

Most of us are familiar with the distinctive smell of the seaside. It is caused by the volatile gas dimethyl sulfide, $(CH_3)_2S$. Dimethyl sulfide (DMS) is produced from a compound called dimethylsulfoniopropionate (DMSP) that many microscopic marine algae contain. When the algae die, certain ocean bacteria convert DMSP into DMS, which is then released into the atmosphere.

▲ **Figure 14** Conversion of DMSP into DMS in the oceans.

▲ **Figure 15** Microscopic marine algae – some types produce dimethyl sulfide that is released into the atmosphere.

DMS molecules are thought to be one of the main sources of cloud condensation nuclei – small particles in the atmosphere that bind water molecules together and so encourage the formation of clouds. Since clouds can reflect solar radiation back into space, DMS molecules may play an important role in climate regulation.

In 2007, it was reported that certain genes had been identified that appear to affect how ocean bacteria process DMSP molecules. It appears that one particular gene modifies the DMSP molecule so that it breaks down into DMS, whereas a different gene causes the bacteria to produce other sulfur-containing compounds that then enter the marine food web. Finding out which bacteria contain one or both of these genes will be an important step in understanding how marine sulfur is cycled. There may subsequently be some potential in helping to reduce global warming.

DMS has also been implicated in a problem we have known about for some time. The adverse effects of acid rain (or more correctly *acid deposition*) have been experienced in many parts of the world, and compounds containing oxidised sulfur are among the handful of chemicals involved.

Assignment 5

a Draw a full structural formula for dimethyl sulfide. What shape would you expect the molecule to adopt?

b Sulfur is in Group 6 of the Periodic Table, along with oxygen.

- **i** What do we call the series of compounds related to dimethyl sulfide in which the sulfur atom is replaced by an oxygen atom?
- **ii** Dimethyl sulfide, like its oxygen-containing relative, is volatile. Use your knowledge of intermolecular bonds to explain why.

c $(CH_3)_2SO$ and SO_2 are two of the compounds produced in the atmosphere from dimethyl sulfide. Explain why formation of these products corresponds to oxidation of the sulfur.

In the 1960s, scientists were trying to find out more about acid deposition by measuring the quantities of sulfur that circulated around the land, the oceans and the atmosphere. They found that their sums didn't add up … there was a missing link in the sulfur cycle (see Figure 16).

It is now thought that DMS is the missing compound in the cycle. Oxidation of DMS can take place in the atmosphere or by certain ocean bacteria that feed on DMS to form acidic sulfur compounds.

A North Sea survey showed just how significant algal production of dimethyl sulfide can be. Maximum activity occurs in April and May along the coast from Germany to France and probably accounts for about 25% of all acidic pollution over Europe at that time. So we have to accept that some acidic emissions are beyond our control! Acidification of the oceans could potentially lead to some disastrous consequences, as you will learn in section **O4**.

The story of dimethyl sulfide is an example of how our understanding of global processes grows as we learn more about the interrelationship between life, the oceans and the atmosphere. In the next section you will look at another element, carbon, and the vital role of the oceans in determining the amount of carbon dioxide in the atmosphere.

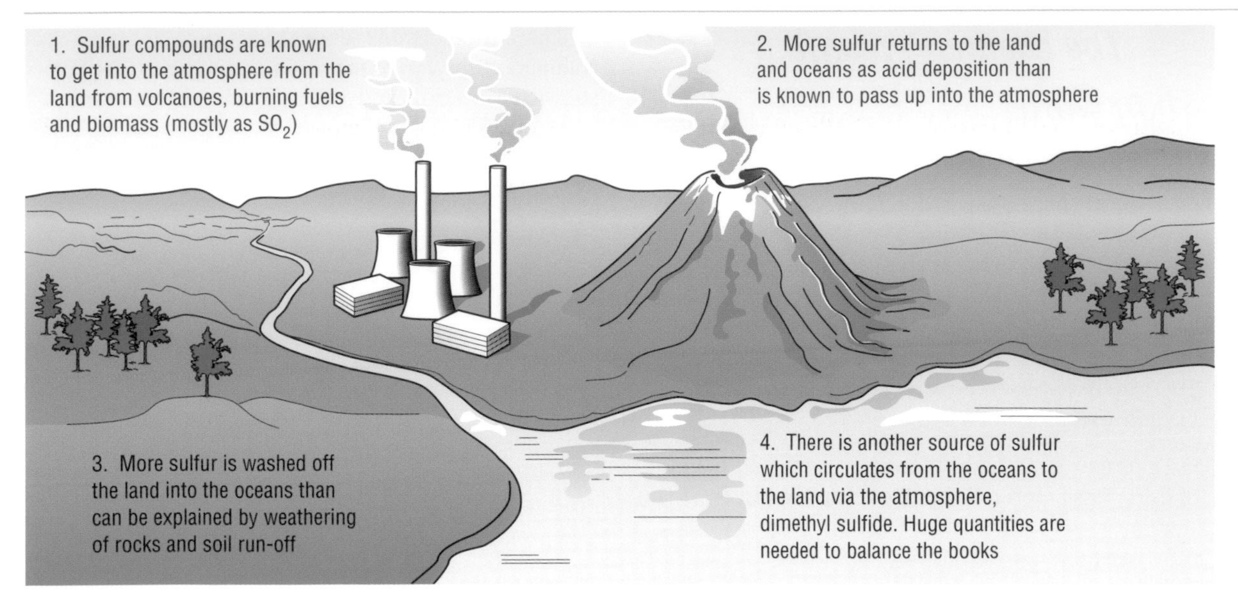

Figure 16 The missing link in the sulfur cycle.

At this point you will find it useful to look back at **Chemical Ideas 8.1** to review acid–base reactions and to learn more about **conjugate acids** and **conjugate bases**. **Chemical Ideas 8.2** tells you more about acids and bases and the ionisation of water (K_w). You will also learn how to calculate the pH of solutions of **strong acids** and **strong bases.**

In **Activity O3** you will carry out a simple practical that will help you understand the logarithmic nature of the pH scale.

GREEN FOR LIFE

You have probably noticed that the sea is a different colour depending on where you are. The colour depends partly on the weather and light conditions, but also on what's in the sea.

Figure 17 Is a blue sea really beautiful?

The sea contains microscopic organisms called phytoplankton that, similar to plants on land, contain chlorophyll and obtain their energy through photosynthesis. Phytoplankton are a crucial link in the global food web and are also responsible for much of the oxygen that is present in the Earth's atmosphere.

Chlorophyll absorbs blue and red light – which is why plants appear green (see **Chemical Ideas 6.7**). The colour of the sea gives a measure of how many phytoplankton are present – green water indicates lots of phytoplankton, blue water means that very few are present. Scientists use this information to create worldwide maps of chlorophyll from satellite images of ocean colour. This provides information on how well these microscopic 'plants' in the oceans are growing. However, it's not always that easy – rivers wash mud and dead vegetation into the sea and this can also produce a green colouration in sea water close to land masses. Care must be taken to avoid drawing false conclusions from these images.

O4 The oceans – a safe carbon store?

Getting rid of carbon dioxide

Some scientists have predicted that the carbon dioxide concentration in the atmosphere could rise to over 650 ppm by 2100 (see **The Atmosphere** storyline). To keep the carbon dioxide concentration stable, even at 650 ppm (it was 383 ppm in 2007), the emissions of the gas worldwide will need to be reduced drastically.

There are basically three ways to reduce the build-up of carbon dioxide in the atmosphere:

- use alternative methods to produce energy instead of using fossil fuels (see the **Developing Fuels** storyline)
- use fossil fuels more efficiently
- capture and store carbon dioxide.

The last of the three methods listed is probably the most revolutionary and *could* involve:

- turning CO_2 into useful products (as yet unknown!)
- growing more trees and increasing the organic content of soils
- storing the gas in deep natural trenches on the sea floor, where the pressure will cause it to liquefy
- injecting the gas onto the sea floor – at a depth of, say, 3500 metres where it will form a lake of liquid carbon dioxide that should remain undisturbed.

There is a danger that storing carbon dioxide in the deep ocean *may* disturb the natural environment. Even so, this may be less damaging than allowing carbon dioxide to be absorbed naturally in the surface waters of the ocean. To weigh up the advantages and disadvantages of any proposed method for reducing atmospheric concentrations of this greenhouse gas, we need to understand how carbon dioxide interacts with, and cycles around, the atmosphere and the oceans.

Before going further you may find it useful to remind yourself about chemical equilibrium and the dissolving of carbon dioxide in the oceans by reviewing **Chemical Ideas 7.1** and **The Atmosphere** storyline.

Storing carbon dioxide

Imagine you are planning a couple of hours on the beach in the middle of your summer holiday. You have a choice – to take cans of fizzy drink with you, or to leave them in the fridge to drink when you return. From experience you decide to leave the cans in the fridge – you know that more of the fizz stays in a fizzy drink if you keep it cool. If you leave a can standing in the Sun, the gas will come frothing out explosively when you open the top.

The carbon dioxide in the drink is like all gases – its solubility *decreases* as temperature increases. Table 1 shows this for carbon dioxide, oxygen and nitrogen.

Table 1 Variation of the solubilities of some gases with temperature (at atmospheric pressure).

Temperature/°C	Solubility/mg per 100 g water		
	CO_2	O_2	N_2
0	338	7.01	2.88
10	235	5.47	2.28
20	173	4.48	1.89
30	131	3.82	1.65
40	105	3.35	1.46
50	86	3.02	1.35

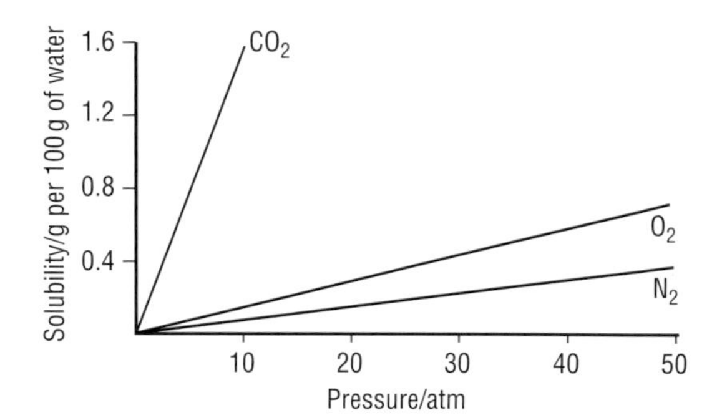

Figure 18 Variation of the solubilities of some gases with pressure at 298 K.

Carbon dioxide is put into fizzy drinks under considerable pressure – about 14 atmospheres in most cases. That's why there is such a 'pop' when you open the can. The high pressure makes more gas dissolve, as Figure 18 shows.

A high pressure and a low temperature help to keep a drink fizzy – these conditions also encourage carbon dioxide to dissolve in the oceans. This takes the gas out of the atmosphere and helps to maintain a stable environment on Earth.

Elegant equilibria

The exchange of carbon dioxide between the atmosphere and oceans takes place quickly – more quickly than if you left some water in a beaker to come to equilibrium. Like so many environmental processes, we cannot explain this in terms of physical factors alone. Uptake of carbon dioxide by the oceans is speeded up by the action of marine life, as shown in Figure 19.

Table 1 and Figure 18 show that carbon dioxide is more soluble in water than oxygen or nitrogen is. CO_2 molecules contain polar C=O bonds that can form hydrogen bonds with water molecules.

$$CO_2(g) \rightleftharpoons CO_2(aq) \qquad \text{(reaction 1)}$$

In addition, some of the carbon dioxide molecules react chemically with water and are removed from the

Assignment 6

Perrier® water is officially described as 'naturally carbonated, natural mineral water' – which means that the fizz comes from carbon dioxide produced naturally underground, rather than from chemically manufactured gas added from a cylinder. Lots of people prefer the taste of bottled mineral water to their tap water, and many prefer the naturally carbonated drink to the other kind of fizzy water.

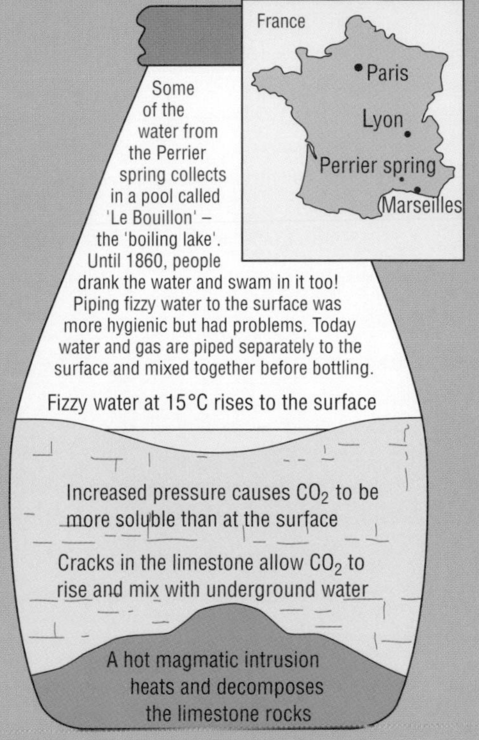

a Write an equation for the thermal decomposition of calcium carbonate.

b The pressure below the Perrier spring is 443 kPa (4.4 atm). How many times more soluble is carbon dioxide at this pressure than at atmospheric pressure (see Figure 18)?

c Why do you think the Perrier pool was called the 'boiling lake'?

d The Perrier company had problems when they tried piping the carbonated water out of the ground. Describe one difficulty they would have encountered and how it would be avoided by piping the gas and water separately.

equilibrium. More carbon dioxide therefore dissolves to maintain the equilibrium position:

$$CO_2(aq) + H_2O(l) \rightleftharpoons H^+(aq) + HCO_3^-(aq)$$ (reaction 2)

$$HCO_3^-(aq) \rightleftharpoons H^+(aq) + CO_3^{2-}(aq)$$ (reaction 3)

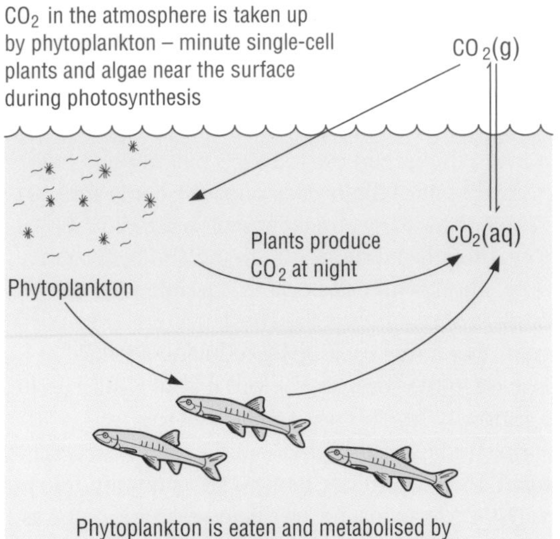

▲ **Figure 19** Phytoplankton increase the rate at which carbon dioxide dissolves in the oceans – they play a key role in removing carbon dioxide from the atmosphere.

The reaction with water produces a mixture that contains mainly hydrogencarbonate ions (HCO_3^-) and H^+ ions, together with some carbonate ions (CO_3^{2-}).

Much of the excess carbon dioxide we release into the atmosphere from the combustion of fuels is absorbed by the oceans. Estimates vary, but it seems likely that 35–50% is removed in this way. The oceans continue to 'soak up' carbon dioxide because surface water, rich in CO_2, is constantly being removed and stored away for hundreds of years in the deep ocean. The maximum amount of carbon dioxide is removed because this takes place in cold regions, where CO_2 is most soluble. So currents, chemistry and marine life together make up a very efficient CO_2 removal system.

Assignment 7

The oceans can keep on taking carbon dioxide out of the atmosphere because the dissolving of CO_2 takes place in what chemists call an *open system*. This means that material can enter or leave the system and so prevent equilibrium ever being established.

In this example, $CO_2(aq)$ is removed to the deep oceans, so more $CO_2(g)$ dissolves to replace it. Drying washing on a clothes line and beer going flat in an open bottle are also examples of open systems that are never allowed to reach equilibrium. Explain what happens in these two examples.

Life on Earth

The Earth's early life forms evolved in the oceans – and that's where they stayed throughout most of the Earth's history. The planet was nearly 4 billion years old and living things had existed for 3 billion years before life moved out of the oceans onto the land (Table 2).

The Earth's early atmosphere consisted mainly of carbon dioxide, ammonia, methane and hydrogen sulfide. Photosynthesis became possible with the evolution of cyanobacteria. These bacteria produced oxygen, but it was used up by reducing agents dissolved in the sea water before it could build up in the atmosphere. This was just as well for the cyanobacteria, because they cannot tolerate oxygen. Instead, they use sulfate ions (SO_4^{2-}) and nitrate(V) ions (NO_3^-) as oxidising agents in respiration. It was only later that organisms could use the free oxygen that was dissolved in the sea water or that had built up in the atmosphere.

The Earth's atmosphere has changed dramatically since the early days of life – reducing agents such as methane and acidic gases such as carbon dioxide have been largely replaced by a neutral, oxidising mixture of nitrogen and oxygen. If life had been forced to evolve on land in contact with the air, the primitive organisms would have become extinct. Instead they have been protected by their watery environment, which has altered remarkably little over billions of years.

Assignment 8

Molecules such as 2-hydroxypropanoic acid and ethanol were probably present in small quantities in the early oceans. An example of an oxidation using sulfate ions, such as might have been carried out by a primitive bacterium, is

$$2CH_3CH(OH)COOH(aq) + SO_4^{2-}(aq) \rightarrow 2CH_3COOH(aq) + S^{2-}(aq) + 2CO_2(g) + 2H_2O(l)$$

a Explain why this equation represents a redox reaction.

b Calculate the enthalpy change in the oxidation of 2 moles of 2-hydroxypropanoic acid by sulfate ions. (Assume that the enthalpy changes of formation of the aqueous solutions of the acids are: 2-hydroxypropanoic acid, $\Delta H_f^{\ominus} = -694\,kJ\,mol^{-1}$; ethanoic acid, $\Delta H_f^{\ominus} = -486\,kJ\,mol^{-1}$. Use the **Data Sheets** to find the other information you need.)

c **i** Write an equation for the complete oxidation of 2-hydroxypropanoic acid by oxygen to produce carbon dioxide and water.

ii Calculate the enthalpy change for the reaction in part **c, i.**

d Use your calculations to explain which is more efficient – the aerobic respiration or anaerobic respiration of 2-hydroxypropanoic acid.

Table 2 The origins of life on Earth.

Million years ago	Life-form changes	Atmosphere changes
4600	origin of Earth	first atmosphere lost from hot planet
3800	oldest sedimentary rocks organisms probably present	atmosphere consists of CO_2, NH_3, CH_4 and H_2S
3500	beginning of fossil record bacteria leave fossil remains	life probably uses energy released during fermentation, or from oxidation of organic molecules by SO_4^{2-} ions and NO_3^- ions
2800	first photosynthetic bacteria (cyanobacteria or blue-green bacteria)	oxygen produced for 1 million years it is used up in the oxidation of S^{2-} and Fe^{2+} ions no ozone is able to form in the atmosphere so life on the surface is impossible owing to ultraviolet radiation bacteria live in the top 200 m of the ocean, screened from ultraviolet radiation by water
2000	first significant $CaCO_3$ deposits formed from shells of marine organisms	removal of two greenhouse gases (CO_2 to form $CaCO_3$, and CH_4 by reaction with O_2) leads to major ice age
1800	evolution of cyanobacteria	oxygen begins to build up in the atmosphere
1500	first green algae – plants with chlorophyll in their cells	oxygen levels continue to build up
800	first sea animals	1% oxygen in atmosphere formation of ozone begins
400	first land plants	10% oxygen in atmosphere ozone layer protects plant cells from ultraviolet radiation
300	first land animals	21% oxygen in atmosphere

Keeping things steady

The ability of the oceans to withstand external changes has been essential for the unbroken evolution of life. For example, the **pH** of the oceans has remained close to 8 for millions of years. Why were the oceans not much more acidic when the atmosphere contained 35% CO_2 – that's a thousand times its present level?

One reason is that a solution of carbon dioxide in water is a **weak acid** – these react incompletely with water. If we represent the acid by the formula HA, we can show the reaction with water by the equation

$$HA + H_2O \rightleftharpoons H_3O^+ + A^-$$

H_3O^+ ions (called **oxonium ions**) make the solution acidic. The position of equilibrium is well over to the left-hand side and only a fraction of the acid added to the water reacts to produce oxonium ions. So the solution is not as acidic as it *could* be if all the acid had reacted.

The equation can be simplified to

$$HA(aq) \rightleftharpoons H^+(aq) + A^-(aq)$$

by leaving out the water, which is present in excess.

You might find it useful here to review equilibrium constants in terms of concentrations in **Chemical Ideas 7.2**, before going on to study **Chemical Ideas 8.2**, which tells you more about solutions of weak acids.

You can compare the properties of strong and weak acids in **Activity O4.1**.

Activity O4.2 gives you the opportunity to develop your understanding of acid strength.

For carbon dioxide, only a small proportion of the $CO_2(aq)$ molecules react to form hydrogen ions and hydrogencarbonate ions:

$$CO_2(aq) + H_2O(l) \rightleftharpoons H^+(aq) + HCO_3^-(aq) \quad \text{(reaction 2)}$$

If a small amount of alkali is added to a solution of a weak acid such as aqueous carbon dioxide, some of the $H^+(aq)$ ions are removed. But this causes more $CO_2(aq)$ to react with water to restore the position of equilibrium and so replace the $H^+(aq)$ ions. In this way a solution of carbon dioxide can maintain a constant pH even when a small amount of alkali is added.

CARBONIC ACID

Sometimes, to emphasise the link with the general equation for weak acids, an aqueous solution of carbon dioxide is represented by the formula $H_2CO_3(aq)$. It is as if the reaction

$$CO_2(aq) + H_2O(l) \rightarrow H_2CO_3(aq)$$

had occurred. This is why some people use the name 'carbonic acid' for a solution of carbon dioxide.

What happens if the ocean becomes more acidic? When the proportion of carbon dioxide in the atmosphere, for example, was much higher, the equilibrium

$$CO_2(g) \rightleftharpoons CO_2(aq) \quad \text{(reaction 1)}$$

would ensure that the concentration of $CO_2(aq)$ was also higher. This in turn would result in the equilibrium in reaction 2 moving to the right and generating more H^+ (aq) ions:

$$CO_2(aq) + H_2O(l) \rightleftharpoons H^+(aq) + HCO_3^-(aq) \quad \text{(reaction 2)}$$

You would therefore predict that the pH of the ocean would have been lower (more acidic). But the ocean could resist even this change, because it is an example of a **buffer solution** – one that remains within a narrow range of pH despite the addition of acid or alkali.

It would help to read **Chemical Ideas 8.3**, which will introduce you to the theory of buffer solutions.

You can investigate some buffer and non–buffer solutions in **Activity O4.3**.

The commonest type of buffer solution is made up from a weak acid with one of its salts dissolved in it. The weak acid acts as a reservoir of $H^+(aq)$ ions. These can react with OH^- ions that are added and so prevent the solution becoming more alkaline. The anions from the salt act as bases – they can 'soak up' additions of $H^+(aq)$ ions and keep the solution from becoming acidic.

Simple buffers rely on the shifting back and forth of the equilibrium

$$HA(aq) \rightleftharpoons H^+(aq) + A^-(aq)$$

The weak acid is represented by HA and the anions in the salt of this weak acid are represented by A^-.

For the ocean to resist an increase in acidity, there must be a supply of $HCO_3^-(aq)$ ions (corresponding to the $A^-(aq)$ ions from the salt of the weak acid in a normal buffer solution). One source of HCO_3^- ions is the material that dissolves in river water and then flows into the sea. But there are also other processes involving seashells, chalk and limestone that can provide an almost limitless supply of hydrogencarbonate ions.

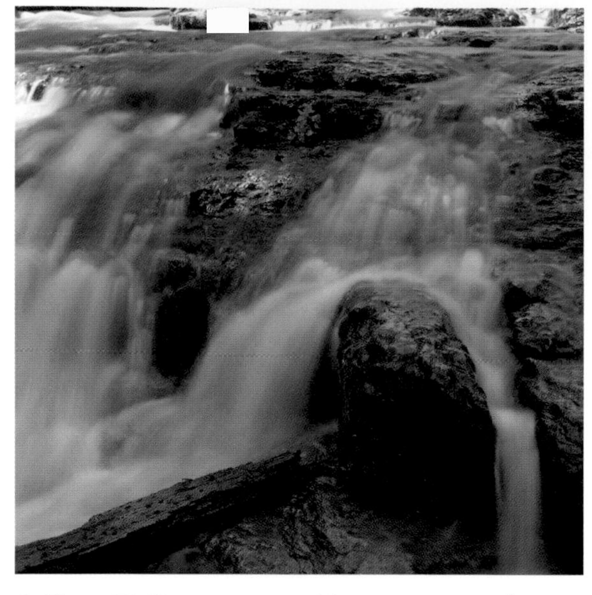

▲ **Figure 20** River waters provide a constant supply of HCO_3^- ions to the oceans from the weathering of limestone rocks.

▲ **Figure 21** Seashells, an important component in pH regulation in the oceans.

Sinking shells

How are seashells involved in the reactions that influence the solubility of carbon dioxide in the oceans? The three reactions that have been discussed so far are linked together – the product of reaction 1 is the reactant in 2, and so on.

$$CO_2(g) \rightleftharpoons CO_2(aq) \qquad \text{(reaction 1)}$$

$$CO_2(aq) + H_2O(l) \rightleftharpoons H^+(aq) + HCO_3^-(aq) \qquad \text{(reaction 2)}$$

$$HCO_3^-(aq) \rightleftharpoons H^+(aq) + CO_3^{2-}(aq) \qquad \text{(reaction 3)}$$

These three equations can be added together to produce just one equation (reaction 4) that shows how the reactants in reaction 1 lead to the products of reaction 3:

$$CO_2(g) + H_2O(l) \rightleftharpoons 2H^+(aq) + CO_3^{2-}(aq) \qquad \text{(reaction 4)}$$

Remember, the reaction does not happen as simply as this, but this equation should make the next part of the story clearer.

Le Chatelier's principle tells us that any way of removing H^+ or CO_3^{2-} ions from solution (reaction 4) will cause more CO_2 to dissolve. Removing H^+ ions by adding a base is one way of doing this. You should be familiar with this process – carbon dioxide is an acidic gas and it dissolves well in alkaline solutions – that's why alkalis such as sodium hydroxide or calcium hydroxide are used to absorb CO_2.

Making the sea alkaline is not a very feasible way of encouraging the oceans to take up carbon dioxide! However, many marine organisms build protective shells composed of insoluble calcium carbonate, using CO_3^{2-} ions in the sea water. The building of these shells provides a route for mopping up carbon dioxide and keeping the composition of our atmosphere constant.

Billions of years ago, the Earth's atmosphere contained very much more carbon dioxide than it does now – probably about 35% CO_2 by volume. Once the process of photosynthesis had evolved, marine life had plenty of raw materials to work on in the form of carbon dioxide and water. Shell production flourished – limestone and chalk rocks are the remains of the shells of marine organisms that lived at that time and changed carbon dioxide from the atmosphere into solid calcium carbonate.

▲ **Figure 22** The chalk cliffs of the Seven Sisters are the legacy of marine organisms that lived billions of years ago.

▲ **Figure 23** Machair in North Uist in the Outer Hebrides. Machair sand is 80–90% calcium carbonate from crushed shells – it is driven by waves and wind over grass and peatland, neutralising acid from the peat. Machair has one of the rarest habitats in Europe.

Calcium carbonate is a good material for shellfish to use for protection at the surface of the oceans. It does not dissolve in sea water – but it does dissolve, very slightly, in pure water. It is an example of a *sparingly* soluble solid – the dissolving of sparingly soluble solids is controlled by equilibria such as

$$CaCO_3(s) \rightleftharpoons Ca^{2+}(aq) + CO_3^{2-}(aq) \qquad \text{(reaction 5)}$$

in which the ions in the *saturated solution* are in dynamic equilibrium with the undissolved solid present.

Calcium carbonate is a safe material from which to build seashells because the concentrations of $Ca^{2+}(aq)$ ions and $CO_3^{2-}(aq)$ ions are already high enough at the surface of the sea for the calcium carbonate in the shells to be effectively insoluble (the equilibrium in reaction 5 lies well over to the left). Remember, though, that the shells are in equilibrium with the ions in sea water, and there will be a constant exchange of Ca^{2+} and CO_3^{2-} between the two.

But things are different deeper in the ocean, where the pressure is higher and the temperature is lower – under these conditions calcium carbonate is more soluble. There is also a continuous downward drift of material from above. It's like a perpetual snowstorm – in fact the falling material is called marine snow. It contains the remains of dead organisms and the waste products from live creatures. Most of the organic material, such as tissue, is consumed or decomposed higher up, but some reaches the deeper water where bacteria break it down to produce carbon dioxide. The shells fall intact, but then react with the extra carbon dioxide and dissolve. These processes are summarised in Figure 24.

There are no shells on the deep ocean floor – they've all dissolved. The creatures that live there cannot use calcium carbonate for a protective coating.

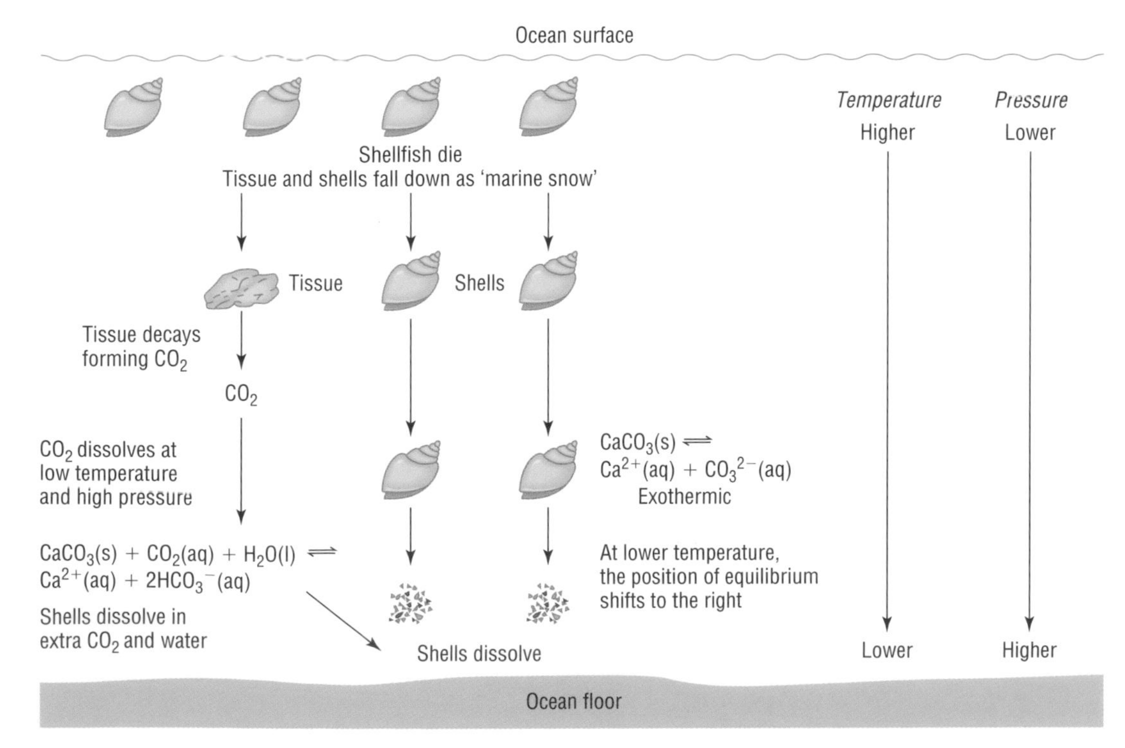

▲ **Figure 24** Dissolving of shells on the deep ocean floor.

The calcium carbonate deposits that built up to form our limestone hills could not, therefore, have formed in deep water. They must have been laid down when our landmass was in shallower seas. The abundance of life also suggests that it was warm, tropical water. Evidence like this helps scientists piece together the distant history of the Earth, and helps us explain how the continents have drifted and how the climate has changed throughout time.

Assignment 9

a Use Hess's law and enthalpy changes of formation in the **Data Sheets** to calculate the standard enthalpy change for the process

$$CaCO_3(s) \rightarrow Ca^{2+}(aq) + CO_3^{2-}(aq)$$

b Explain in terms of equilibria how calcium carbonate production encourages more carbon dioxide to dissolve from the atmosphere.

c Although the dissolving of calcium carbonate

$$CaCO_3(s) \rightarrow Ca^{2+}(aq) + CO_3^{2-}(aq)$$

is a slightly exothermic process, it is accompanied by a large decrease in entropy. We often assume that dissolving is accompanied by an increase in the entropy of the chemicals, but in this case the entropy change for the system, ΔS_{sys}, is large and negative. Explain why you think dissolving might lead to such a large entropy decrease in this situation.
You may need to review **Chemical Ideas 4.3** about entropy and the direction of change to help you to answer part **c**.

Figure 25 summarises the processes that prevent the oceans from becoming acidic. The shells, chalk and limestone in the seas provide the reservoir of anions needed to prevent changes in acidity.

What would happen if the carbon dioxide in the atmosphere rose to the high level of several billion years ago? Solid calcium carbonate would dissolve to produce the carbonate and hydrogencarbonate ions needed to remove the extra H^+ ions. Few CO_3^{2-} ions would remain – most of the carbon would be in the form of HCO_3^- and dissolved CO_2. The sea would be like a mixture of Perrier water and bicarbonate of soda – the shells and white cliffs would disappear!

Assignment 10

Some scientists have predicted that acidification by carbon dioxide will cause the surface waters of the ocean to drop in pH by around 0.1 units from 8.1 to 8.0 in the next 100 years. What would be the change in $H^+(aq)$ ion concentration if this did occur?

▲ **Figure 26** Turret polyp. The world's coral reefs, already damaged by record sea temperatures, are threatened by rising carbon dioxide concentrations in the atmosphere and the oceans. Tiny reef-dwelling creatures called coral polyps produce calcium carbonate, but this is more difficult when the concentration of carbon dioxide in the sea water increases.

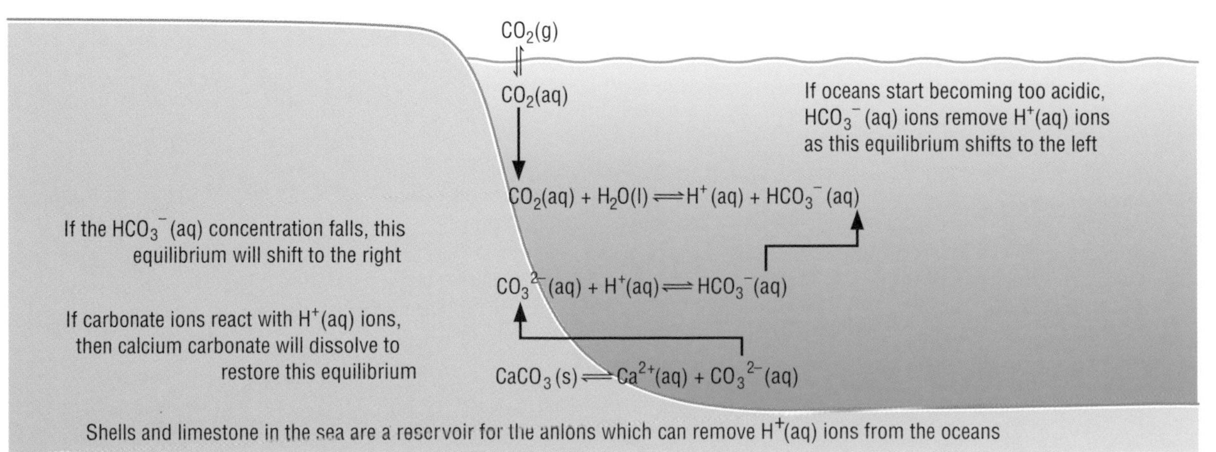

▲ **Figure 25** Buffering action in the oceans – why the oceans do not become more acidic.

Limestone deposits could not form if the atmosphere contained 35% CO_2. The earliest sedimentary rocks are *silicaceous* (consist predominantly of silicon dioxide). The first significant calcium carbonate deposits formed only about 2 billion years ago, over halfway through the Earth's lifetime. Much of the atmospheric carbon dioxide had been used up by then – so much so that its greenhouse effect had diminished and a major ice age had set in.

The $CO_2/CaCO_3$ system is made up of linked equilibria that respond rapidly to change. In the longer term, ion exchange between H^+ ions in the water and Na^+ or K^+ ions in clay sediments provides another very powerful pH control mechanism.

This process can take place only at the bottom of the ocean, where sea water and sediment are in contact. Deep ocean water circulates slowly, perhaps taking 1000 years to complete one cycle. The ion-exchange equilibria may be important over millions of years, but in the surface water, and throughout the oceans on a shorter timescale, it is the $CO_2/CaCO_3$ system that keeps the pH of the ocean stable.

Assignment 11

Review the suggested methods for reducing atmospheric carbon dioxide levels given at the start of this section on page 134. Have your views on the advantages and disadvantages of each method changed now that you know a little more about the complex chemistry involved?

Can you suggest any other methods for reducing carbon dioxide levels?

O5 *The global central heating system*

With the exception of tidal power, all our energy comes ultimately from nuclear sources. On Earth, we can mine radioactive compounds and produce from them fuel for nuclear power stations. Energy from the nuclear processes that go on inside the Earth can be tapped at suitable locations as geothermal energy.

But most of our energy comes from outside our planet – from the great nuclear furnace in the Sun. The Sun's rays heat the Earth directly, and through this drive the winds and waves – solar cells convert sunlight into electricity. Photosynthesis uses sunlight to build up fuels – some, such as wood, can be used almost immediately; others have been changed over millions of years into coal, oil and gas.

When the Sun's rays reach the Earth they can be:

- reflected
- absorbed by the atmosphere
- absorbed by the Earth's surface.

Roughly half the energy we receive from the Sun is absorbed by the land and the oceans, and this causes the Earth's surface to warm up (Figure 27). You saw in **The Atmosphere** module of the course that the Earth in turn radiates energy back into space (see section **A6**).

If the Earth *was* a dry lump of rock with no atmosphere, each part of its surface would soon settle down to a situation in which the energy received from the Sun would, on average, be balanced by energy lost through radiation (see Figure 28). The tropics would be much warmer and the poles even colder than they are – and the Earth would be far less hospitable to life.

But the Earth is surrounded by water and gas – both fluids. Temperature differences set up currents in the oceans and atmosphere that spread out the heating effect of the Sun more evenly. Just like warm air from a radiator spreads around a room, currents in the sea

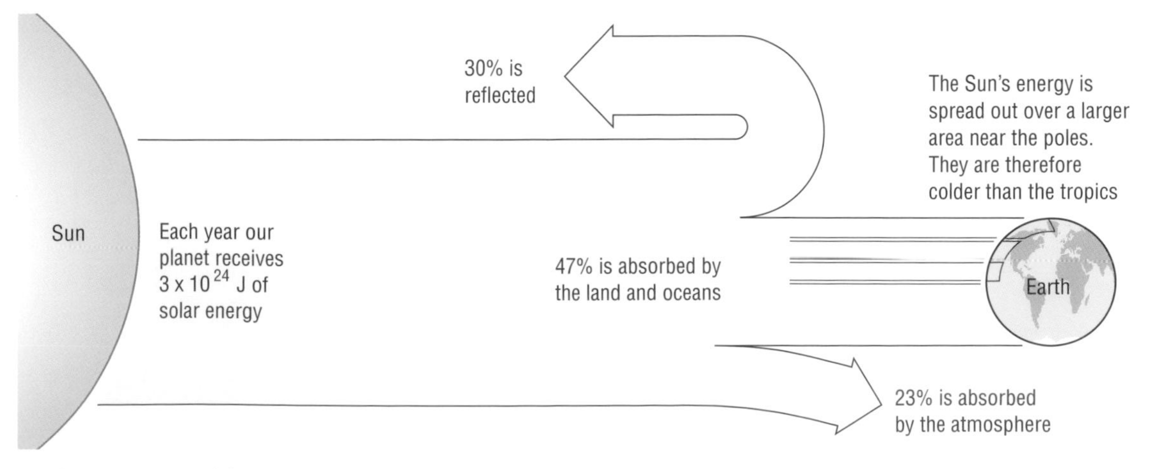

▲ **Figure 27** Fate of the Sun's energy reaching the Earth.

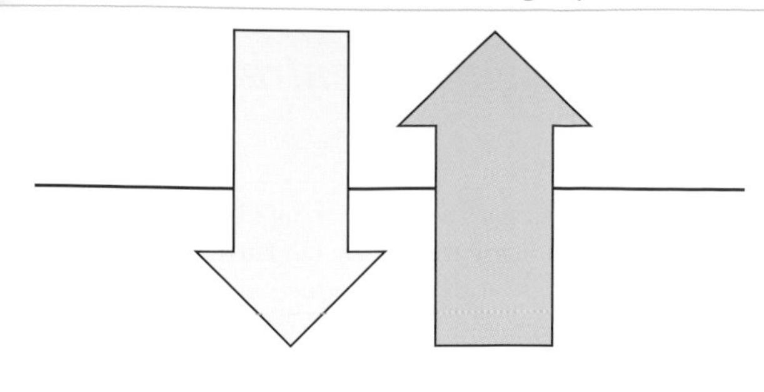

▲ **Figure 28** The fate of the Sun's energy reaching a planet with neither atmosphere nor oceans.

and air take thermal energy from the tropics to the colder regions of the Earth.

In fact, the ocean–atmosphere system is even more effective at spreading out energy. Warm water can do more than circulate – it can evaporate. Energy is taken in when water evaporates, so the situation will be reversed and energy must be released when water condenses. The tropics are cooled by evaporation, and currents in the atmosphere carry the water vapour to colder, high-latitude regions where condensation releases energy (Figure 29).

High-latitude regions receive more energy than is provided by the Sun alone – they are wetter, but warmer. Figure 30 shows the balance of condensation and evaporation around the world.

In the North Atlantic region, the winds and warm water currents flow from SW to NE. Northern Europe, including the UK, is warmed by energy that has been transported from the tropics and the Caribbean. In winter, as much as 25% of our thermal energy may come this way. Eastern North America does not receive this energy. So winters are much more pleasant in Lisbon, Portugal (latitude 38°N) than in Boston, US (latitude 42°N) (Figure 31).

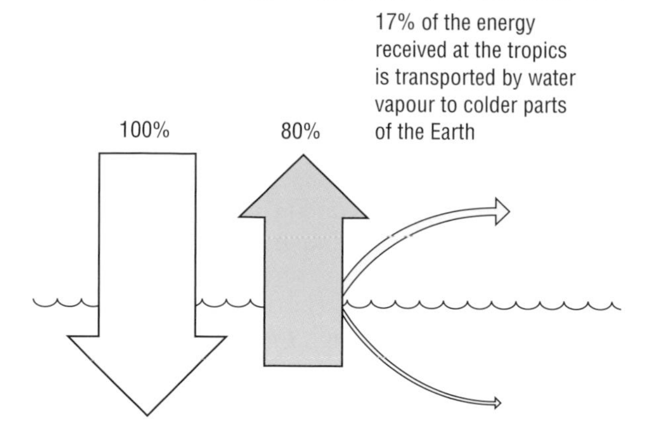

▲ **Figure 29** At the tropics of a planet with an atmosphere and oceans, about 20% of the incoming solar energy is transported to colder regions.

In **Chemical Ideas 4.3** you were introduced to the concept of entropy. **Chemical Ideas 4.4** helps you understand the behaviour of molecules in solids, liquids and gases, and explains why changes occur in terms of entropy.

Energy in the clouds

The molecules in liquid water and water vapour differ in one important aspect. In the liquid, attractive forces between the molecules – **intermolecular bonds** – keep the molecules quite close together. In water vapour, the molecules are much further apart and move about freely.

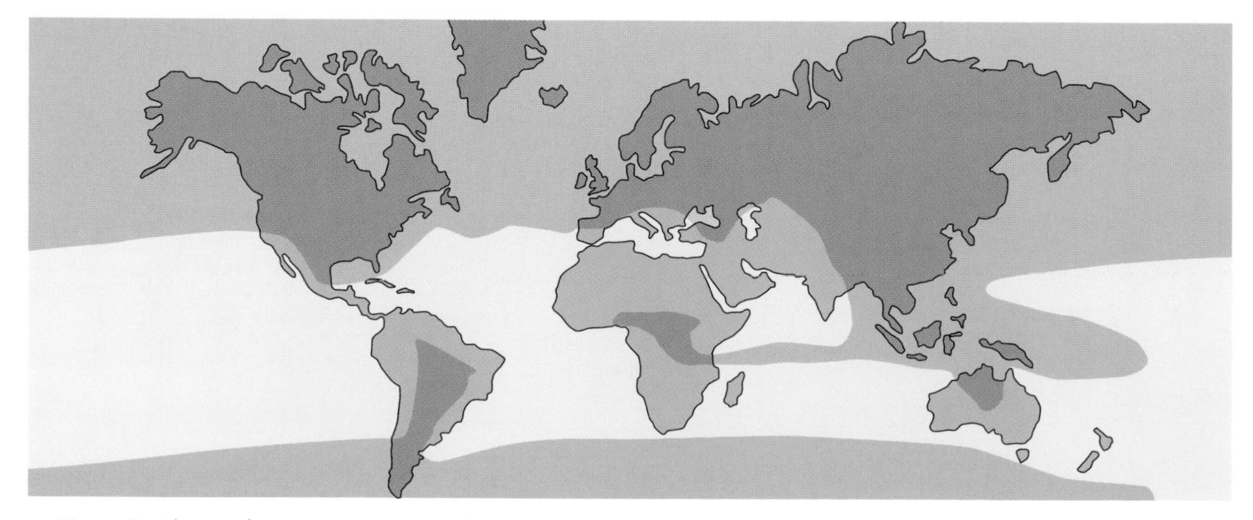

▲ **Figure 30** The condensation–evaporation balance of the Earth – dark shaded areas denote regions where condensation exceeds evaporation.

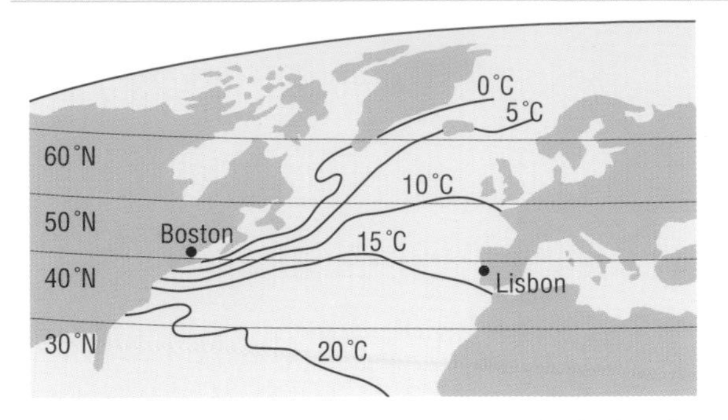

▲ **Figure 31** Average surface temperatures in the North Atlantic in February. This diagram is based on data collected between 1971 and 2000 – more recent data have shown fluctuations in surface temperature as a consequence of sea-ice melting off Greenland.

▲ **Figure 32** Energy is taken in from the surroundings when sea water evaporates, and is released when the water vapour condenses in clouds.

You can remind yourself about hydrogen bonding by reviewing **Chemical Ideas 5.4**. The unique properties of water are explained in **Chemical Ideas 5.5**.

Activity O5.1 provides you with an opportunity to investigate the enthalpy change of vaporisation of water in more detail.

Assignment 12

The positions of three Canadian cities are marked on the map below. Explain the patterns in their winter and summer temperatures.

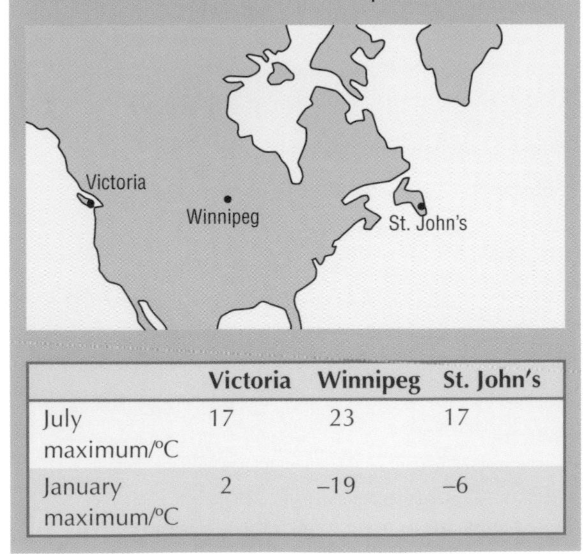

	Victoria	Winnipeg	St. John's
July maximum/°C	17	23	17
January maximum/°C	2	−19	−6

When water evaporates, changing from liquid to vapour, the intermolecular bonds must be overcome – a process that takes in energy. The enthalpy change of vaporisation, ΔH_{vap}, is a measure of this energy.

$$H_2O(l) \rightarrow H_2O(g)$$

evaporation – an *endothermic* process

In the reverse process, condensation, molecules come together again, intermolecular bonds re-form and an equal quantity of energy is released. In other words, condensation is *exothermic* and the enthalpy change is $-\Delta H_{vap}$.

$$H_2O(g) \rightarrow H_2O(l)$$

condensation – an *exothermic* process

Figure 33 illustrates the global water cycle – it summarises the main processes by which water circulates around the world.

Notice that more water evaporates from the oceans than is directly returned to them as precipitation (i.e. rain and snow). Each year, 40×10^{15} kg of water vapour produced from the sea falls as precipitation over the land. The process makes the land wetter and keeps the rivers flowing – it also makes the land warmer.

So evaporation and condensation of water affect the temperature of different parts of the Earth in two ways:

- by transferring energy from low latitudes to high latitudes
- by warming the land through condensation of water that comes from the oceans.

Another world

The UK may seem cold enough and wet enough most of the time, but imagine a world with propanone seas. The enthalpy changes of vaporisation of propanone, water and some other liquids are given in Table 3 (notice that kJ kg^{-1} units are used).

Table 3 Some enthalpy changes of vaporisation.

Substance	Formula	ΔH_{vap}/kJ kg^{-1}
water	H_2O	+2260
ethanol	C_2H_5OH	+840
propanone	C_3H_6O	+520
hexane	C_6H_{14}	+330
mercury	Hg	+300

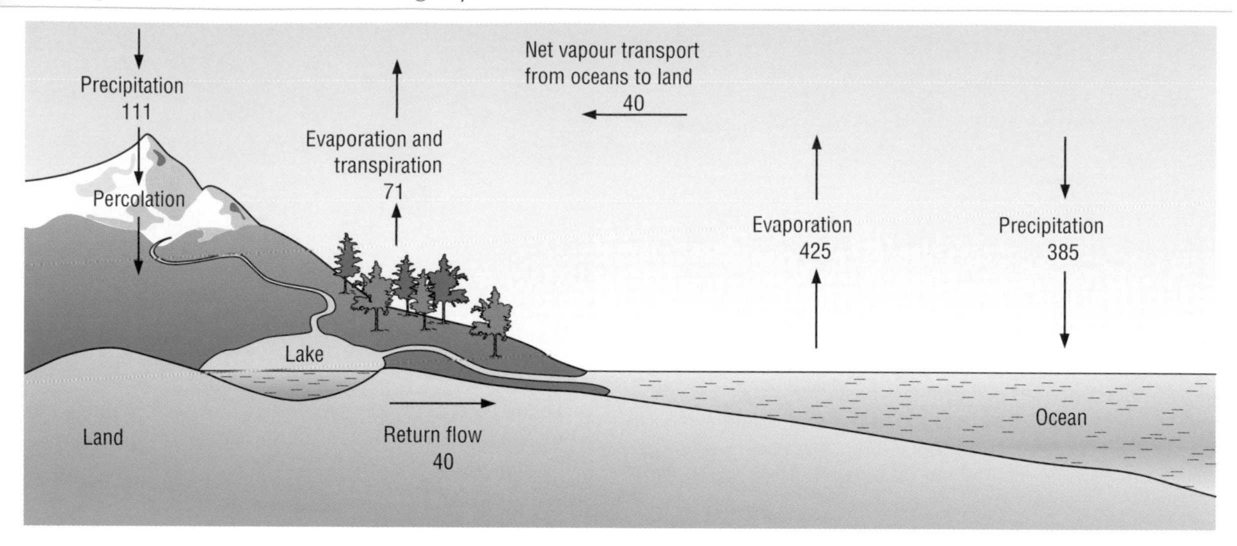

▲ **Figure 33** Global water cycle – the figures represent quantities moved in 10^{15} kg per year.

If our present rain were replaced by the same mass of 'propanone-rain' it would release only about one-quarter of the energy. For enough 'propanone-rain' to fall to keep our temperature the same, there would have to be about four times more rain. Then there would be other problems – an increased fire risk, for example!

Thus, the properties of water make it an ideal liquid for spreading the Sun's energy around the world. Without it, the pattern of evolution and human development would have been very different.

Assignment 13

a Use the data in Figure 33 and Table 3 to calculate the thermal energy released each year when water vapour carried from the oceans condenses over the land.

b The output of a typical power station is about 2000 MW, in other words about 6×10^{16} J per year. Approximately how many power stations would be needed to produce the same energy as that transferred from the oceans to the land by evaporation and condensation?

Warm water from the west

Western Europe is doubly fortunate with its climate (see Figure 31). Winters are kept mild not only by the precipitation of rain, but also by an 'ocean conveyor belt' bringing warm water from the tropics. The 'conveyor belt' is a surface current called the North Atlantic Drift. Surface currents are driven by prevailing winds and they alter course when they are deflected by large land masses (Figure 34).

We can find out how good a substance is at storing thermal energy by looking at its specific heat capacity (c_p). This is a measure of how much energy we have to put into 1 g of the substance to raise its temperature by 1 K. Looked at the other way, it tells us how much energy we can take out of a substance as it cools down by 1 K.

Water has a large specific heat capacity – it will release quite a lot of energy without cooling down too much. Its specific heat capacity is given in Table 4, along with values for some other substances.

Table 4 Some specific heat capacities.

Substance	Specific heat capacity/J g^{-1} K^{-1}
water	4.18
ethanol	2.44
hexane	2.27
propanone	2.18
granite	0.82
copper	0.39
mercury	0.14

Water is one of the best liquids for transporting energy – a Gulf Stream in a 'propanone sea' would provide us with only half as much warmth.

Activity O5.2 provides an interesting demonstration of the heat capacities of air and water.

Heavy water?

On its way across the North Atlantic, the Gulf Stream meets two currents of cold water – one flows down the eastern side of Greenland, the other flows past the Labrador coast of north-east Canada (see Figure 35). The currents are fed by melted ice and snow from the Greenland ice-sheet, so their salinity is low. The Gulf Stream has a relatively high salinity because it is a warm current – water evaporates from it leaving the salt behind.

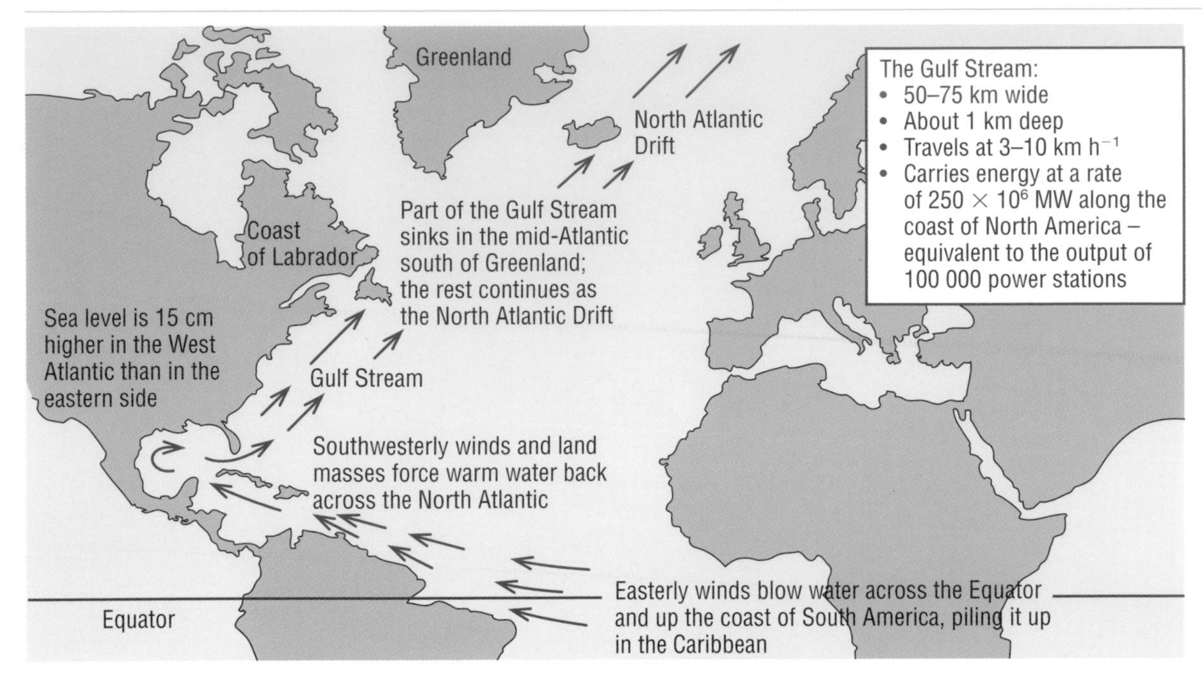

▲ **Figure 34** Ocean currents in the Atlantic.

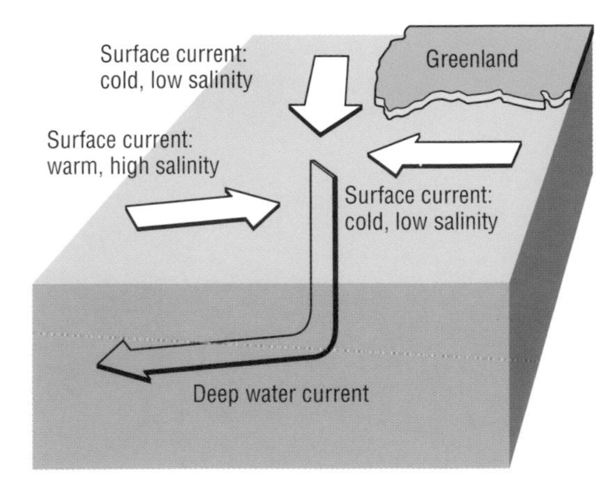

▲ **Figure 35** Deep-water cold current in the Atlantic Ocean.

The density of pure water varies with temperature in a complicated way, but so long as it doesn't go below 4 °C we can say that water is denser when it is colder. Salty water is also denser than pure water. When it meets the East Greenland and Labrador currents, the water in the Gulf Stream becomes much colder. The cold East Greenland and Labrador currents become saltier. The result of cooling the salty Gulf Stream and making the two cold currents saltier is a body of water that sinks because it is denser than the currents that feed it. The landmass of Greenland deflects the sinking water to the south, to produce a deep-water cold current that follows the same route as the Gulf Stream but in the opposite direction (see Figure 35).

Another deep-water current is generated in the Antarctic. The temperature is much colder there than in Greenland, so this current does not come from melted ice – instead it results from sea water freezing under the ice shelves. Salt is not taken up into the ice when this happens – it stays behind, so the residual water becomes saltier and sinks.

Activity O5.3 looks at the effect of freezing solutions of different concentrations.

You can consolidate your understanding of energy, entropy and equilibrium by working through **Activity O5.4**.

You might also like to consult **Chemical Ideas 4.4**.

The two deep-water currents meet in the south Atlantic, where the geography of the Earth and its rotation cause them to flow eastward. The water slowly rises back to the surface in the Indian Ocean and Pacific Ocean and returns in surface currents to replenish the Atlantic, as shown in Figure 36 (page 146).

The deep-ocean currents transport huge volumes of water – 20 times more than all the world's rivers combined – and they move slowly – water that sinks may take over 1000 years to resurface. Materials that are dissolved in the sinking water (e.g. carbon dioxide) are also removed for a long time.

Scientists are learning more about the ocean circulatory system and the oceans' role in controlling the Earth's heat-energy balance. But we need to know much more before these factors can be confidently

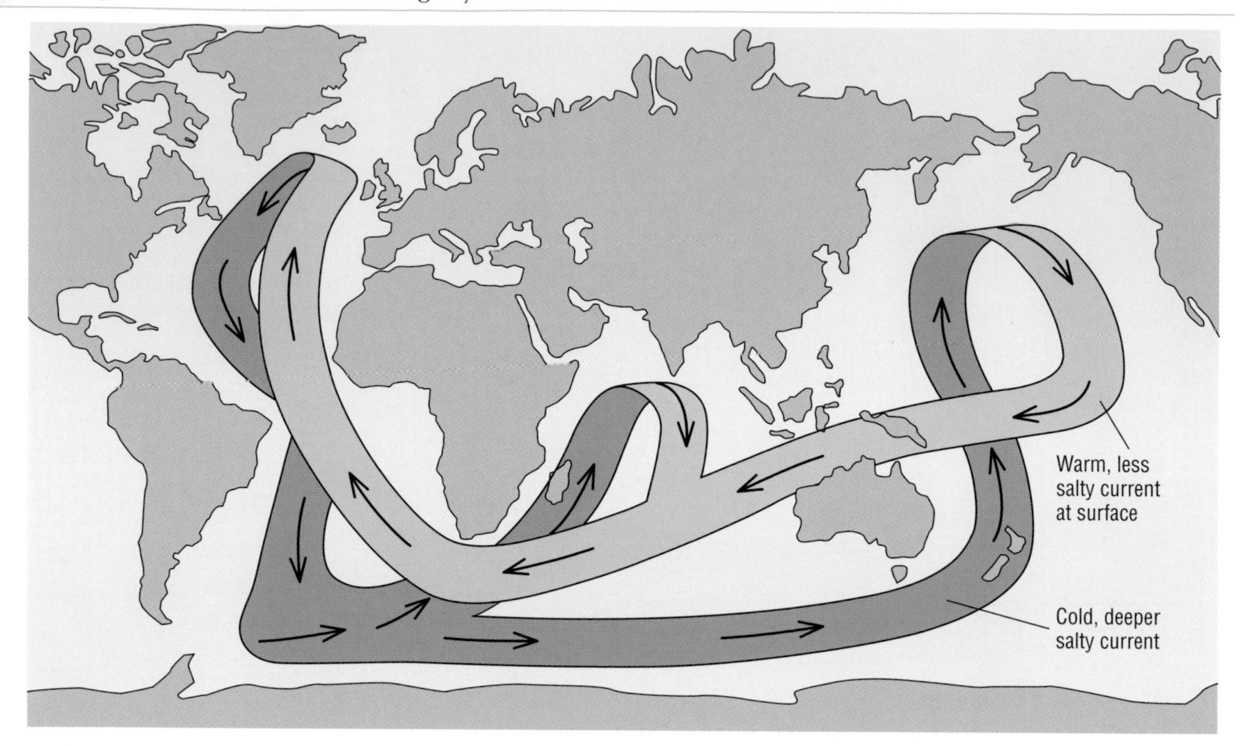

▲ **Figure 36** Deep-ocean currents take cold salty water from the Atlantic to the Indian Ocean and Pacific Ocean on a giant 'conveyor belt'.

included in global climate models – models that explain our present climate and predict how it may change in the future. We know that ocean currents are important, but we cannot yet predict with any certainty how they may change and, therefore, how our climate may be affected if the Earth warms up.

On and off

It seems possible that the deep-ocean current was shut down during the last ice age, which was at its coldest about 18 000 years ago. The flow of warm surface water also stopped, making lands around the north Atlantic cooler by an extra 6–8 °C. The currents were re-established at the end of the ice age as the Earth warmed up. But the warming caused the ice that covered much of North America to melt and flow out along the St Lawrence river. The water in the north Atlantic became much less salty and it could not sink. The deep-ocean current and, therefore, the warm Gulf Stream were switched off again. The ice took nearly 1000 years to melt and during that time northern Europe stayed cold while the rest of the world warmed up.

The ice has gone from North America, but what might happen if global warming caused much more Greenland ice to melt? (In the early stages of global warming this is thought to be more likely than the melting of Antarctic ice.) Low-salinity water might pour into the north Atlantic once more, this time into the

▲ **Figure 37** Melting ice.

East Greenland and Labrador currents. The ocean circulatory system might shut down again and northern Europe might become colder.

It is important to keep saying *might* because, until we have more knowledge, we cannot make accurate predictions. However, we are sure that the consequences of global warming are likely to be very uneven, with a probable increase in more extreme weather.

OCEANS AND CLIMATE CHANGE

NEW SCIENTIST
SCIENCE

30 NOVEMBER 1996

Will a sea change turn up the heat?

Fred Pearce

GLOBAL warming could be happening much faster than climate researchers had feared, a study warns this week. Rising temperatures could reduce the oceans' ability to absorb carbon dioxide by as much as 50 per cent, leaving the greenhouse gas in the atmosphere to heat the Earth further.

Until now, climate models such as those used by the Intergovernmental Panel on Climate Change (IPCC) have assumed that the oceans' capacity to remove CO_2 from the atmosphere will stay constant as the world warms. But Jorge Sarmiento and Corinne Le Quéré of Princeton University in New Jersey question this assumption. Their model, which predicts climate events over the next 350 years, suggests that conditions could be radically different.

"This really is a startling finding," says Sarmiento. "Warmer oceans will be more stratified, causing the

▲ **Figure 38** *New Scientist* article.

This headline in the *New Scientist* introduces another concern about global warming. As the atmospheric temperature increases, so does the temperature of the ocean surface. Less carbon dioxide will dissolve in the warmer water. One estimate is that, unless we reduce the rate at which CO_2 is building up in the atmosphere, up to 50% less gas will dissolve in the oceans, with the consequence that the enhanced greenhouse effect will lead to further global warming.

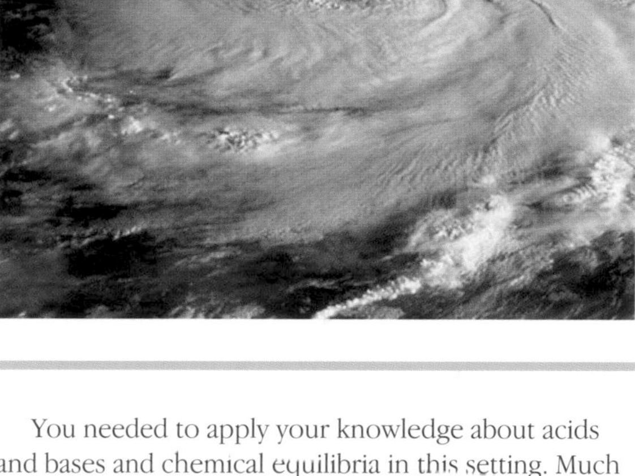

► **Figure 39** Stormy weather ahead?

O6 *Summary*

This module has given you a glimpse of just a few of the processes going on in the oceans – you also learned more about solutions, dissolving and enthalpy changes involving ionic compounds.

The oceans play a major role in absorbing and redistributing the energy the Earth receives from the Sun and keeping the planet hospitable to life. Water is an ideal substance to carry out this role because of its unique properties. This led you to revisit ideas on hydrogen bonding and its effect on the properties of water.

In this context, you extended your ideas about entropy and saw how the entropy change for a process determines whether or not it occurs spontaneously.

You have seen that carbon dioxide appears to exert a profound influence on the Earth's climate. Life has evolved in the oceans, and these materials have become inextricably linked to the life cycles of many creatures. Buffering in the oceans is largely controlled by chemical reactions involving carbon dioxide.

You needed to apply your knowledge about acids and bases and chemical equilibria in this setting. Much of this was an extension of earlier work covered in **The Atmosphere** and **What's in a Medicine?**, but you then went on to study weak acids, the pH scale and buffer solutions in more detail.

One of the big challenges scientists face is to understand how human activities are likely to affect the global environment. What is becoming clear is that global conditions are determined by the way in which the oceans, the atmosphere, the land and life interact with one another. This is an enormous task involving all branches of science – as this module has shown, chemistry has a major part to play.

Activity O6 will help you to summarise what you have learned in this module.

MD

MEDICINES BY DESIGN

Why a module on 'Medicines by Design'?

This module describes some examples from an area of chemistry that has had a major influence on the quality of our lives. As we have increased our understanding of the ways in which pharmacologically active compounds interact with the human body, so chemists have been able to design medicines that are more effective and have fewer undesirable side effects than earlier remedies.

The module begins with a look at ethanol, how it interacts with the human body and how infrared spectroscopy is used to measure alcohol concentration in the breath. It then looks at a group of medicines called 'statins' – these are in common use in Western society to reduce the risk of heart disease and strokes by reducing the level of cholesterol in the body. The mechanism for statin action in the body is then examined. The module concludes with a study of how penicillin inhibits a bacterial enzyme.

The chemical reactions that are used to synthesise and modify medicines are drawn almost exclusively from the field of organic chemistry, and this module serves as a good way of consolidating your knowledge of the organic reactions you have encountered in the course.

Synthetic chemists use spectroscopic techniques to determine the structure of the compounds they make. You will learn how proton nuclear magnetic resonance can help in this task. You will also use your knowledge of mass spectrometry, infrared and ultraviolet/visible spectroscopy to elucidate the structure of organic compounds.

Overview of chemical principles

In this module you will learn more about ideas introduced in earlier modules in this course:

- pharmaceutical chemistry (**What's in a Medicine?**)
- the importance of molecular shape and molecular recognition in biological activity (**The Thread of Life**)
- enzymes (**The Thread of Life**)
- the interpretation of spectroscopic data (**Polymer Revolution**, **What's in a Medicine?**, **The Thread of Life** and **Colour by Design**)
- interpretation of gas–liquid chromatography data (**Colour by Design**)
- mass spectrometry (**Elements of Life** and **What's in a Medicine?**)
- reactions of organic functional groups and the classification of organic reactions (several modules)
- isomerism (**Developing Fuels, Polymer Revolution** and **The Thread of Life**).

You will also learn new ideas about:

- the interaction of biologically active molecules with active sites
- the role of chemists in designing and making new medicines
- the synthesis of organic compounds
- nuclear magnetic resonance.

MD

MEDICINES BY DESIGN

MD1 *Alcohol can be a problem*

Alcoholic drinks make many people feel better for a short time – they help them to relax or to cope with stress. Alcohol can make them feel happier and relieve tension, anxiety or boredom. These effects are all outcomes of a single aspect of the behaviour of ethanol molecules in the body – they depress the activity of the central nervous system.

This depression of nervous activity has important short-term effects. Drinking alcohol reduces vigilance, slows reaction times and impairs judgement. It is largely because of these effects that so many laws have been introduced to control the use of alcohol – this is particularly important in relation to driving.

Measuring alcohol in the blood

Blood alcohol concentration (BAC) is closely related to the extent of the effects of alcoholic drinks:

$$\text{BAC} = \text{mg of ethanol per } 100\,\text{cm}^3 \text{ of blood}$$

The concentration of ethanol in the blood rises for some time after taking a drink, as the alcohol is absorbed. Then it slowly decreases as the ethanol is excreted or metabolised.

'Excuse me Sir, ...'

80 mg per $100\,\text{cm}^3$ of blood is the legal limit for BAC when driving a motor vehicle. A roadside blood test is impractical when a motorist suspected of drink-driving is stopped by the police. A quick BAC estimate is needed in order to decide whether or not to take things further.

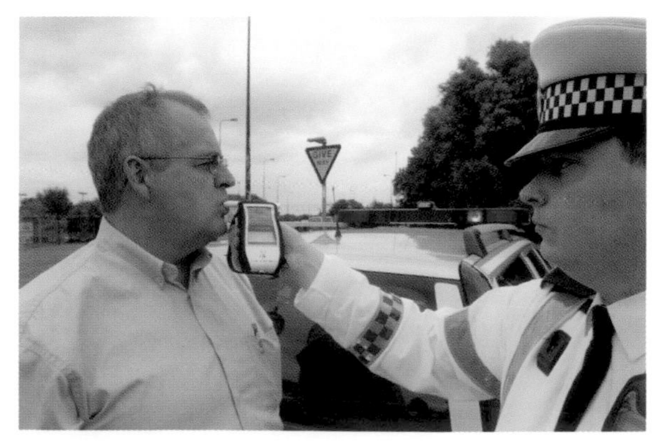

▲ **Figure 1** Testing for drink-driving.

Ethanol (the alcohol present in alcoholic drinks) is almost the only commonly used drug that is sufficiently volatile to pass from the blood to the air in the lungs. The distribution

$$C_2H_5OH\text{ (blood)} \rightleftharpoons C_2H_5OH(g)$$

is an example of chemical equilibrium, and it is governed by an equilibrium constant (K_c) that has a fixed value at a particular temperature.

In the human body, K_c for this process is 4.35×10^{-4} at body temperature. A measurement of the ethanol concentration in the breath (BrAC) therefore gives an indication of the BAC.

Assignment 1

a Write down the expression for K_c for the distribution of ethanol between the blood and the air in the lungs.

b The position of this equilibrium depends on temperature and pressure. These conditions will be the same in all drink-drive suspects. What values will they have?

c In the human body, K_c for this process has a value of 4.35×10^{-4} under these conditions. What is the BrAC (in mg of ethanol per $100\,\text{cm}^3$ of air) that corresponds to the 80 mg per $100\,\text{cm}^3$ legal limit for BAC?

One of the earliest successful methods of BAC detection – the 'breathalyser' – was based on a familiar chemical reaction – orange crystals of potassium dichromate(VI) turn green when they react with ethanol. The extent to which the crystals in a tube change colour is an indication of the BAC.

Assignment 2

a What causes the green colour in a breathalyser that uses potassium dichromate(VI) crystals?

b Using oxidation states for chromium, explain what happens to the dichromate(VI) ions in the breathalyser when indicating an over-the-limit result.

c Draw the structural formulae for two possible organic products formed from ethanol in the breathalyser reaction.

▲ **Figure 2** The Lion Alcolmeter contains a fuel cell – it has replaced the traditional breathalyser for roadside tests.

Oxidation of ethanol is a redox reaction and involves electron transfer. An alternative way of finding the ethanol concentration is therefore to measure the voltage of an electrochemical cell that incorporates the reaction. The Lion Alcolmeter is a cell designed to do just this.

An electrochemical cell in which a fuel (such as ethanol) is oxidised is sometimes called a *fuel cell* – the fuel is being oxidised to produce electrical energy rather than being burned to produce heat.

In the Lion Alcolmeter the cell consists of a porous glass plate in which phosphoric acid is absorbed. The plate is sandwiched between coatings of platinum or silver, which act as electrodes.

Oxygen is reduced to water at one electrode and ethanol is oxidised to ethanoic acid at the other. The instrument is calibrated by passing different concentrations of ethanol in air through the instrument and measuring the resulting cell voltage.

The instrument is set so that a green light comes on when there is no alcohol in the breath, an amber light when a small amount is present, amber and red when close to the limit, and red above the limit.

Assignment 3

Write half-equations for the reactions at each electrode of a Lion Alcolmeter – remember to include water molecules and/or hydrogen ions where necessary to balance the equations.

The science behind the Lion Alcolmeter is fairly straightforward, but the technology needed to develop it is more challenging. The instrument must be:

- small enough for hand-held use beside the road
- rugged
- safe and easy to use
- reusable with a short turn-round time
- reliable – the consequences of the test can be serious for the person being tested.

Then it has to be marketed and sold.

No need to go to the station …

If a roadside test indicates a high level of alcohol, the driver will usually be taken to a police station for a more accurate determination of the BAC.

At the police station, most suspects undergo a second type of breath test, where the ethanol concentration is measured using infrared spectroscopy (Figure 3). The driver is asked to breathe continuously into the cell of an i.r. spectrometer and the intensity of radiation absorbed at $2950\,cm^{-1}$ is measured. If the driver is incapable of doing the test or suffers from a breathing problem such as asthma, then blood or urine samples are taken by a doctor. Two samples are taken – one for the forensic service and one for the driver, who can use it to have the analysis checked independently.

▲ **Figure 3** The Lion Intoxilyzer analyses ethanol in a suspect's breath by absorption of infrared radiation.

However, a new type of Lion Intoxilyzer has been developed – a portable infrared analyser weighing less than 7 kg that runs on a 12 V supply. It can be carried in a police car or van and be used at the roadside to provide an immediate printout of the BAC. Subject to changes in legislation, it is envisaged that these printouts will be acceptable as evidence in court.

▲ **Figure 4** The portable Lion Intoxilyzer analyses the ethanol in a suspect's breath at the roadside.

You can revise your knowledge of infrared absorption spectroscopy in **Chemical Ideas 6.4**.

Assignment 4

The infrared absorption spectrum of ethanol is shown in Figure 5.

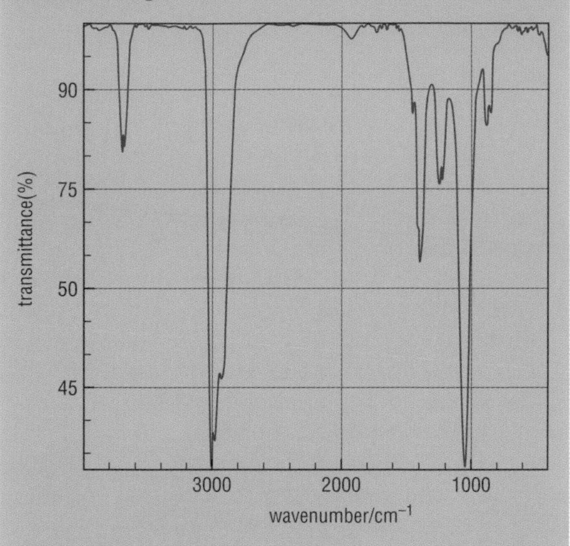

▲ **Figure 5** Infrared absorption spectrum of ethanol in the gas phase.

The ethanol concentration in the breath of a drink-driving suspect can be determined by measuring the intensity of one of the absorption bands in the ethanol spectrum. The absorption at 2950 cm^{-1} is used in the Lion Intoxilyzer 6000 UK machine shown in Figure 3 and the Lion Intoxilyzer 8000 shown in Figure 4. This gives rise to strong infrared absorptions centred at about 3800 cm^{-1}, 3600 cm^{-1}, 3200 cm^{-1} and 1600 cm^{-1}.

a Compare the positions of the water absorptions in an infrared spectrum with those of ethanol shown in Figure 5. Suggest why the 2950 cm^{-1} band is chosen for ethanol detection.

b Explain why ethanol and water both give rise to infrared absorptions in the same region of the spectrum.

c Look at the table of characteristic infrared absorptions in the **Data Sheets**. What bond is responsible for the 2950 cm^{-1} absorption in ethanol?

d People who suffer from diabetes or those who are on certain types of diet commonly produce propanone vapour in their breath. Suggest why infrared breath testing of such a person might appear to give a positive result, even though the person had not drunk any alcohol.

MD2 *Statins for all*

Heart disease is Britain's biggest killer, accounting for over 30% of all deaths. Treatment costs in the EU are estimated to be in the region of 192 billion euros in 2008. Since the 1950s, scientists have provided overwhelming evidence that an increased blood cholesterol level is linked to an increased probability of getting heart disease – the higher your blood cholesterol level, the higher your chances of having a heart attack.

CHOLESTEROL

Cholesterol is a lipid, a type of fat, found in the cell membranes of all tissues. It is needed by the body for several important purposes:

- as a building block for cell membranes
- to insulate nerve fibres so that chemical messages can travel
- for synthesising hormones.

It is a white, waxy solid, insoluble in water, with a melting range of 148–150 °C.

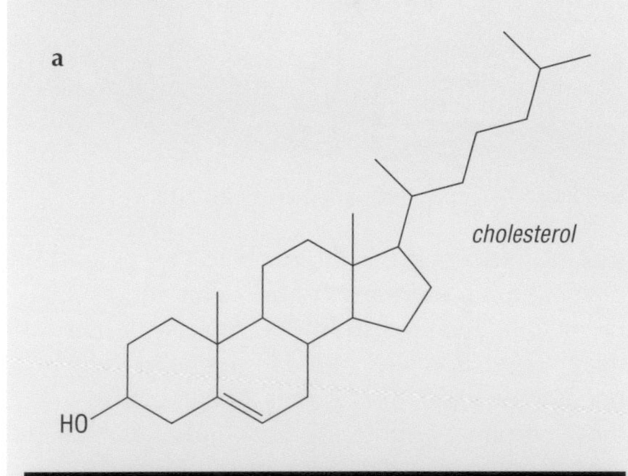

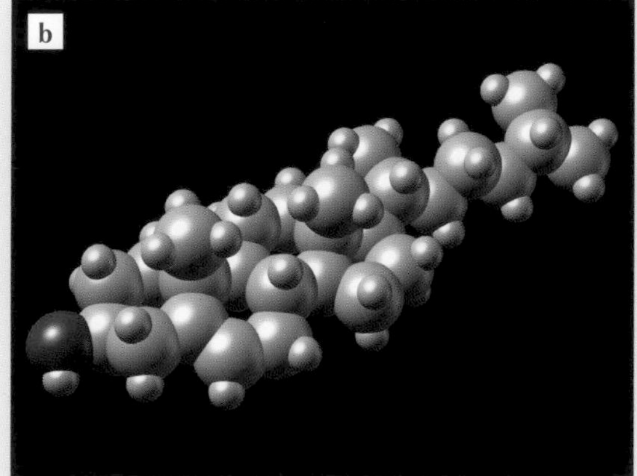

▲ **Figure 6** **a** Structure of cholesterol; **b** space-filling model of the cholesterol molecule.

Why cholesterol is a problem

There are two types of cholesterol found in the body – 'bad' or low-density lipoprotein cholesterol (LDL), and 'good' or high-density lipoprotein cholesterol (HDL).

LDL carries cholesterol around the body in the bloodstream and deposits it in arteries, where it can cause narrowing and eventually blockage (see Figure 7).

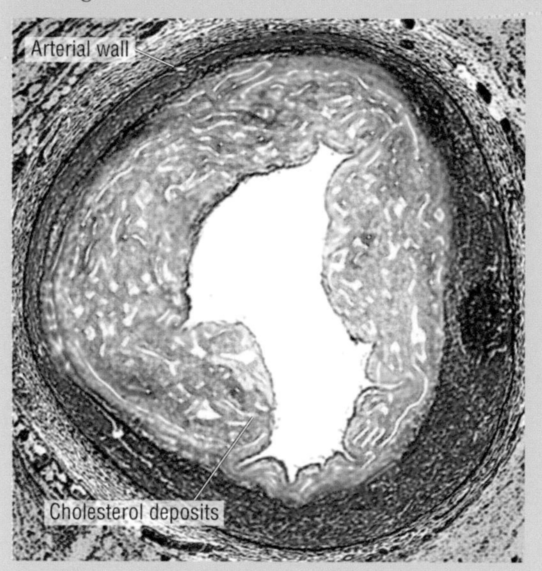

▲ **Figure 7** Cholesterol deposits in an artery.

It is this process that contributes to the development of diseases of the heart and its connecting blood vessels. For example, when the arteries supplying blood to the heart itself become blocked, the result is a heart attack. If these coronary arteries are only partially blocked then insufficient oxygen may reach the heart, resulting in chest pain.

HDL is described as 'good' because it collects cholesterol and transports it to the liver to be broken down. For this reason, it is not just your total cholesterol level that is significant (LDL + HDL) but the ratio of HDL to the total – a higher ratio is more beneficial.

Your cholesterol levels are affected by the amount of dietary cholesterol eaten. Two common sources of cholesterol are full-fat dairy products and animal fats. However, most of the cholesterol in your blood system is made by the body itself.

Assignment 5

In order to answer these questions you might need to revise **Chemical Ideas 3.5, 5.3** and **5.4.**

a Name the functional groups present in cholesterol.

b Identify all the chiral carbon atoms present in cholesterol.

c Explain in terms of intermolecular bonds why cholesterol has a relatively low melting point and is insoluble in water.

It is little wonder that scientists have tried to make cholesterol-reducing drugs – economic and health reasons go hand-in-hand for pharmaceutical companies. The discovery of the first drug of its type for reducing cholesterol levels in the blood, lovastatin, heralded a new era in the treatment of heart disease. Today there are a number of different drugs based on lovastatin – these are referred to collectively as 'statins'.

In the summer of 2007, a UK government adviser suggested that statins should be offered to all men over 50 and women over 60 as an effective 'shortcut' to prevent heart disease. Later on that year *The Times* newspaper published an article with the headline 'The pill of life: statin benefits last a decade' – the opening paragraphs of the article are shown in the 'The pill of life' purple box on page 153.

Before you look at the discovery and development of statins in more detail, you need to have a more general appreciation of how scientists discover new medicines in order to see where statins fit into the overall picture.

Discovering a new medicine

When scientists set out to design a new medicine, they focus their attention on two areas:

- developing a biological understanding of the disease or condition they want to cure
- finding a 'lead compound' – not a compound of the element lead, but one that provides a *lead*, showing some promise and giving some clues.

Sometimes, an idea for a lead compound comes from research into the chemical processes that go on in the body. Medicines have also been developed from lead compounds that are the active ingredients in traditional remedies. For example, in **What's in a Medicine?** you learned how aspirin was developed from a natural remedy made from willow bark and how drug trialling takes place. Many ethical issues need to be addressed when trialling drugs. First, a detailed protocol is developed before trialling can start – this sets out the purpose of the research and explains how the trial will be conducted and the results analysed. All human participants in clinical trials confirm their willingness to

THE PILL OF LIFE

The pill of life: statin benefits last a decade

***Nigel Hawkes*, Health Editor**

People who take cholesterol-lowering drugs are protected from heart disease and premature death years after they stop taking them, a major study has shown.

New research into statins – the world's biggest-selling medication – offers dramatic evidence of their long-lasting ability to halt and even reverse the progression of heart disease.

The study, involving 6,500 men, found that those who took statins were still showing benefits of the drugs ten years after they had finished taking them. The chances of suffering a fatal heart attack over the period dropped by more than 25 per cent, the scientists found, while there was no evidence of unexpected side-effects.

This remarkable result will increase pressure on GPs to prescribe statins to an even greater number of middle-aged people with raised cholesterol levels ...

Statins are currently taken daily by an estimated three million Britons to tackle high chloresterol ...

Statin prescriptions have risen by 150 per cent in England in the past five years. The trial raises the question of whether they should be given to an even wider group, including younger people in whom heart disease has yet to get a start.

The Times, 11th October 2007

participate in drug trials after being informed of aspects of the trial that are relevant to their decision to participate. In order to reduce the use of animals in drug testing, chemists and biochemists may work on developing alternative techniques, such as using isolated tissues and cell cultures, rather than using animals.

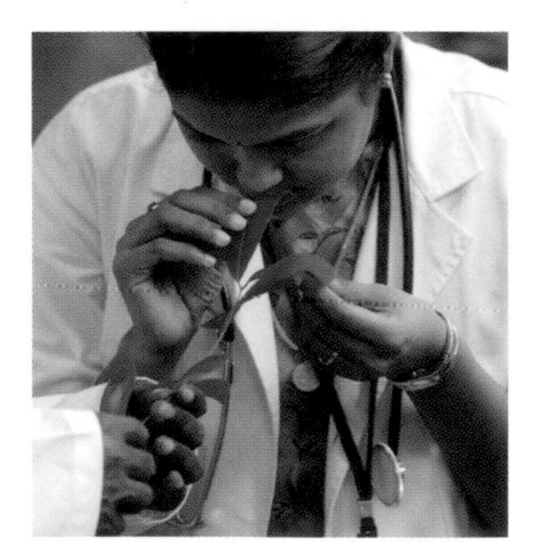

▲ **Figure 8** Traditional medicines from plants can provide lead compounds for new medicines.

Some naturally occurring lead compounds have been discovered by accident. A classic example is the discovery of penicillin – you can learn more about this in section **MD4**.

Some medicines have been found from random screening – looking for activity in a very wide range of compounds, some of which might already be used to treat other things medically.

However, the early years of the twenty-first century have seen pharmaceutical companies increasingly using the new technology of *chemical informatics* in the process of drug discovery. This involves storing vast amounts of chemical information about compounds in computer databases – the information is then searched and analysed to help to identify suitable lead compounds for effective drugs.

Despite the increasing success of medicine design based on scientific understanding of the chemical processes that go on in the body, only one in about 10 000 of the compounds that are synthesised survive today's rigorous testing procedures and become commercially available for medical use.

The discovery of lovastatin

The first discovery of a statin was by the Japanese biochemist Dr Akira Endo, who was working on fungi and cholesterol. From his initial studies in the 1970s he hypothesised that fungi used chemicals to ward off parasites by preventing the synthesis of cholesterol.

He studied over 6000 compounds found in the cell walls of fungi. Only three of these compounds showed an effect, one of which, *mevastatin*, was the first member of the statin group of drugs. Clinical trials (see section **WM8**) showed that mevastatin had a very good inhibitory effect on cholesterol production. However, extensive and long-term toxicity tests showed that the higher doses were far too toxic for it to be given to humans.

In 1978, scientists at Merck Research Laboratories, using Dr Endo's research, isolated a new natural product from the fermentation of the fungus *Aspergillus terreus*. Their new chemical had a powerful effect in inhibiting the synthesis of cholesterol in fungi, but was much less toxic than mevastatin – it became known as *lovastatin*. In August 1987 this was first statin to be marketed as a prescription drug in the USA.

Dr Akiro Endo was honoured academically for his research and for the discovery of mevastatin. The financial investment in research and development, together with the eventual profit from the drug, was the domain of the large pharmaceutical companies who then made the product available commercially.

Both mevastatin and lovastatin are natural products and can be obtained by fermentation of certain fungi – they have very similar structures. In order to simplify things, the stereochemistry is not indicated at this stage.

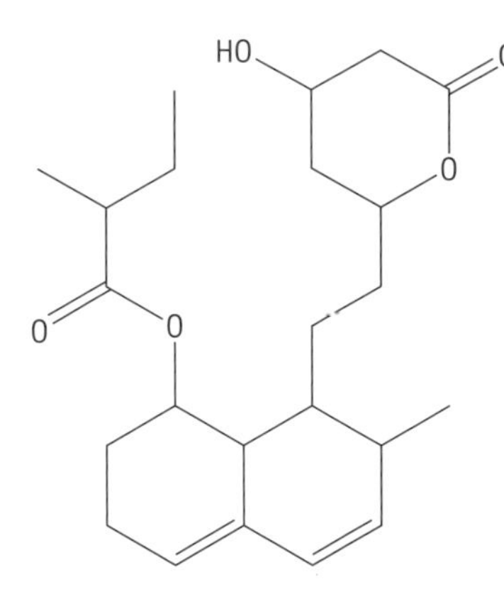

mevastatin

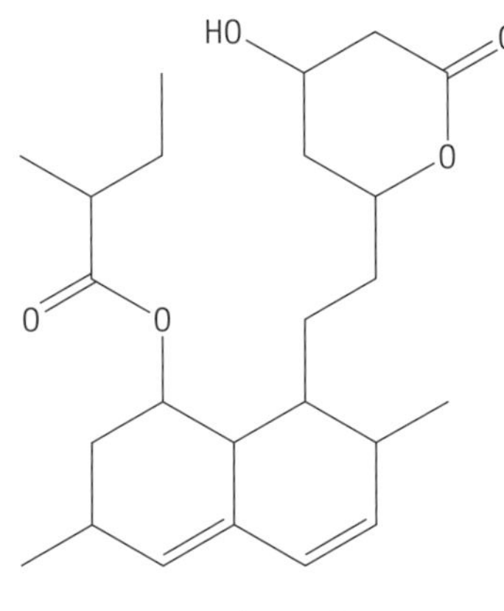

lovastatin

Lovastatin is found in the wild oyster mushroom, *Pleurotus ostreatus*, prized for its edibility and now cultivated worldwide (see Figure 9).

Chemists elucidated the structures of mevastatin and lovastatin using a variety of instrumental techniques, including nuclear magnetic spectroscopy.

▲ **Figure 9** Fruiting body of the oyster mushroom.

In **Activity MD2.1** you can perform some simple test-tube reactions to identify various functional groups.

Professional chemists use a variety of spectroscopic tests to identify not only the functional groups present in a molecule, but its actual full 3-dimensional structure. In **Chemical Ideas 6.6** you can learn about nuclear magnetic resonance spectroscopy and how it can give information about the different environments in which hydrogen atoms are to be found within chemical compounds, so helping to elucidate molecular structure.

You can revise earlier work on the use of other instrumental techniques in determining the structure of organic compounds by reading **Chemical Ideas 6.4** and **6.5**.

Activity MD2.2 will help you appreciate how the various spectroscopic techniques you have met can be used together to determine a compound's structure.

Assignment 6

a Name the functional groups present in both mevastatin and lovastatin.

b How would the infrared spectra of these statins show the presence of the functional groups that you named in part **a**?

c Describe how these two statins are different.

d Describe one way that nuclear magnetic resonance spectroscopy would distinguish between the two statins.

Mevastatin and lovastatin can be named systematically – though they may look complex, a series of well-defined rules makes it relatively easy to give them their correct unique names.

Activity MD2.3 will help you to revise the naming of compounds containing the functional groups met during your chemistry studies.

Research indicates that statins have anti-inflammatory properties and have shown some promise in treating multiple sclerosis, an inflammatory autoimmune disease. Ongoing research may well provide confirmation of the significance of these indications.

MD3 Designer statins

In order to design more effective statins, scientists had to first work out how lovastatin prevents cholesterol being synthesised. The synthesis of cholesterol in the body is a multi step process and the rate-determining step (see **Chemical Ideas 10.4**) involves an enzyme called HMG-CoA reductase (HMGR). This enzyme converts the substrate HMG-CoA into the carboxylate ion of mevalonic acid (see Figure 10). The reaction takes place in the liver. The HMGA-CoA substrate enters the enzyme's active site, where it forms intermolectlar bonds with functional groups in the active site. The reaction proceeds via the mevaldyl CoA transition-state intermediate shown to the right. Once the reaction is complete the product molecule diffuses away from the active site, and the reaction can then occur again with a new substrate molecule.

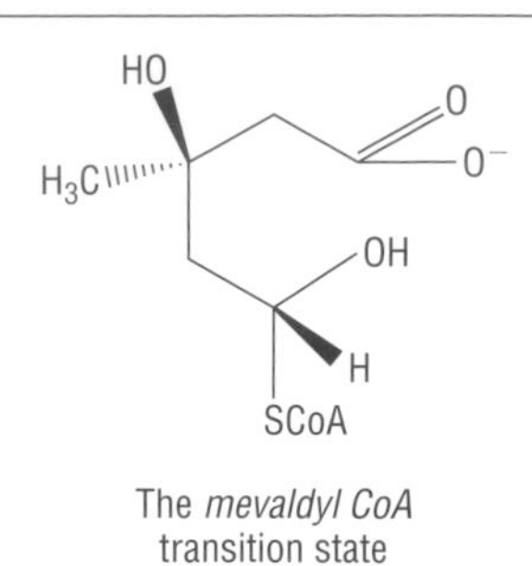

The *mevaldyl CoA* transition state

Lovastatin and mevastatin prevent this reaction from happening. How do they do this? If they react in a similar way to the substrate HMGA-CoA, they should have similar structural features it. The fragment of the molecule that binds to the active site of an enzyme is called a *pharmacophore*. It is this set of structural features in a molecule that interact at a receptor site

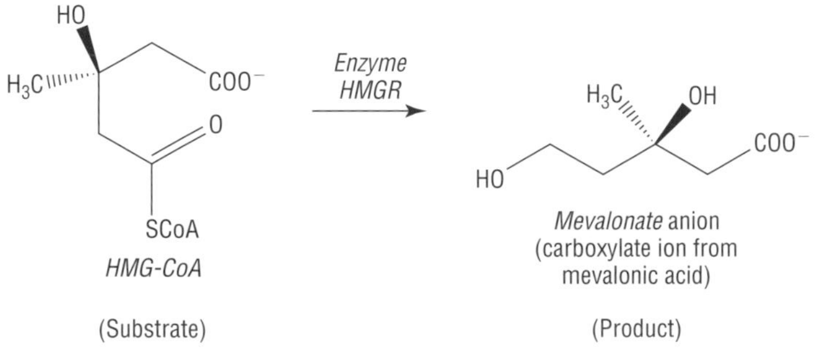

▲ **Figure 10** Rate-determining step in the formation of cholesterol.

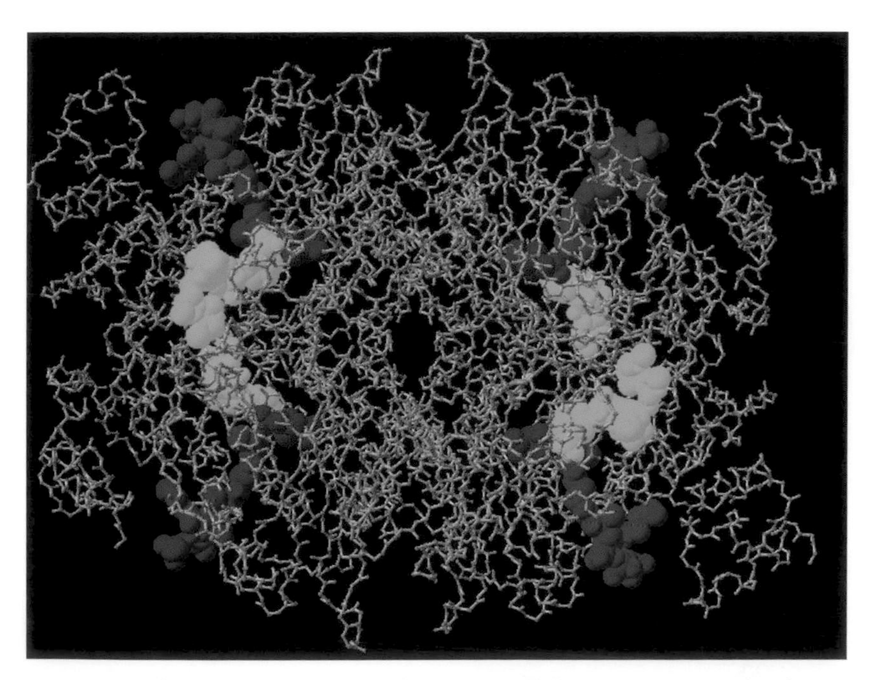

▲ **Figure 11** Using computer software programs to model enzymes and their interactions with substrate molecules is vital in designing modern medicines. This shows a computer-generated graphic of HMGR forming complexes with HMG-CoA substrate molecules – the red coloured parts show the HMG pharmacophore.

that is responsible for that molecule's biological activity. The statin pharmacophore should be structurally similar to the structure of HMGA-CoA. Looking at the structures of mevastain and lovastatin, this does not seem to be the case. However, in the body they are hydrolysed to form a carboxylate anion and the statin pharmacophore can now be identified. Figure 12 shows the part of the statin molecule that is hydrolysed.

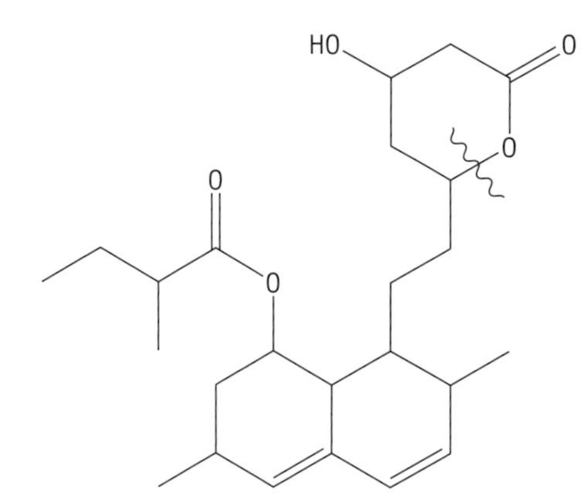

▲ **Figure 12** The bond within the statin structure that is hydrolysed is indicated by a wavy line.

Assignment 7

a Redraw Figure 12 showing the structure of the anion produced after hydrolysis.

b The statin pharmacacophore has been identified as

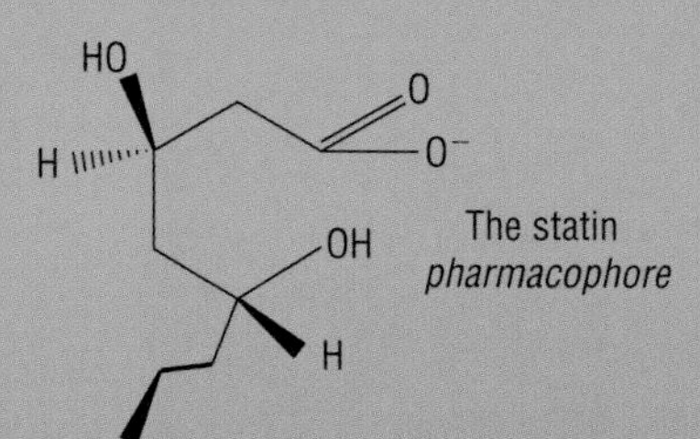

Compare the structure of the statin pharmacophore with that of HMGA-CoA for similarities.

c The active site of the HMGA enzyme includes $-NH_3^+$ and $-COO^-$ groups . Draw diagrams to show:

i how these ionic groups on the active site interact with HMGA-CoA

ii how these ionic groups interact with the statin pharmacophore.

In each case, clearly label and identify the type of intermolecular bond involved.

The statin molecules bind strongly to the enzyme's active site – it's a case of *molecular recognition*. The statin pharmacophore fits precisely into the active site of the enzyme, and functional groups on both are correctly positioned to interact. This strong bonding of the statin molecule to the enzyme's active site prevents HMG-CoA molecules from doing so – this is called *competitive inhibition*.

Molecular recognition was introduced in section **TL4** in terms of its importance with regard to enzyme activity. The size, shape and orientation of both the statin molecule and the active site of the enzyme, together with their ability to form bonds, are all factors that need to be taken into account for successful molecular recognition. This inhibition of the HMG-CoA enzyme reduces the rate of cholesterol synthesis in the liver, which leads to a decrease in blood cholesterol levels.

Molecular shape and bond angles are crucial factors in molecular recognition – you can revise this work in **Chemical Ideas 3.2**.

Chemical Ideas 12.1 allows you to revise how to represent molecules in 3-D.

Armed with this knowledge, chemists decided that the ideal statin should have the following properties:

- a high affinity for the enzyme's active site
- selective absorption into liver cells – synthesis of cholesterol in non-liver cells is necessary for normal cell functions
- a relatively long period of effectiveness.

Initially, lovastatin extracted from fermentation broth was modified using simple chemical reactions to create new statins. These statins, such as pravastatin and simvastatin, were semi-synthetic and were called type 1 statins:

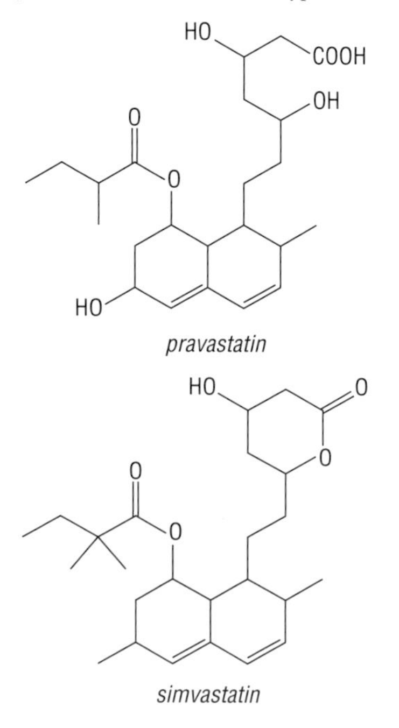

Later, chemists discovered how to make fully synthetic statins using a series of chemical reactions – the synthesis of lovastatin requires about eight steps.

The new statins are fully synthetic and have larger groups linked to the statin pharmacophore. These statins are called type 2 statins. In Figure 13 two examples of type 2 statin are compared with lovastatin, a type 1 statin.

One of the main differences between the two types is the replacement of the butyryl group of type 1 statins with the fluorophenyl group of type 2 statins. This group is responsible for additional polar interactions that result in stronger bonding to the HMGR enzyme.

When synthesising statins it is important that chemists ensure that they make the correct isomer, otherwise the compound would be inactive.

Activity MD3.1 will help you revise the different types of structural and stereoisomerism you have met during your chemistry studies. This activity revises **Chemical Ideas 3.3**, **3.4** and **3.5**.

You can remind yourself how to represent three-dimensional structures by reading **Chemical Ideas 12.1**.

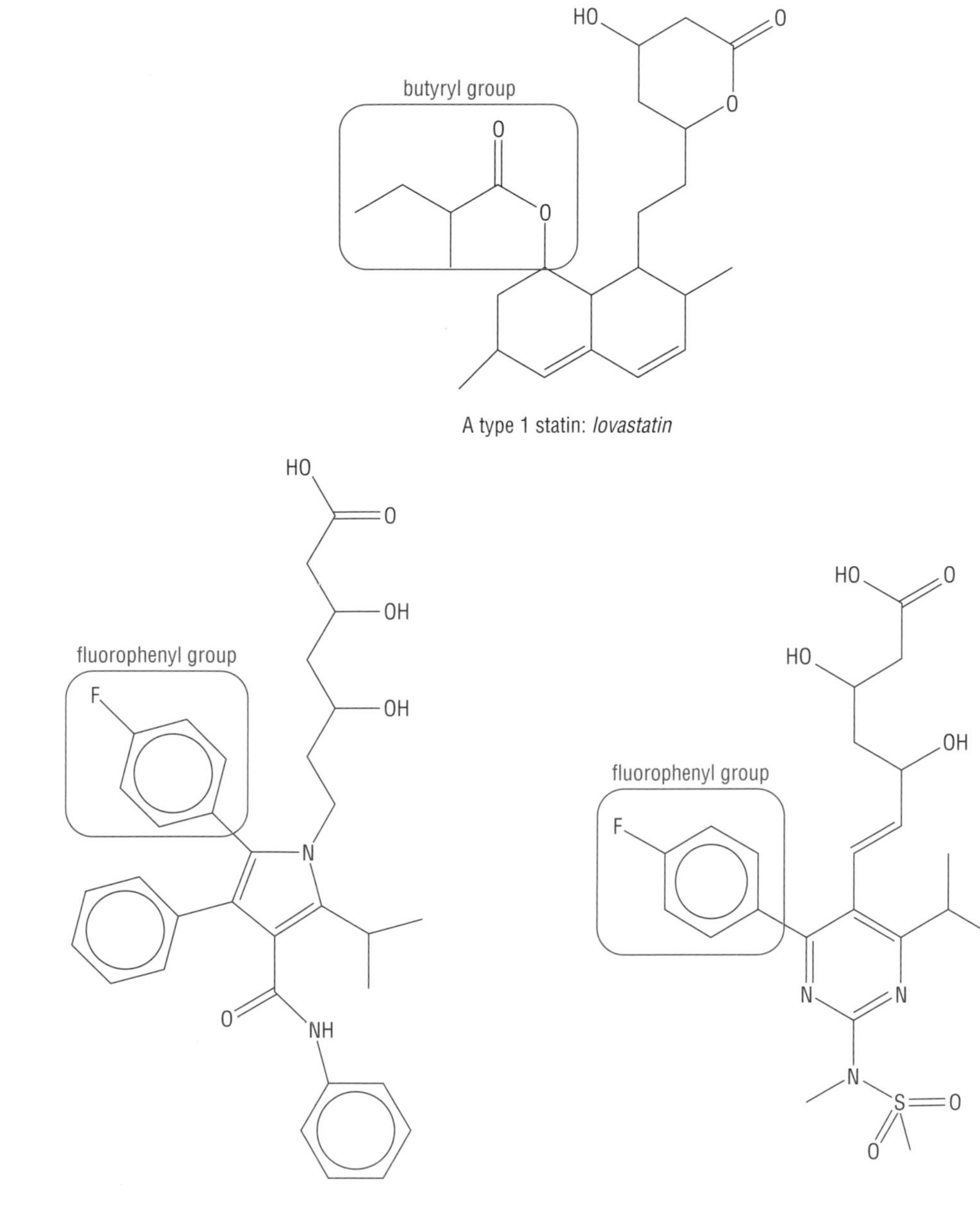

▲ **Figure 13** The general structures of type 1 and type 2 statins.

Assignment 8

Figure 14 shows how rosuvastatin binds to the enzyme HMGR.

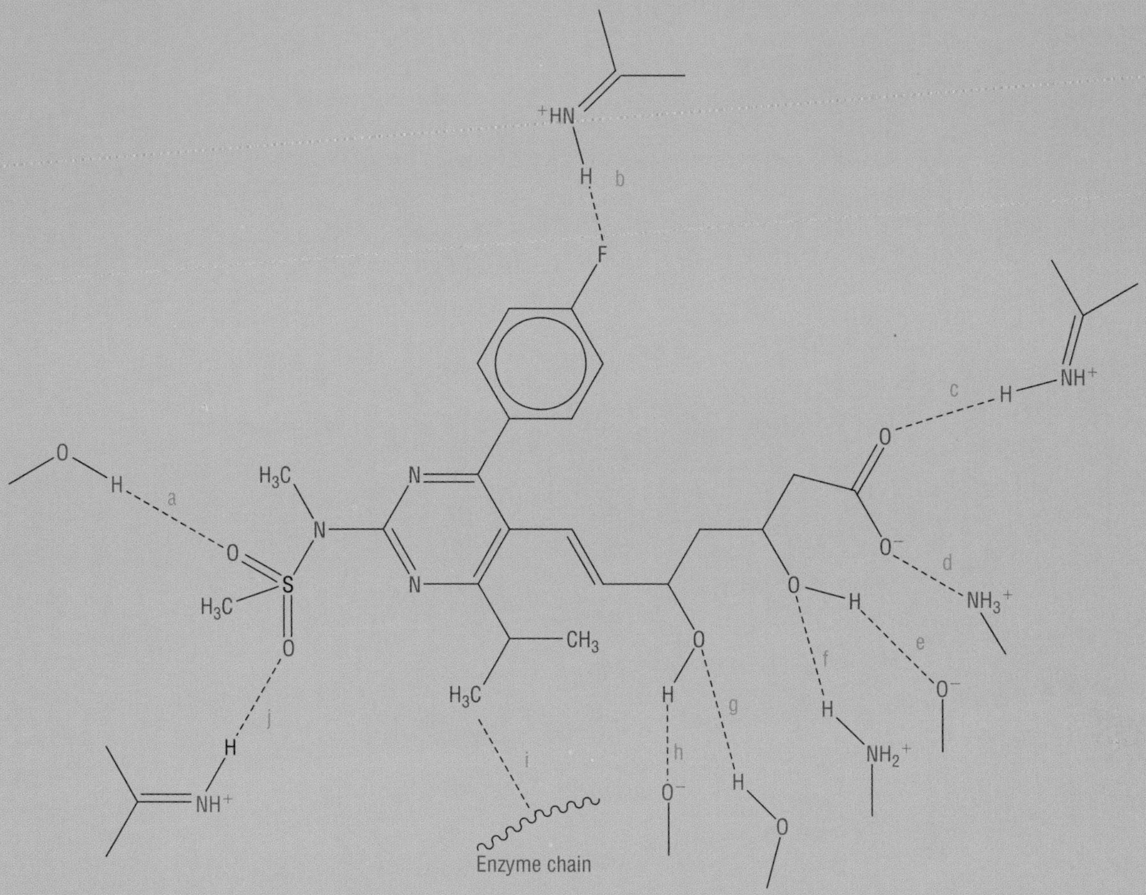

▲ **Figure 14** Representation of how rosuvastatin binds to the enzyme HMGR. The red coloured atoms are part of the rosuvastatin structure, whereas the blue atoms are part of the active site of the enzyme.

a Identify the different types of intermolecular bonds labelled *a–j* and show as dashed lines.

b The active site is stereoselective and only one of the optical isomers of rosuvastatin will bind to the enzyme. Identify the chiral carbon atoms present in rosuvastatin.

c Explain why the arrangement of atoms/groups of atoms around these carbon atoms is important for binding the statin to the enzyme.

d Rosuvastatin also shows *E/Z* isomerism.

i What structural feature is responsible for this?

ii Explain why only one of the possible *E/Z* isomers is active.

Many of the possible active statins designed do not occur naturally – chemists had to make them synthetically. Making a large number of compounds, each involving a number of individual reaction steps, is no mean feat. Today's chemists are able to use a technique called *combinatorial chemistry*, which has reduced enormously the time needed to produce a large number of new compounds.

In **Activity MD3.2** you can use your knowledge of organic chemistry to build up a 'tool-kit' of organic reactions.

Activity MD3.3 will help you to classify these reactions by the type of mechanism involved.

Chemical Ideas 14.2 will help you with these two activities.

You can find out about some of the factors involved in planning a synthesis by reading **Chemical Ideas 14.1**.

COMBINATORIAL CHEMISTRY

This is the rapid synthesis of a large number of similar but structurally different molecules. Using such techniques, a pharmaceutical company can easily produce over 100 000 distinct compounds per year. First, researchers create a 'virtual library' of perhaps thousands or even millions of 'virtual' compounds. These are constructed by combining a chosen *pharmacophore* from a *lead* compound with all the reactants that are thought appropriate. A small fraction of these is then selected for synthesis, based on the following:

- Knowledge about how the lead compound is absorbed by the body and distributed within the body, how it is broken down and how it is then finally eliminated from the body.
- Calculations using computational chemistry using theoretical calculations, and sometimes experimental results, to calculate the structure, shape and properties of a molecule – this is often called *molecular modelling.*
- Quantitative knowledge about how chemical structure affects biological activity – this is an area of *chemical informatics.*

The selected syntheses are then carried out using robots – machines that can rapidly perform all the required chemical processes, previously carried out by chemists in a laboratory, in the correct sequence (see Figure 15). Robots, of course, do not work a set number of hours per week, but can work every hour of every day.

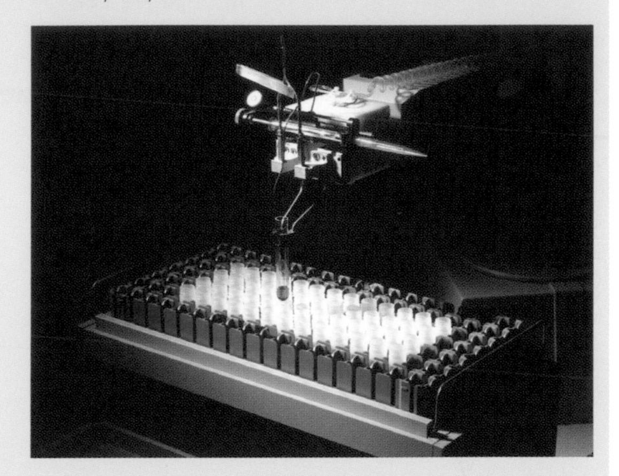

▲ **Figure 15** Chemical robot that can perform multiple syntheses.

PETER'S STORY

As you will realise by now, devising synthetic routes for the preparation of organic compounds is fundamental to the successful development and production of many medicines. The interview with Peter O'Brien, right and overleaf, provides an insight into the work of an organic synthesis research group. Peter is a professor in the Department of Chemistry at the University of York and leader of a research group.

▲ **Figure 16** Peter (right) and a member of his research group in the Department of Chemistry at the University of York.

Tell me about your research group.

'At any one time, my group usually consists of six researchers, made up of post-doctoral researchers (post-docs) and PhD students. The post-docs have successfully completed their PhD and are experienced researchers. The PhD research students are expected to carry out original research and report on it in a thesis after three/four years. Each thesis is examined and the student interviewed (in an oral examination called a *viva*) by a distinguished chemist from another university along with another member of the University of York Chemistry Department.'

And what is your background?

'I did A levels in chemistry, maths and physics. Then I did my first degree at Cambridge. During two of the summer vacations I worked in two different pharmaceutical companies. I saw at first hand the importance of organic chemistry to industry and to the world at large. In the final year of my undergraduate course I did a research project which involved making new organic molecules. This excited and fascinated me, so I stayed on at Cambridge and did a PhD in synthetic organic chemistry. On completing my PhD, I moved to York as a post-doc and, after three months, was appointed as a lecturer in 1996. I was promoted to professor in 2007.'

How do you decide what research problems to tackle?

'It is curiosity-driven. From my knowledge of the literature on organic synthesis and certain branches of medicine and the pharmaceutical industry, I identify an aspect of organic synthesis that has not been done before and that could be of interest to the pharmaceutical industry.

'It is actually quite easy to produce new compounds – it is much more challenging and satisfying to develop a more elegant and simple laboratory method to prepare a compound that is already known to be potentially useful. I am particularly interested in problems associated with the synthesis of chiral molecules.'

How are the problems broken down for different members of the group?

'Each of the research students must have their own clearly defined project, because the award of the PhD degree has to be based on their own work. But we try to have people working on related areas and, in particular, on topics related to the work of a more experienced researcher so that they can support each other. I supervise all projects, but as people gain experience I become more of a consultant.'

What does the actual research involve?

'Organic synthesis involves three main stages. The first is the reaction stage. We carry out reactions at different temperatures, from 120 to −78 °C. Often reactions are carried out under nitrogen, which provides an inert atmosphere – this requires special techniques. Sometimes reactions are done in simple round-bottomed flasks and sometimes in more complicated equipment.

'During the second stage the product is purified. This can take a long time. Recrystallisation, distillation and column chromatography are the main techniques used. Proton n.m.r. spectroscopy is the principal method we use to judge the purity of a product. If the spectrum shows peaks that we would not expect from the compound, then the compound contains impurities and needs to be purified further.

'The third stage is the analysis of the product. This is where we collect evidence that the compound is what we say it is. For confirmation of the identity of a product we need n.m.r, i.r. and mass spectra. We also need a combustion analysis that provides the percentages of carbon, hydrogen and nitrogen in the compound and so confirms its empirical formula.'

What happens next?

'The most satisfying part of the whole process is the publication of our research findings in specialist journals. These publicise our contribution to new knowledge and the research students get to see their names in print.'

So how does all of this relate to the work of the pharmaceutical industry?

'In my research group, we develop new methods for the synthesis of chiral organic molecules, especially those containing nitrogen. Chiral molecules contain a carbon atom bonded to four different groups and therefore they exist in two isomeric forms, called enantiomers. A common problem with chiral compounds is that it is difficult to synthesise one enantiomer rather than a mixture of the two. We try to develop ways of making just one enantiomer.

'Many of the new medicines developed by pharmaceutical companies are chiral nitrogen-containing compounds. Thus, any new methods that can be used to make such compounds are of interest to the pharmaceutical industry. We work in collaboration with them to develop our new methods. For example, they provide us with a grant to support the research, which I use to buy equipment and chemicals, and to give a scholarship to one of the research students. Our ultimate aim would be that one of our new methods is used to prepare a medicine of the future.'

MD4 *Targeting bacteria*

Some medicines inhibit the action of enzymes present in bacteria. As a result, the bacterial cells do not grow and divide, and infection is prevented. Of these medicines, penicillin is one of the most widely prescribed.

Penicillin first became available during the Second World War. Its use saved the lives of many injured people who would otherwise have died from bacterial infections – it was hailed as a 'miracle cure'.

The discovery of penicillin occurred as a result of a chance observation. It proved to be an extremely valuable lead compound and set the scene for the discovery and development of a vast range of *antibiotics.* These are compounds that selectively destroy disease-causing bacteria.

The production of penicillins on a large scale is an early example of biotechnology.

TYPES OF BACTERIA

Bacteria are classified according to the chemical composition of their cell walls. The cell wall confers a particular shape to each type (Figure 17). Bacteria multiply by cell division, which occurs approximately every 20 minutes.

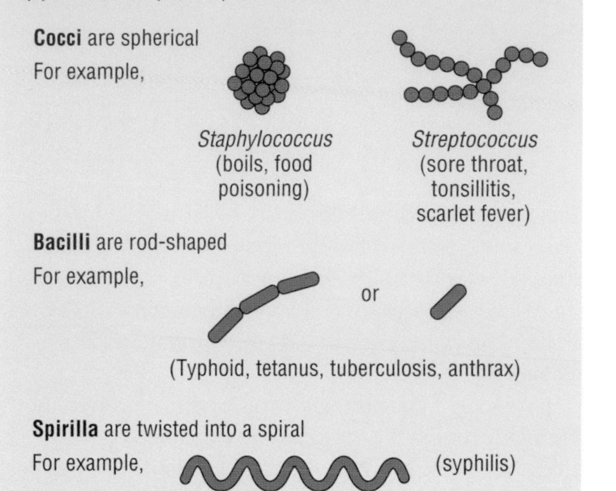

▲ **Figure 17** Types of bacteria and their effects on the body.

Most bacteria are harmless – some are essential to us, such as those in the intestine that aid digestion. A minority, called *pathogens*, can cause infections of, for example, the lungs (pneumonia), brain (meningitis), heart (endocarditis) or bloodstream (septicaemia).

▲ **Figure 18** Penicillin mould on bread.

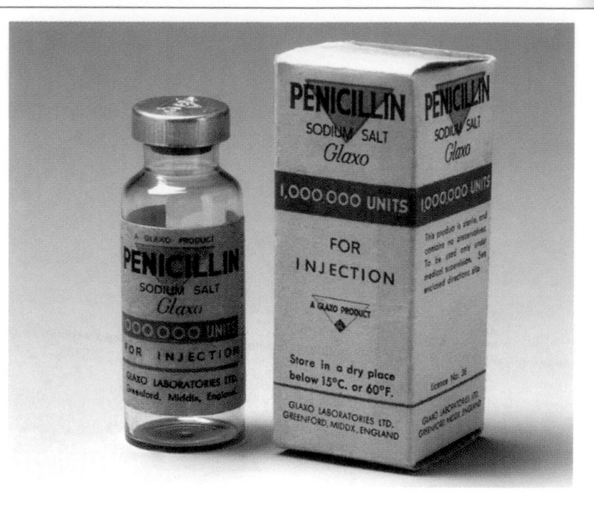

▲ **Figure 19** Early sample of penicillin.

The miracle cure

Natural materials have been used in many forms to treat infections in the past – the *Old Testament*, for example, describes the use of fungi and moulds to treat infected wounds.

In 1928, these remedies began to be shown to have a scientific basis when Alexander Fleming noticed that a mould (*Penicillium notatum*) produced a substance that appeared to inhibit bacterial growth.

Accidentally, one of Fleming's experiments with bacteria had become contaminated with a mould. On return from a holiday, Fleming noticed that bacterial growth was restricted in the areas where the mould had developed. He deduced that the mould had affected the bacteria by producing chemicals that he called *penicillins* – after the name of the mould.

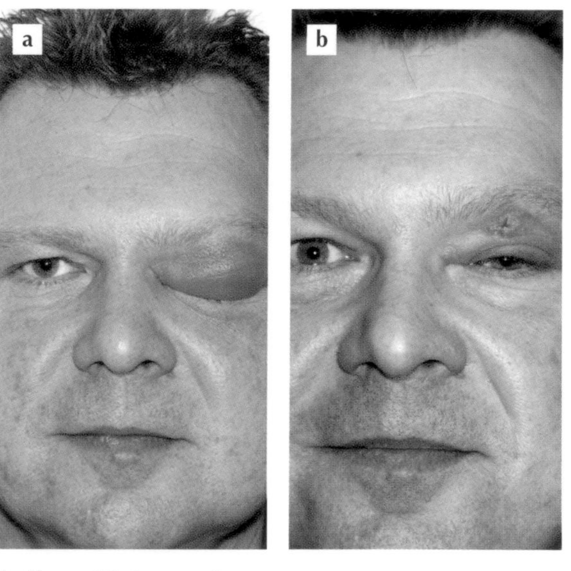

▲ **Figure 20** An eye abcess
a before teatment with antibiotics
b part way through the treatment.

Fleming also showed that the culture appeared to be non-toxic to animals. However, he did not go further to see if penicillin was effective against infections in the animals. This is not surprising, because the age of antiobiotics had not begun and penicillin was considered only for local use as an antiseptic.

Interest was revived in May 1940 when Howard Florey and Ernst Chain, in Oxford, showed that penicillin injections in mice were effective against a lethal streptococcal infection. Penicillin was later introduced into the clinic for use in humans, with dramatic consequences.

The antiobiotic era had dawned. Fleming, Florey and Chain shared a Nobel Prize in Medicine in 1945 for their work.

During the 1940s, penicillin was extracted in bulk from mould cultures, both in the UK and the US, and it became widely available.

Getting moulds to do the work

At first, scientists relied entirely on the fermentation of moulds to make penicillins. The mould *Penicillium notatum* makes several antibiotic compounds as products of its natural metabolism. These were isolated and called, for example, penicillin F, G, K and X.

Researchers found that the mould can be encouraged to produce just one penicillin, and that the type of penicillin produced can be altered, by changing the nutrient on which the mould is grown.

For example, growing the mould with compound I as the nutrient gives just penicillin G – this was the first 'miracle cure':

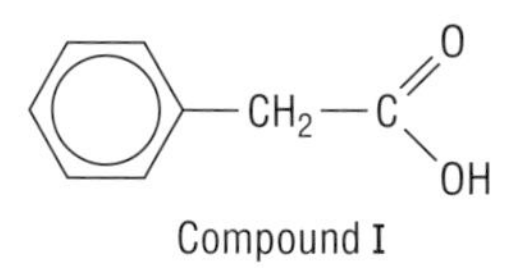

Compound I

Unfortunately, penicillin G is active against only a limited variety of bacteria. Also, stomach acidity causes it to lose its activity, so it must be given by injection. Growing the mould with compound II gives the improved penicillin V, which is not so susceptible to attack by acids and can be taken orally:

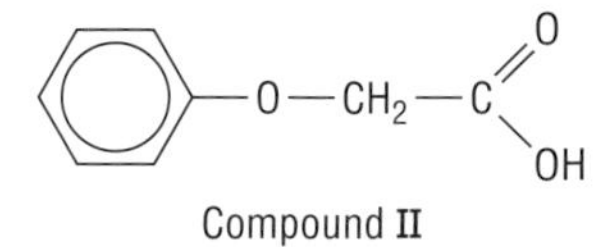

Compound II

Improving on nature

When the National Health Service was set up in 1948, Beecham (which became SmithKline Beecham (SKB) and subsequently GlaxoSmithKline in 2000), known then as the 'pills and potions' company, decided to look more closely at the preparation of antibiotic compounds.

A major breakthrough came in the late 1950s. While trying to make a new penicillin, their scientists found the penicillin 'nucleus' – the core structural framework of the penicillin molecule. They isolated this compound, called 6-aminopenicillanic acid (6-APA), and showed that it has this structure:

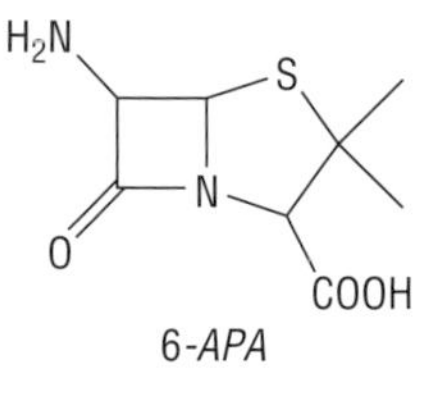

6-*APA*

This was an important discovery because it helped to make sense of the different types of penicillin. It also enabled scientists to make new penicillins partially by chemical techniques (i.e. semi-synthetically) without relying completely on moulds to do the necessary reactions.

6-APA itself has little effect on bacterial growth, but if you add an extra 'side chain' to the amino ($-NH_2$) group you have a disease-curing penicillin. You can see the structure of a penicillin in Figure 21. The nutrient compounds I and II react with 6-APA to give different side chains.

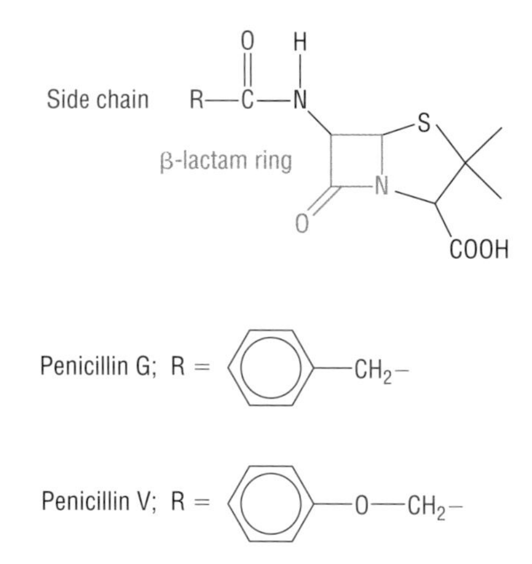

▲ **Figure 21** Structure of a penicillin – all penicillins have the same basic structure; only the R group varies.

Research now focused on chemical reactions that would attach different side chains to 6-APA. An **acylation** reaction is used, as shown in Figure 22.

In **Activity MD4.1** you can make a semi-synthetic penicillin and test its antibacterial activity.

The different penicillins produced have different properties and ranges of antibacterial activity. A doctor decides which of the many penicillins available is the best one to use against the particular infection being treated.

Making a so-called semi-synthetic penicillin involves making two biologically inactive molecules and clipping them together to make the active compound.

One of the molecules is synthesised in the laboratory – this is the side chain in the form of an acyl chloride. The other is 6-APA. This is obtained by treating naturally produced penicillin G with an enzyme to hydrolyse the amide linkage in the side chain.

THE β-LACTAM RING

Penicillin contains a fused-ring system containing a nitrogen atom and a sulfur atom. The four-membered ring contains a *cyclic amide* group and is called a **β-lactam ring.**

The simplest β-lactam ring is

$H_2C — CH_2$
$O{=}C — NH$

You can think of it as arising from the β-amino acid

$$\overset{\beta}{} \qquad \overset{\alpha}{}$$
$$H_2N — CH_2 — CH_2 — COOH$$

by the $-NH_2$ group bending round and condensing with the –COOH group on the other end of the molecule.

β-lactam rings are very sensitive to acids and alkalis – they react readily to give open-chain compounds. This is one of the reasons why it was so difficult to isolate and purify the first penicillins.

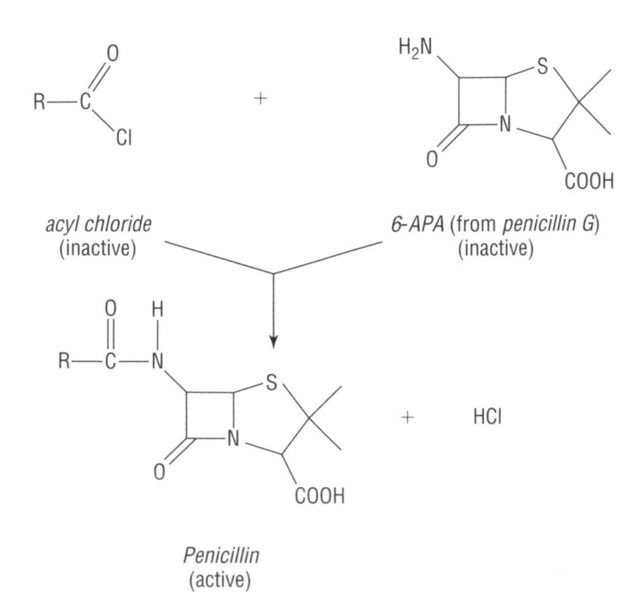

▲ **Figure 22** Preparation of a semi-synthetic penicillin.

How penicillin works

Penicillin does not normally attack bacteria that are fully grown or in a resting state. Instead, it stops the growth of new bacteria by inhibiting the action of an enzyme responsible for constructing the cell wall.

A bacterium is protected by a rigid cell wall that is built up from a network of polysaccharide chains joined by polypeptide cross-links (see Figure 23).

Assignment 9

Use the 'tool-kit' of organic reactions you built up in **Activity MD3.2** to devise a synthesis for phenylpenicillin

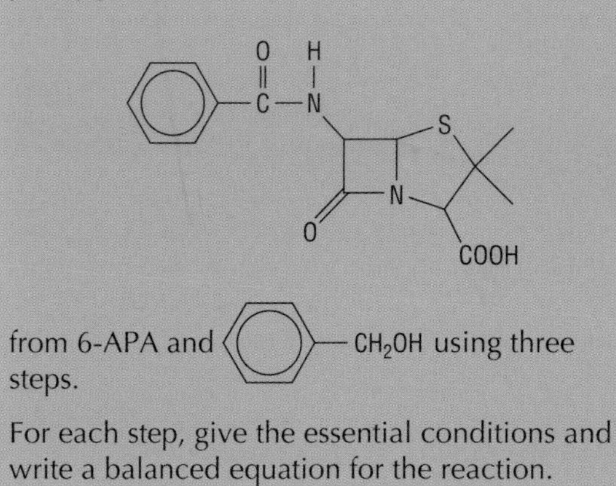

from 6-APA and [benzene ring]—CH_2OH using three steps.

For each step, give the essential conditions and write a balanced equation for the reaction.

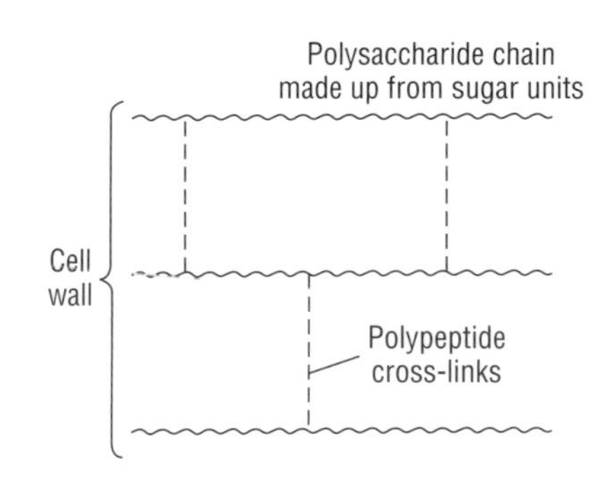

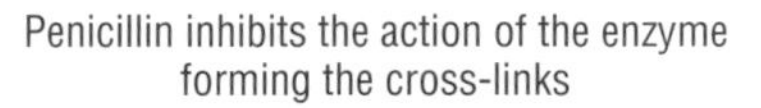

▲ **Figure 23** Simplified diagram of a bacterial cell wall.

Making a so-called *semi-synthetic* penicillin involves making two biologically inactive molecules and clipping them together to make the active compound

One of the molecules is synthesised in the laboratory. This is the side chain in the form of an acyl chloride. The other is *6-APA*. This is obtained by treating naturally produced *penicillin G* whth an enzyme to hydrolyse the amide linkage in the side chain.

The shape of the penicillin molecule resembles that of the crucial amino acids used in the cross-linking process. The penicillin reacts with the enzyme that makes the cross-links and inhibits its action – this results in weakening of the cell wall.

The enzyme never satisfactorily completes the wall, and eventually the contents of the cell burst out and the bacterium dies.

The structure of the cell wall in bacteria is unique and different from the cell walls in plants and the cell membranes in animals. One of the amino acids used to make the polypeptide cross-links is D-alanine. Only its isomer, L-alanine, is found in humans. As a result, penicillin is not toxic to humans and animals generally. It selectively targets bacteria.

Nature fights back

Sometimes, penicillin-resistant bacteria cause serious outbreaks of infections in hospitals, especially in surgical wards. These bacteria produce an enzyme called β-lactamase that attacks the four-membered β-lactam ring and makes the penicillin inactive.

Chemists have responded by developing new side chains that make the penicillin molecule less susceptible to attack by this enzyme. Such compounds include *methicillin* and *flucloxacillin*.

In **Activity MD4.2** you can make models of some penicillins, and investigate how the structure of the side chain affects their antibacterial activity.

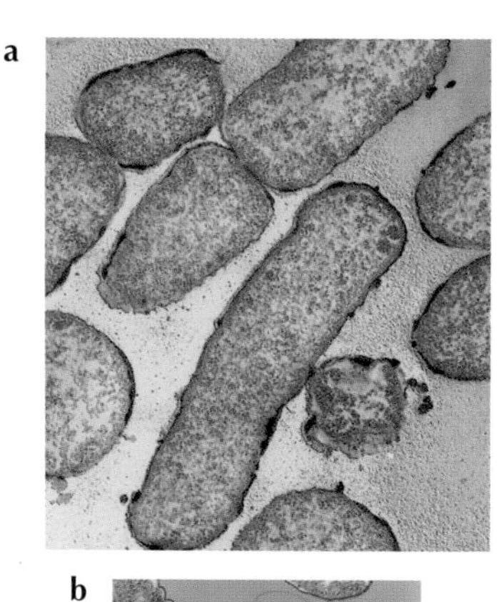

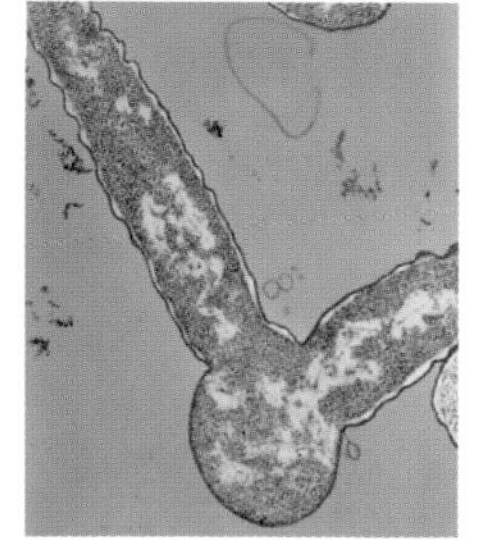

▲ **Figure 24** *E. coli* being treated with amoxycillin: **a** before and **b** after the dose. The cell wall is unable to form normally and the bacterial cells appear unusually long and thin.

Assignment 10

New strains of bacteria appear as a result of mutations – these are spontaneous changes in the genetic material of the cell. They occur at random and are passed on when cells divide.

a Explain how penicillin-resistant strains of bacteria develop at the expense of susceptible ones.

b When you are prescribed a course of antibiotics, you should always finish the medicine – even if you feel better. Suggest a reason for this.

A group of scientists working at Beecham Research Laboratories (now GlaxoSmithKline) tried an alternative approach to overcoming bacterial resistance. In addition to the search for penicillins with new side chains, they reasoned that a compound might be found that would inhibit the action of the β-lactamase enzyme, leaving the penicillin free to carry out its normal antibacterial action.

After investigating many possible compounds, they came up with a natural product called *clavulanic acid*. It has a structure similar to that of a penicillin molecule and is recognised by the β-lactamase enzyme:

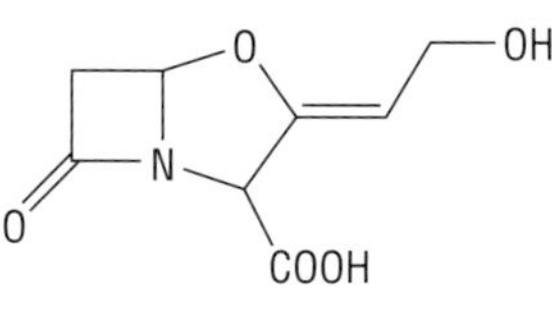

clavulanic acid

Clavulanic acid alone has virtually no antibiotic activity, but when mixed with previously ineffective penicillins it can be used against pathogenic bacteria that produce β-lactamase, with excellent results.

The story of penicillin doesn't end here. Work continues to produce an ever-widening range of 'magic bullets' – compounds that are targeted at specific bacteria in particular tissues of the body. Because bacteria are constantly mutating, these compounds are in constant danger of becoming obsolete. Ways have to be found (like using the β-lactamase inhibitor, clavulanic acid) to prolong their usefulness by overcoming resistance.

Assignment 11

Compare the structure of clavulanic acid with that of the penicillin shown in Figure 21.

a What features, common to both molecules, might the active site of the β-lactamase enzyme recognise?

b Suggest how clavulanic acid might inhibit the action of the β-lactamase enzyme.

▲ **Figure 25** A worker inspecting the fermentation tanks in the production hall of a pharmaceutical factory.

MD5 *Summary*

This module has centred around the search for efficient medicines that are increasingly selective in the way they work. Once scientists understand the way in which a compound interacts with the human body, or with a bacterial cell, they can design new substances that fit selectively onto a target site and bring about a desired effect. This means that medicines can be more effective and have fewer side effects.

The module began with a study of ethanol and how it interacts with the human body. It also looked at how infrared spectroscopy is used to measure alcohol concentration in the breath. The module then went on to consider the problems associated with high levels of certain types of cholesterol and the discovery of statins. Central to the work was the concept of molecular recognition. The shape and size of a biologically active molecule are crucial to its action. Its structure and precise shape must be known, because certain groups may need to be in specific positions to bind the active site of an enzyme. Computer-generated graphics help chemists to investigate the interactions involved and are the basis of modern drug design.

Synthetic organic chemists play an important role in preparing new compounds and modifying existing ones. So alongside the work on the module, the accompanying chemical ideas introduced you to organic synthesis and allowed you to revise the various functional groups and organic reactions that you have encountered throughout the course. These were organised into a 'toolkit' of reactions that allow you to write schemes for converting one organic compound into another.

You were introduced to nuclear magnetic spectroscopy and used it, plus a range of other instrumental techniques, to determine the structure of organic compounds.

Activity MD5 will help you to check your knowledge and understanding of this module.

INDEX

Note: **CS** – Chemical Storylines A2;
CI – Chemical Ideas.

The winning formula for Salters Advanced Chemistry

Developed in exclusive partnership with OCR, Heinemann's complete Salters Advanced Chemistry series will support and inspire you to reach your potential.

STUDENT BOOKS

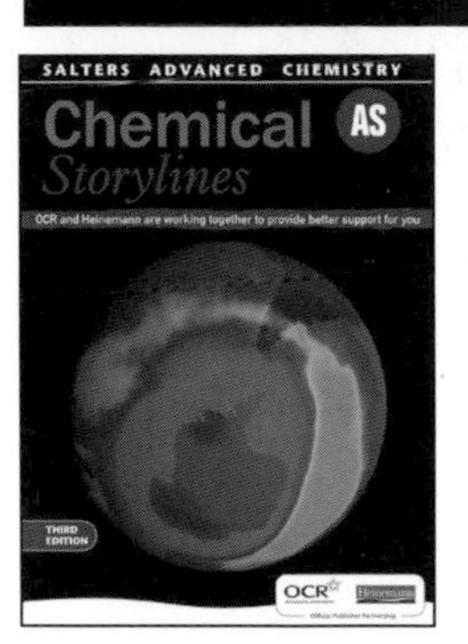

Chemical Storylines AS
(3rd Edition)
978 0 435631 47 5

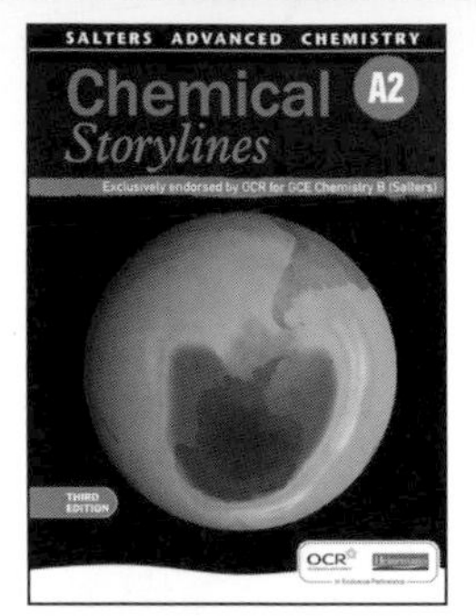

Chemical Storylines A2
(3rd Edition)
978 0 435631 48 2

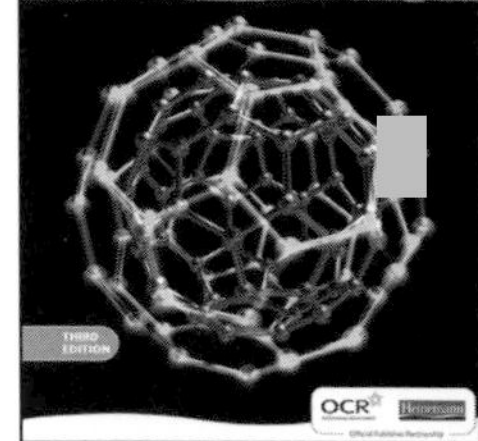

Chemical Ideas
(3rd Edition)
978 0 435631 49 9

REVISION GUIDES

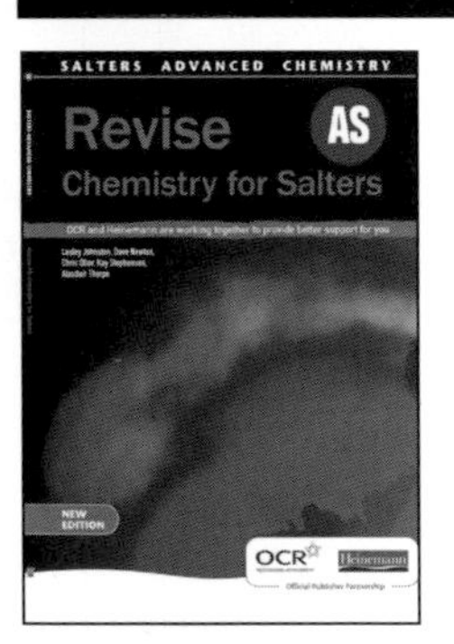

Revise AS Chemistry
for Salters (New Edition)
978 0 435631 54 3

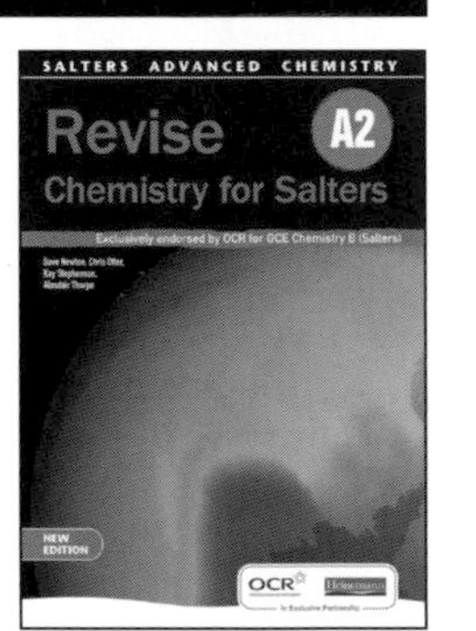

Revise A2 Chemistry
for Salters (New Edition)
978 0 435631 55 0

TEACHER SUPPORT PACKS

Support Pack AS
(3rd Edition)
978 0 435631 50 5

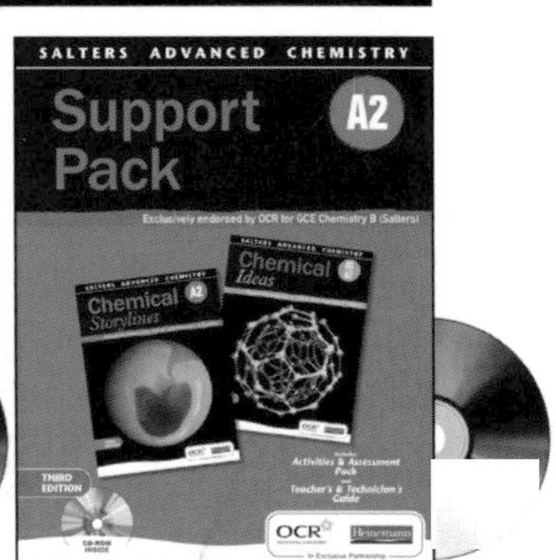

Support Pack A2
(3rd Edition)
978 0 435631 51 2

INTERACTIVE PRESENTATIONS CD-ROMS

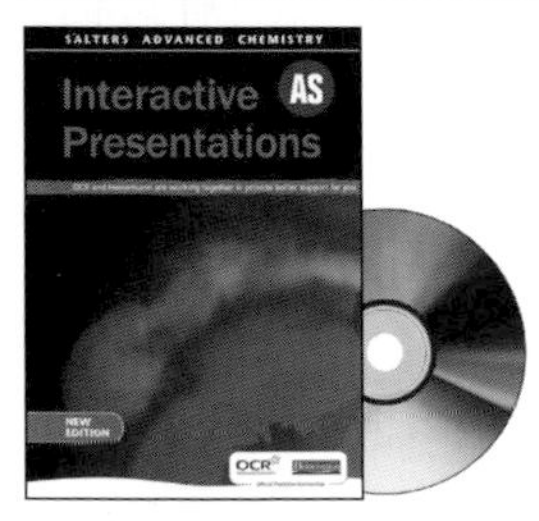

Interactive Presentations AS
(New Edition)
978 0 435631 52 9

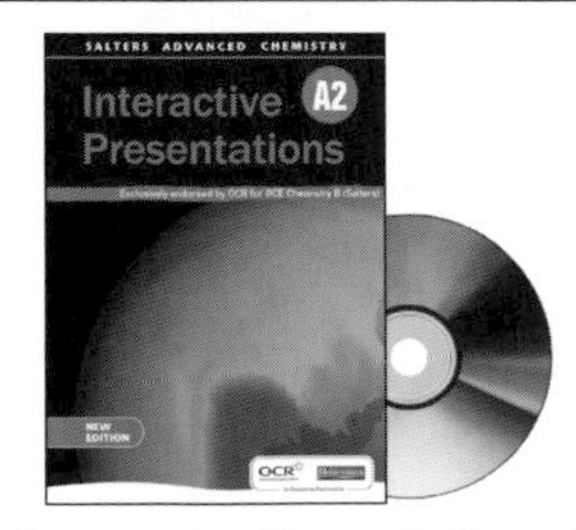

Interactive Presentations A2
(New Edition)
978 0 435631 53 6

(T) 0845 630 33 33
(F) 0845 630 77 77
(E) myorders@pearson.com
(W) www.heinemann.co.uk

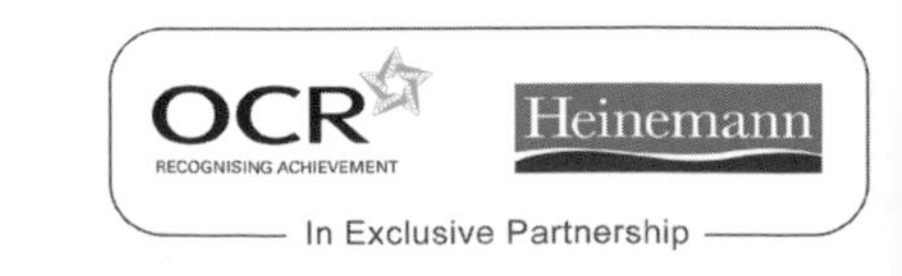